IV International Young Researchers Conferences "Youth, Science, Solutions: Ideas and Prospects" (YSSIP-2017)

MATEC Web of Conferences Volume 143 (2018)

Tomsk, Russia
25 - 27 October 2017

Editors:

N.O. Kopanitsa
T.Y. Ovsyannikova
V.S. Plevkov
A.A. Klopotov
L.A. Teplyakova
P.A. Radchenko

ISBN: 978-1-5108-5580-9

Printed from e-media with permission by:

Curran Associates, Inc.
57 Morehouse Lane
Red Hook, NY 12571

Some format issues inherent in the e-media version may also appear in this print version.

Printed by Curran Associates, Inc. (2018)

For additional information, please contact EDP Sciences – Web of Conferences at the address below.

EDP Sciences – Web of Conferences
17, Avenue du Hoggar
Parc d'Activité de Courtabœuf
BP 112
F-91944 Les Ulis Cedex A
France

Phone: +33 (0) 1 69 18 75 75
Fax: +33 (0) 1 69 28 84 91

contact-edps@webofconferences.org

Additional copies of this publication are available from:

Curran Associates, Inc.
57 Morehouse Lane
Red Hook, NY 12571 USA
Phone: 845-758-0400
Fax: 845-758-2633
Email: curran@proceedings.com
Web: www.proceedings.com

TABLE OF CONTENTS

Preface
IV International Young Researchers Conference "Youth, Science, Solutions: Ideas and Prospects" (YSSIP-2017)

IV International Young Researchers Conference "Youth, Science, Solutions: Ideas and Prospects" (YSSIP-2017) was held on 25-27 October 2017 at Tomsk State University of Architecture and Building (Tomsk, Russia).

The IV International Young Researchers Conference "Youth, Science, Solutions: Ideas and Prospects" has a cross-disciplinary nature and covers a wide range of relevant research topics, such as investigation of novel building materials, their structure, properties and production technologies; modern robust buildings and structures; materials physics, ultrafine-grained materials and nanostructures; energy and resource efficiency; economics; efficient use of resources; urban development, etc.

The Conference is featured by promotion of research activities among young scientists, post-graduates and students through discussions of problems and results of topical research. The number of participants of the Conference reached 200 researchers from 21 scientific and educational institutions, 110 contributed oral presentations were given.

Chairman of the Program and Scientific Editorial Board expresses gratitude to members of the organizational committee and the secretariat for the Conference organization and all of the staff who assisted in providing efficient Conference operation. Organizers acknowledge participants for their important contribution in the form of their oral reports and the present proceedings.

Editor of the Proceedings Volume,
Prof. Natalia O. Kopanitsa,
Dr. Tech. Sc., Prof.
Tomsk, December, 2017

edp sciences

MATEC Web of Conferences

Statement of Peer review

In submitting conference proceedings to MATEC Web of Conferences, I certify to the Publisher that I adhere to the **Policy on Publishing Integrity** of the journal in order to safeguard good scientific practice in publishing.

1. All articles have been subjected to peer review administered by the proceedings editors.
2. Reviews have been conducted by expert referees, who have been requested to provide unbiased and constructive comments aimed, whenever possible, at improving the work.
3. Proceedings editors have taken all reasonable steps to ensure the quality of the materials they publish and their decision to accept or reject a paper for publication has been based only on the merits of the work and the relevance to the journal.

Title, date and place of the conference

IV International Young Researchers Conference "Youth, Science, Solutions: Ideas and Prospects" (YSSIP-2017)
October 25 – 27, 2017
Tomsk, Russia

Proceedings editor(s):

1. Natalia O. Kopanitsa, Prof. Dr. (Chief Editor)
2. Tatiana Y. Ovsyannikova, Prof. Dr.
3. Vasilii S. Plevkov, Prof. Dr.
4. Anatoly A. Klopotov, Prof. Dr.
5. Luydmila A. Teplyakova, Prof., Dr
6. Pavel A. Radchenko, Assoc. Prof., Cand. of Sc.

Date and editor's signature / Natalia Kopanitsa, December 7, 2017

17, avenue du Hoggar - PA de Courtaboeuf – BP 112 - 91944 Les Ulis Cedex A (France)
Tel.: 33 (0)1 69 18 75 75 - Fax: 33(0)1 69 07 45 17 – www.edpsciences.org

Assessment of structural reliability of precast concrete buildings

Alexandr Koyankin[1,*], and *Valeriy* Mitasov[2]

[1]Siberian Federal University, 660041 Krasnoyarsk, Russia
[2]Novosibirsk State University of Architecture and Civil Engineering, 630008 Novosibirsk, Russia

Abstract. Precast housing construction is currently being under rapid development, however, reliability of building structures made from precast reinforced concrete cannot be assessed rationally due to insufficient research data on that subject. In this regard, experimental and numerical studies were conducted to assess structural reliability of precast buildings as described in the given paper. Experimental studies of full-scale and model samples were conducted; numerical studies were held based on finite element models using "Lira" software. The objects under study included fragment of flooring of a building under construction, full-size fragment of flooring, full-scale models of precast cross-beams-to-columns joints and joints between hollow-core floor slabs and precast and cast-in-place cross-beams. Conducted research enabled to perform an objective assessment of structural reliability of precast buildings.

Introduction

Precast construction is rapidly developing all over the world due to its ability of combining the advantages of precast and cast-in-place housing construction at the highest possible minimizing of its drawbacks. In particular, well-known Saret construction system suggested by French builders in the middle of the XXth century is being used by construction companies of the world (France, Russia, Czech Republic, Great Britain, French-speaking countries of Africa, etc.). The system underwent some structural changes depending on natural and climatic, seismic, regulatory and other peculiar features of the regions [1-8].

Popularity of precast housing construction defines the relevance of conducting research devoted to specific aspects of precast structures deformation. Scientists from many countries conducted various studies; e.g. the frame of B1.020.1-7 series (Russian grade) was multiply subjected to field tests on buildings under construction [9-11]. The study [12] describes experimental testing of load bearing capacity of the standard slabs along normal and oblique section with regard to their special work by supporting on three sides or flat loading of flooring. Professor A.S. Semchenkov conducted experimental study of RADIUSS frame [13, 14] that proved its effectiveness. Testing of precast flooring with wide cross-beams was described in [15].

The analysis of previous research demonstrated that the amount of research, namely experimental ones, is insufficient for assessing structural reliability of precast frame made

* Corresponding author: koyankinaa@mail.ru

with precast cross-beams and hollow-core slabs supported on them. Therefore, the present paper is aimed at assessing structural reliability of precast frame of Saret type.

Field tests were conducted on flooring segment and fragment of precast frame; testing of full-scale models of precast cross-beams-to-columns joints was performed; hollow-core floor slabs to precast and cast-in-place cross-beams joints were tested; numerical simulation was conducted on models of slab to flooring joints and cross-beam to column joints.

Experimental

Researchers of Institute of Engineering and Construction, Siberian Federal University (Krasnoyarsk, Russia) together with the research group of Novosibirsk State University of Architecture and Civil Engineering (Novosibirsk, Russia) conducted a series of experimental and numerical studies aimed at defining the stress-strain state of the frame mentioned above [16, 17]. Part of the flooring mesh on the first floor of a building under construction No. 1 of the II Residential area Pokrovskii in Krasnoyarsk was subjected to field tests; mesh size for testing was equal to 6.0x7.2 m in plan view (Fig. 1). The testing included four-stage loading of the mentioned fragment of flooring by covering it with ceramic bricks up to the loading of 400 kg/m^2. In these conditions we could observe fixation of flooring deflection and possible shear of cross-beams with reference to columns.

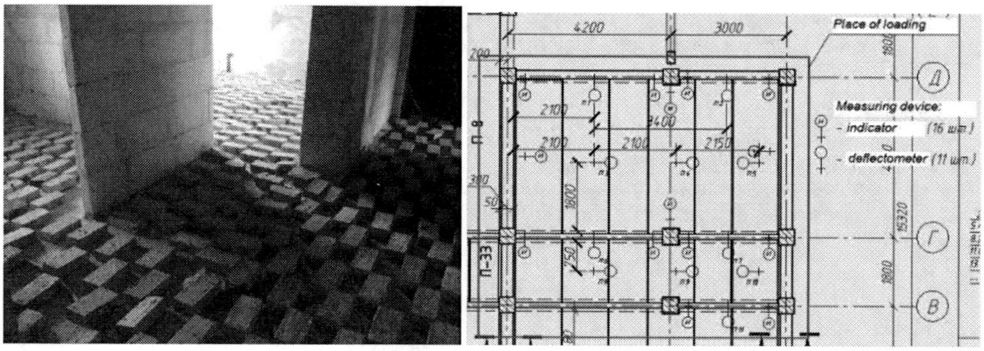

Fig. 1. Fragment of flooring of a building under construction.

Field tests were conducted on a fragment of precast frame sized 6430x4940x2800(h) mm, mounted and tested at the factory "Monolitresurs", LLC based in Krasnoyarsk (Fig. 2). The loading was conducted within four stages, by covering it with solid foundation blocks and precast cross-beams. The maximum load on the balcony was 6.06 kN/m^2 on the floor area 0.8 m wide from the balcony edge; the load acting on the flooring was 5.60 kN/m^2. Flooring deflections of the structural elements were defined by means of deflectometers and indicators, while the strain was recorded using universal multi-channel measuring-recording device TEPEM-4.0 with displacement sensors DLP-10.

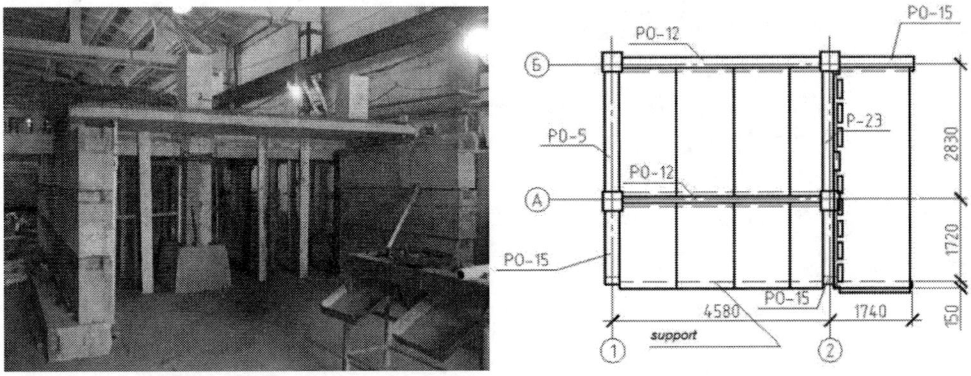

Fig. 2. Fragment of precast frame.

Moreover, four full-scale models of joints between precast cross-beam and column were tested in the Structures Testing Laboratory of Siberian Federal University (Fig. 3). Overall dimensions of models enabled to obtain M and Q stresses combination in the joint corresponding to the stress state of the joint in 6 m spans and larger. Samples were tested using specially developed test bench mounted in the "streams" of strong floor of the testing laboratory. Meanwhile, the loading of test samples mounted directly on the strong floor was provided by two 100-tons jacks DU100P150 (Russian grade) placed on the edges of samples consoles and transmitting top-down load. Vertical displacements were measured by indicators and deflectometers, while deformations – by the measuring system MMTC-64 with resistance strain gauges KF5P1-50-120-A-12 (concrete) and KF5P1-10-200-A-12 (reinforcement). The width of cracks opening was defined by the microscope MPB-3. In addition to experimental research using Lira software complex numerical simulation of the above mentioned models was conducted using solid finite elements for concrete and rod finite elements for reinforcement.

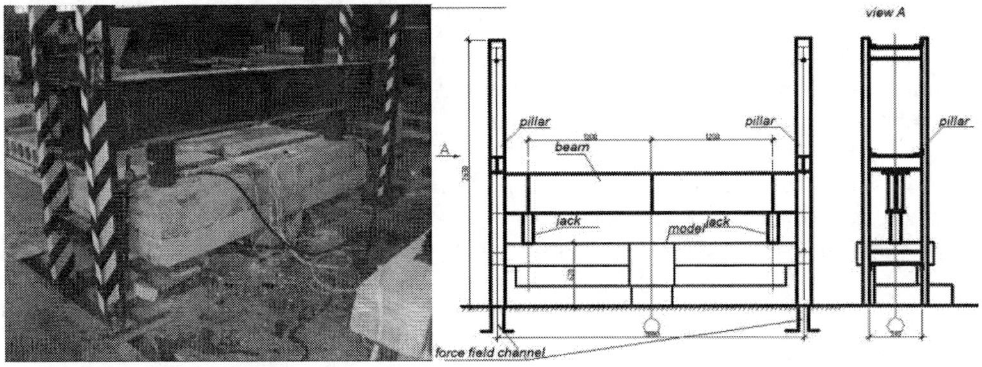

Fig. 3. Testing of cross-beam to column joints.

Another testing included three joints between hollow-core floor slab and precast and cast-in-place cross-beam (Fig. 4). Testing was held using test bench mounted in the "streams" of power floor of the testing laboratory. The load was applied top-down by two 100-tons jacks DU100P150, placed on the edges of samples consoles and transmitting vertical load. Deflections were defined by deflectometers and indicators, while strain was defined by the measuring system MMTC-64. Numerical studies were conducted in a similar way to finite element models of cross-beam to column joints.

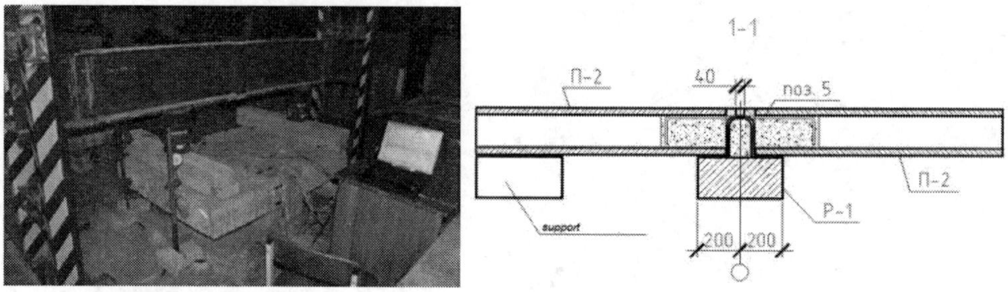

Fig. 4. Testing of joints between hollow-core slab and cross-beam.

Results

Structural solution of precast frame under study provides the necessary load bearing capacity as required by technical prescriptions, also the required stiffness and crack resistance to the full extent. However, experimental data highlight the necessity to perform a number of structural technical solutions enabling to provide larger structural reliability of precast frame and further reduction of construction costs.

Detailed analysis of the data obtained from the studies makes it possible to conclude that the rigidity of structure of precast flooring is provided with measurable reserve. This became possible due to quite high bending stiffness and, at the same time, light weight of hollow-core floor slab, as well as due to the fact that slabs and cross-beams were pre-stressed. Moreover, additional bending stiffness is provided by key joint of hollow-core slab with precast cross-beam, causing their joint deformation.

In all conducted tests deflections did not reach the admissible limit values. Particularly, vertical displacements on field tests on a building under construction and frame fragment hardly exceeded 1 mm (deflection in the center of fragment mesh of pre-fabricated frame was 1.03 mm with the load of 400 kg/m^2). Along with that, the values of deflections obtained by using the calculation methodology suggested by regulatory documents and as a result of numerical studies using Lira software were 0.81 mm and 0.82 mm respectively.

Experimental study of cross-beam to column joints demonstrated the maximum value of deflection 14.4 mm with the load of 41 kN in the joints without hollow-core slabs and 8.5 mm in the joints with fragments of slabs. This testifies that inclusion of hollow-core slabs into combined deformation with cross-beam has a significant influence on bending stiffness (causes its increase) of the cross-beam. As can be seen from numerical studies, the maximum deflections defined on the edges of cross-beams were 11.1 mm with the load of 41 kN, i.e. the difference between the obtained experimental and numerical data (Fig. 5) did not exceed 23 %, thus validating the obtained results.

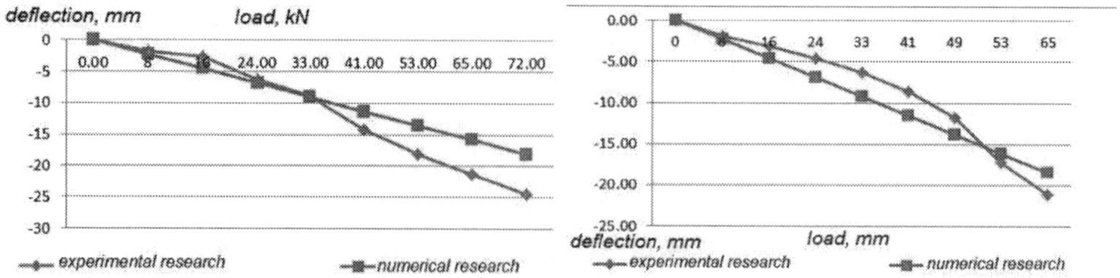

a *b*

Fig. 5. Deflection curve: *a* – samples with fragment of hollow-core slabs; *b* – samples without fragment of hollow-core slabs.

Crack resistance of the precast frame structure corresponds to the requirements of rules and regulations of the Russian Federation. Particularly, during field tests on a building under construction and fragment of the building frame no cracks were observed at the experimental load on the flooring of 4 kN/m^2 and 5.6 kN/m^2 respectively (results of strain-gauge measurement showing that tensile strain of concrete did not exceed the limit value of 0.0001 indirectly proved the absence of cracks).

The absence of cracks during experimental loading was also proved by theoretical calculations demonstrating that in hollow-core slab the cracks must occur with the load of 27.2 kN/m^2 applied to the flooring, while in precast cross-beam (in the solid part on the support) it should be the load of 13.7 kN/m^2. During experimental research of cross-beam to column joints only normal cracks (the width of opening 0.2…0.3 mm, which does not exceed the admissible limit value) were registered in cast-in-place reinforced concrete in cross-beam-to-column joints.

It should also be noted that along with the positive influence of slab to cross-beam key joint on increasing the cross-beam stiffness, also the negative influence of hollow-core slab inclusion into combined deformation with the cross-beam was observed. In this case hollow-core slab starts to bend in lateral non-operating direction with further formation of longitudinal cracks. This was recorded during testing of the cross-beam to column joint.

The picture of crack formation obtained from experiments was also proved by stress isofields obtained during numerical studies (Fig. 6). Particularly, the maximum tensile stresses in precast cross-beam reaching the limit value of 1.05 MPa were concentrated in the place of joining between the cross-beam and the column, in the exact place where cracks formation was observed during experiment. The lower compression zone keeps its load bearing capacity where the maximum tensile stresses did not exceed 12.9 MPa. Moreover, occurrence of longitudinal crack in the hollow-core floor slab which is included into operation of cross-beam under bending due to key joint was proved by numerical studies as appearing of ultimate tensile stresses in the slab in lateral non-operating direction for the slab.

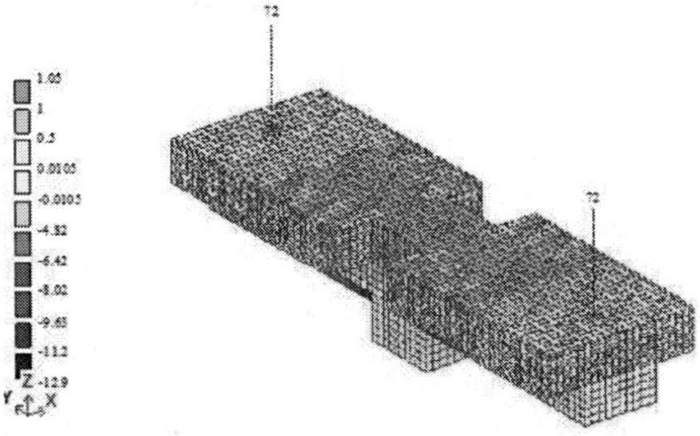

Fig. 6. Stress isofields (MPa) of the sample with fragments of hollow-core slabs.

In spite of the fact that during field tests on a building under construction and fragment of building frame the structures were not subject to complete fracture, still quite high load bearing capacity of precast frame can be inferred according to the following features, as defined from experimental studies:

- the load reached the reference value;
- the signs of failure were not visually observed;
- the value of tensile strain defined in the place of sensors placement did not exceed the admissible limit value equal to 0.002;

- calculation of load-bearing capacity according to standards of the Russian Federation showed that admissible load on the hollow-core slab was 16.8 kN/m^2, while fracture of precast cross-beam requires load of 59.4 kN/m^2 acting on the flooring.

Unlike the field tests, cross-beam to column joints were subject to fracture occurring under the load of 65-73 kN; that testifies on the high load bearing capacity of the joints (load bearing capacity defined in technical regulations is 60 kN). The depletion of load-bearing capacity occurred due to reaching the yield stress of 400 MPa in longitudinal reinforcement, the value was recorded by strain-gauge sensors. Fracture of compression zone of concrete resulting from experiment did not occur, as seen from visual examination, and also from the data of strain-gauge measurement that did not record the ultimate strain of compressed concrete (Fig. 7).

Similar results were shown by numerical studies, where the stresses in longitudinal reinforcement reached the limit values of 400 MPa, while the maximum tensile stresses in concrete were 12.9 MPa.

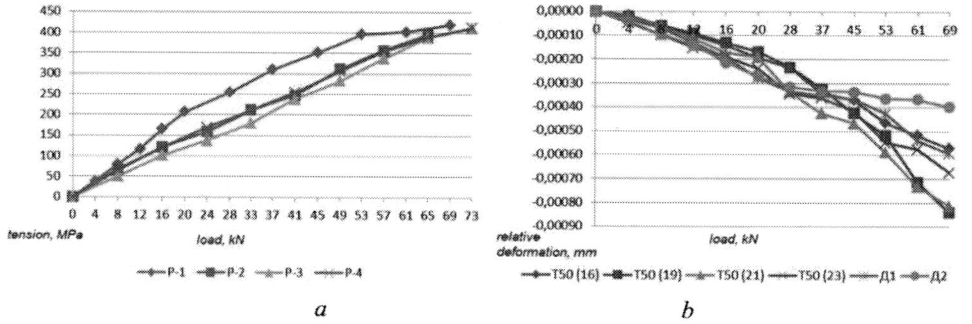

Fig. 7. The readings of strain-gauge sensors: *a* – mean stresses of reinforcement; *b* – relative strain of compressed concrete in the sample P-1.

During the conducted studies of joints between hollow-core slabs and cross-beams brittle failure of the joint connection occurred at the load of 20 – 24 kN resulting from formation of critical crack on the upper tensile surface of the hollow-core slab fragment at the place where solid connector ends. This fact indicates sufficient stiffness of key joint of the hollow-core slab with precast cross-beam. The reason why further withstanding of the load is impossible is absence of reinforcement of the upper slab part. Insignificant opening of the upper part of junction node of cast-in-place and precast reinforced concrete was recorded at the load of 12…16 kN (the last but one stages of loading).

Stresses in longitudinal reinforcement of keys, relative strain of concrete in the upper (tensed) surface of the cross-beam and the slab nearby the joint between hollow-core slab and cast-in-place part of the cross-beam, as well as concrete strain of the lower surface of the slab do not reach admissible standard values. It is well observed from the data of strain gauge sensors, as well as from visual inspection. Numerical simulation of joint connection of the slab with the cross-beam shows the similar fracture pattern. The formation of critical crack is also observed, resulting in fast further fracture in the hollow-core floor slab in the area where concrete key ends, i.e. the place where there is no reinforcement of the upper tensile zone. Ultimate breaking load was 20 kN; this value is 17 % less than the experimental data.

Conclusion

Conducted experimental and numerical research and the data obtained herein qualitatively and significantly enlarge the existing materials on description of the stress-strain state of precast buildings. Moreover, as specified in the objective of the given work the obtained data enables to accurately assess the structural reliability of pre-fabricated frame of Saret type (with precast cross-beams and hollow-core floor slabs). The following conclusions can be made:

1. Precast buildings which are designed and built in accordance with the Saret system correspond to the required parameters of the 1st and 2nd groups of ultimate behavior, as defined by the technical prescriptions;
2. The span between vertical load bearing elements up to 4.5 m provides significant reserve of load bearing capacity, stiffness and crack resistance, thus increasing the structural costs;
3. Results of testing joints of cross-beam with the column and slab with the cross-beam showed that this structural solution has some drawbacks; however, they were not defined during field tests on the buildings under construction due to small size of actual spans (4.5 m) of cross-beams and slabs and overestimated stiffness parameters of horizontal load-bearing elements. Still, civil designers and builders will face these drawbacks when trying to enlarge the spans. Rather rigid connection of hollow-core floor slab with the cross-beam can be referred to one of the mentioned drawbacks leading to the following possible negative consequences:

a. occurrence of tensile stresses the slab in the lateral direction during inclusion of the hollow-core slab into the process of cross-beam deformation, then resulting in formation of longitudinal cracks in it;

b. occurrence of support bending moment in the hollow-core slab in longitudinal direction with further cracks formation in the place of concrete key ending (the slab does not have upper reinforcement).

The following measures can be recommended in order to optimize the structural solution of pre-fabricated frame based on the conducted research:

- construction of flooring with larger span, linking closely the parameters of the existing load-bearing capacity of separate structural elements (slabs and cross-beams) and stresses from external load. Currently, the spans of load bearing horizontal elements up to 4.5 m provide significantly overestimated and at the same time unreasonable strength reserve of slabs and cross-beams (there are cases when hollow-core slabs reinforced with pre-stressed rebars 7Ø12K1500 cover the spans of 1.8 m, and the cross-beams 450 mm high and reinforced with pre-stressed rebars 3-4Ø12K1500 cover the spans of 3.0 m);
- application of upper longitudinal and lateral reinforcement in hollow-core slabs (at least in the support zone) with the aim of withstanding tensile stresses caused by existence of key connection (which is quite rigid).

Experimental and numerical studies show that cast-in-place structures require optimization in order to find the most effective structural solution.

References

1. V.A. Shembakov, *Sborno-monolitnoe karkasnoe domostroenie. Rukovodstvo k prinyatiyu resheniya* [Cast-in place and Precast Frame House-Building. Guidance for Decision-Making] (2005) (in Russian)
2. N.V. Nikitin, P.I. Franov, E.M. Timonin, *Rekomendatsii po proektirovaniyu konstruktsii ploskogo sborno-monolitnogo perekrytiya "Sochi"* [Recommendations

for Structural Design of a Flat Precast-Cast-In-Place Slab "Sochi"] (1975) (in Russian)
3. A.I. Mordich, V.N. Belevich, V.N. Simbirkin, D.I. Navoi, A.N. Mironov, V.P. Raichev, A.I. Chubrik, *Effektivnye konstruktivnye systemy mnogoetazhnykh zhilykh domov I obschestvennykh zdanii (12…25 etazhei) dlya uslovii stroitelstva v Moskve i gorodakh Moskovskoi oblasti, naibolee polno uovletvoryauschie sovremennym marketingovym trebovaniyam* [Effective Structural Systems of Multistory Blocks of Flats and Civil Buildings (12…25 Stories) for Construction Conditions in Moscow and the Cities of Moscow Region, Fulfilling Modern Marketing Demands More Completely] (2002) (in Russian)
4. A.I. Mordich, Sborno-monolitnye i monolitnye karkasy mnogoetazhnykh zdanii s ploskimi raspornymi perekrytiyami. Montazhnye i spetsialnye raboty v stroitelstve [In-cast and Precast Joists and Cast-In-Place Frames of Multi-Storey Buildings with Flat Space Slabs], **8-9** (2001) (in Russian)
5. A.S. Semchenkov, Concrete and Reinforced Concrete, **4** (2008) (in Russian)
6. A.S. Semchenkov, Concrete and Reinforced Concrete, **6** (2010) (in Russian)
7. A.M. Kimberg, *Effektivnaya konstruktivnaya sistema karkasno-panel'nykh zdaniy s natyazheniem armatury v postroechnykh usloviyakh* [Effective Structural System of Frame-Panel Buildings with Tensioning of the Steel in Site Conditions] (TbilZNIIEP Publ., Tbilisi, 1985) (in Russian)
8. B. Koprivitsa, Zhilishchnoe Stroitel`stvo [Housing Construction], **1,** (1984) (in Russian)
9. A.I. Mordich, V.N. Belevich, V.N. Simbirkin, D.I. Navoy, BST – Bulletin of Construction Equipment, **8,** (2004) (in Russian)
10. A.I. Mordich, V.E. Sadokho, I.I. Podlipskaya, N.A. Taratynova, Concrete and Reinforced Concrete, **5,** (1993)
11. A.I. Mordich, S.L. Galkin, Stroitelnaya Nauka i Tekhnika [Construction Science and Technology], **3,** (2011) (in Russian)
12. A.S. Semchenkov, Concrete and Reinforced Concrete, **4,** (1993) (in Russian)
13. A.S. Semchenkov, A.K. Khavkin, B.S. Sokolov, Concrete and Reinforced Concrete, **6**, (2008) (in Russian)
14. A.S. Semchenkov, Concrete and Reinforced Concrete, **1,** (2009)
15. L.N. Zaitsev, V.V. Ivanov, V.L. Zaitseva, Concrete and Reinforced Concrete, **2,** (2009) (in Russian)
16. V.M. Mitasov, A.A. Koyankin, News of higher educational institutions. Construction, **3,** (2014)
17. A.A. Koyankin, V.M. Mitasov, Vestnik MGSU (Scientific and Engineering Journal for Construction and Architecture), **5,** (2015)

Thermal state analysis of energy saving structures of cast-in-place lightweight aggregate concrete walls

Nikolay Tsvetkov[1,*], *Andrei* Khutornoi[1], *Alexandr* Kozlobrodov[1], *Anna* Kolesnikova[1], *Elena* Ivanova[1], and *Artem* Golovko[1]

[1]Tomsk State University of Architecture and Building, 634003 Tomsk, Russia

Abstract. The purpose of this paper is numerical study on the impact of location of cylindrical fillers made of insulating material on the thermal state and thermal protection properties of the corner fragment of cast-in-place lightweight aggregate external wall. Solution of spatial thermal conductivity problem was obtained and the effect of low-thermal-conductivity fillers location inside the corner fragment of a wall was analyzed using the software complex Ansys. The nature of temperature fields distribution was determined both inside the wall and in the inner corner of the structure. The method for temperature increase in the inner corner is suggested and its quantitative evaluation is performed. The effect of thermophysical and geometrical parameters of wall materials on its thermal insulation properties is analyzed.

Introduction

Effective use of energy sources in various branches of national economy is an important strategic task for Russia. In Russian construction industry, 23% of total energy consumed is spent on provision of heating energy to buildings and structures [1-3]. Heating energy loss through external walls in buildings can reach up to 45% depending on their structural peculiarities [4-5]. In this regard, design of new structures with improved thermal insulation properties and justification of the possibility of their effective use is a relevant task.

The issue of construction of buildings that satisfy ever increasing standard requirements regarding energy efficiency is directly connected with the level of thermal protection of external walls, which is greatly dependent on whether effective insulation is installed in the wall structure. External walls are divided into 3 types with regard to location of insulation: walls with insulation located on external surface of the wall; walls with insulation located on inner surface of the wall; and walls with insulation located inside the wall [6].

In the research [7] two types of structures with the same thickness of insulation layer are analyzed: a conventional multilayer facade and a hinged ventilated facade. As the result of the analysis hinged ventilated facade was found to be the most energy efficient in the climate conditions of Saint Petersburg. The works[8-9] suggest the new system of facade insulation by panels with ventilated channels; they also present analysis results for buildings with brick

* Corresponding author: nac.tsuab@yandex.ru

external walls and prove high efficiency of using the suggested system in Siberian climate conditions.

The research [10] provides analysis and development of a numerical model for simulation of unstable thermal behavior, both on the surface and in the corner joint of an external wall; consequently, this enables to define a building's demand in heating energy. Also, in [11] the authors present quantitative assessment of heat transfer through the external walls in the corner fragment.

Since external corner in the most heat-stressed area of the envelope of a building, the research [12] presents the model for calculating heat transfer in the external corner area. As the result of the simulation, the influence of thickness and location of insulation layers on the temperature field was determined; methods for reducing and even eliminating the impact of a heat-stressed element were developed.

In [13], the authors suggest a structural scheme for a cast-in-place external lightweight aggregate concrete wall with regularly located vertical vapor-proof compound formworks with efficient insulation [14]. Numerical and experimental studies on thermal protection properties of a wall were performed. However, regularities of fields distribution on the wall surface considerably differ from such regularities in wall corners [15]. The temperature of internal corner surface, as a rule, is lower than that of the wall surface, which can lead to condensate formation. For that reason, investigation of thermal state peculiarities of such heat-stressed elements of building envelopes, including corner of a building studied in the paper, shall be considered new and have remarkable practical importance. The main contribution of the paper to the previous research in the field consists in the advanced consideration of physics of thermal processes in multilayer structures with heat-stressed elements and in defining new regularities and relations that allow evaluating the impact of geometrical factors and heat transfer conditions on thermal state of a building envelope with regard to its energy efficiency.

Thus, the purpose of this paper is mathematical modeling and numerical study of thermal state of a corner fragment of a cast-in-place external lightweight aggregate concrete wall with various locations of insulating fillers inside the structure, as well as assessment of the effect of cable embedded in the corner.

Physico-mathematical statement of a problem

Research of thermal protection properties of a cast-in-place lightweight aggregate concrete wall is performed for a corner fragment with various locations of insulating fillers inside the structure (Fig. 1).

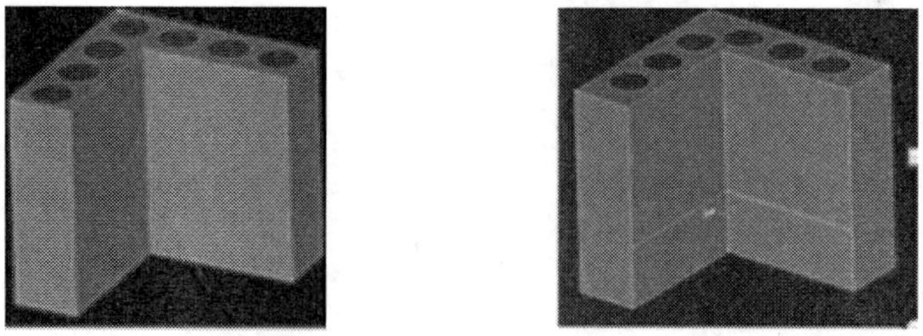

a) *b)*

Fig. 1. Structures of external wall under study: *a* – structure of a wall with a vertical insulating filler located on the line that connects vertices of internal and external angles (hereinafter referred to as variant 1); *b* – structure of a wall without a vertical insulating filler on the line that connects vertices of internal and external angles (hereinafter referred to as variant 2).

For mathematical problem statement, size parameters of envelope structures and their thermophysical characteristics are considered as given, generally, they may depend on the temperature. Geometrical parameters of the fragments are as follows: wall thickness δ_c = 0.5 m, fillers diameter d = 0.3 m. Thermal conductivity coefficients for lightweight aggregate concrete λ = 0.41 W/(m·°C), for insulating fillers made of polyurethane foam λ = 0.04 W/(m·°C). The outdoor temperature t_{otd} and indoor temperature t_{ind} are set, as well as heat transfer coefficients for external and internal surfaces. The indoor and outdoor temperature values are +22 and –40 °C, and heat transfer coefficients for external and internal surfaces, respectively, are α_{otd}= 23 W/(m²·°C) and α_{ind} = 8.7 W/(m²·°C).

Unstable spatial transfer occurring in the structural fragment under study is described in Cartesian coordinate system as a system of nonlinear stationary three-dimensional equations for thermal conductivity [12], their number being equal to the total number of constituent elements (n):

$$\frac{\partial t_i}{\partial \tau} = \frac{\partial}{\partial x}(\lambda_{x,i}\frac{\partial t_i}{\partial x}) + \frac{\partial}{\partial y}(\lambda_{y,i}\frac{\partial t_i}{\partial y}) + \frac{\partial}{\partial z}(\lambda_{z,i}\frac{\partial t_i}{\partial z}), \tag{1}$$

where i = 1,…,n with corresponding initial and boundary conditions.

For initial conditions are taken temperature field that corresponds to the solution of the stationary problem, and uniform temperature field or temperature for each element (layer) of a structure.

$$t_i|_{\tau=0} = t_0(x, y, z); i = 1,..., n; x, y, z \in \Omega, \tag{2}$$

where Ω – closed region of variation of independent spatial variables.

Boundary conditions for all internal and external surfaces are the boundary conditions of type III which account heat exchange between these surfaces and the environment; for example, for internal surfaces such conditions can be put in the following form:

$$-\lambda_{n,m}\left.\frac{\partial t_m}{\partial n}\right|_{n=0} = \alpha_{int}(t_{int} - t_{surf,m}), \tag{3}$$

where n – the direction of normal to the corresponding surface; m– number of the element that contacts with indoor air; $\lambda_{n,m}$ – thermal conductivity coefficient of the material of this element; t_{int}, $t_{surf,m}$ – temperatures of internal air and the contacting surface, respectively.

In the interface of two adjacent elements the boundary conditions of type IV are established, according to which the temperatures and the heat flows rate shall be equal:

$$t_{bound.,m}|_{bound.} = t_{bound.,m+1}|_{bound.}; \tag{4}$$

$$-\lambda_{n,m}\frac{\partial t_m}{\partial n} = -\lambda_{n,m+1}\left.\frac{\partial t_{m+1}}{\partial n}\right|_{bound.}. \tag{5}$$

For all edges of the structure symmetry condition are established, which corresponds to the boundary conditions of type II with the value of heat flow rate equal to zero:

$$q|_{bound.} = 0. \tag{6}$$

Numerical study

Calculation of temperature fields and heat flow rate in the studied structural elements are performed in Cartesian coordinate system with the use of ANSYS Mechanical APDL 17.0.

The calculation results for temperature fields in normal cross-section of the studied structures are presented in Figures 2-5.

Figure 2 shows the calculation results for temperature fields in the cross-section of a corner fragment for the set structural elements.

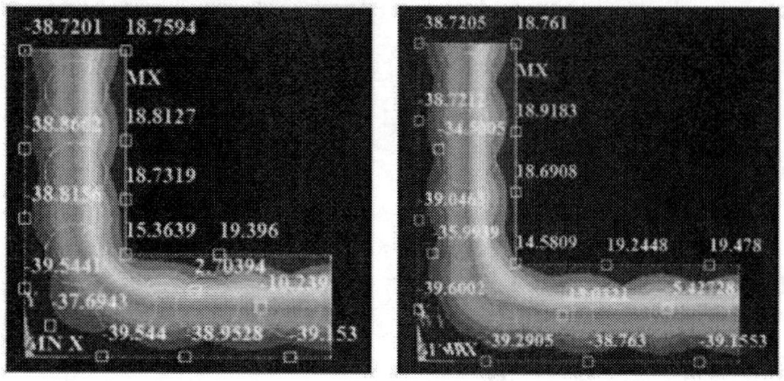

a) *b)*

Fig. 2. Behavior of temperature fields in structures: *a* – wall structure produced according to variant 1; *b* – wall structure produced according to variant 2.

The results of temperature fields calculation presented in Figure 2 show that temperature of the corner of the structure produced according to variant 1 is 0.78 °C higher than that of the structure produced according the variant 2. However, formation of condensate in the wall corner is excluded in both cases with $t_{ind} = +22\ ^{o}\mathrm{C}$ and relative humidity of indoor air φ = 50%, the dew point temperature being $t_{\delta} = +10{,}5\ ^{o}\mathrm{C}$.

Figures 4 and 5 illustrate the calculation results for heat flow rate in the studied fragments of envelope structures.

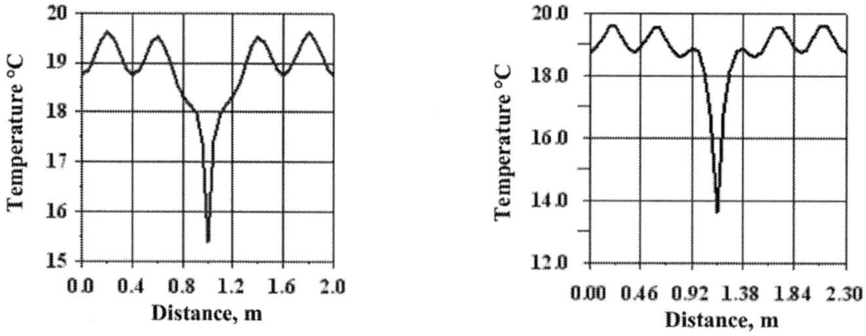

a) *b)*

Fig. 3. Temperature curve on the internal surface of the structure: *a* – wall structure produced according to variant 1; *b* – wall structure produced according to variant 2.

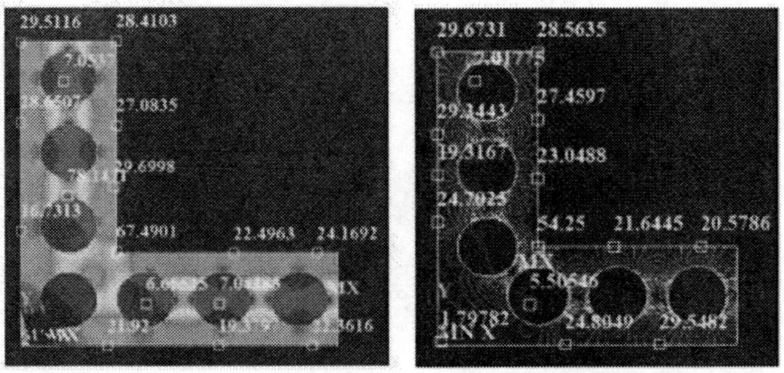

a) *b)*

Fig. 4. Distribution of heat flow rates in structures: *a* – wall structure produced according to variant 1; *b* – wall structure produced according to variant 2.

After analyzing Figures 4 and 5 one can conclude that the highest values of heat flow rates inside both types of structures are observed between the insulating fillers. For the structure produced according to variant 1 the maximum value of heat flow rate in the corner is 67.49 W/m^2, and for the structure of variant 2 - 54.25 W/m^2.

With the increase of thermal conductivity of lightweight aggregate concrete the temperature on the internal surface of external wall reduces, which causes the risk of condensate formation in the corner of the wall (Fig. 6). Thus, with thermal conductivity of lightweight aggregate concrete equal to 0.92 W/(m·°C) the value of wall surface temperature is lower by the average value of 2.7 than that of the wall made of lightweight aggregate concrete with thermal conductivity of 0.41 W/(m·°C) (Fig. 6).

In the corner of the lightweight aggregate concrete with thermal conductivity of 0.92 W/(m·°C) the temperature value is 9.45 °C, which is lower that the dew point temperature when t_{ind}=+22 °C and relative humidity of internal air is φ = 50%. One of the measures to increase the temperature in the corner is the use of heating cable [15]. In Figure 6, the curve 2 characterizes the temperature distribution on the internal surface of the wall with heating cable.

The performed calculations showed that when installing a heating cable with the power of $q_v = 50\,W/\mathrm{m}$ in the lightweight aggregate concrete with the thermal conductivity of 0.92 W/(m·°C), the temperature in the corner increases up to 14 °C (Fig. 6). The cable with the specified power heats the structure up to the distance of about 0.3 m from the corner, further wall surface temperature does not change.

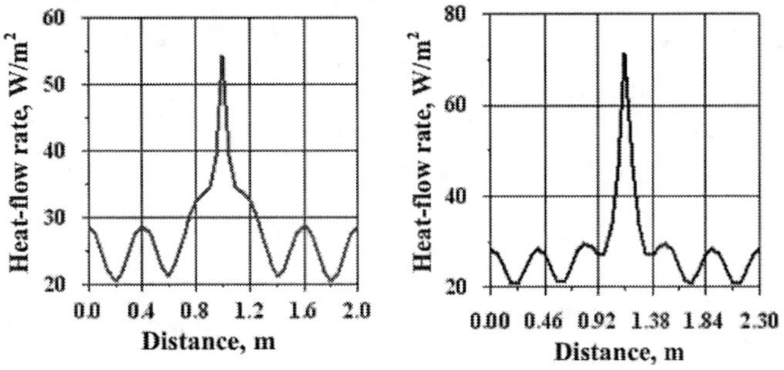

a) *b)*

Fig. 5. Distribution of heat flow rates in structures: *a* – wall structure produced according to variant 1; *b* – wall structure produced according to variant 2.

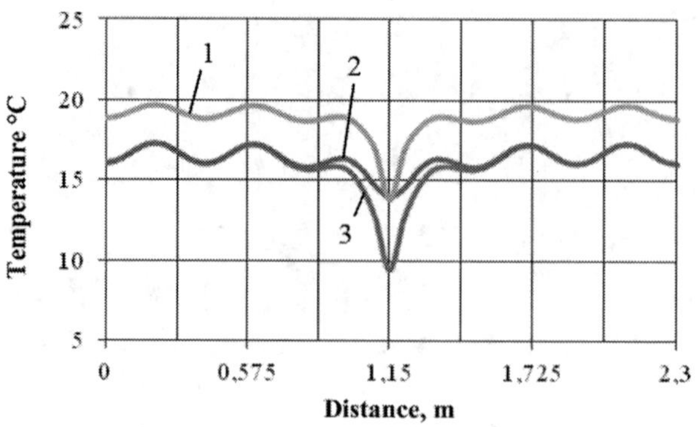

Fig. 6. Temperature curves for the internal surfaces of the structure produced according to variant 1: 1, 3 - curves that characterize distribution of temperatures on the internal surface of the wall without heating cable; 2 - curves that characterize distribution of temperatures on the internal surface of the wall with heating cable; 1 - thermal conductivity of lightweight concrete is 0.41 W/(m·°C); 2, 3 - thermal conductivity of lightweight concrete is 0.92 W/(m·°C).

Conclusion

The obtained research results show that mathematical modeling of thermal state of external envelope structures with the use of modern software packages enables to solve the problem of construction of buildings that satisfy ever increasing standard requirements regarding energy efficiency and to develop recommendations for design of such buildings.

For a more complex analysis of thermal protection properties of envelope structures other heat-stressed structural elements shall be studied except for corner fragments of buildings, such as joints between external wall, flooring plate and balcony plate, as well as window and door frames. It is also necessary to investigate thermal state of various heat-stressed elements under unstable conditions and extreme heat exchange conditions (intermittent heating, fluctuations of outdoor air temperature), and to study the issues of heat resistance.

The research was performed with the financial support of the Russian Foundation for Basic Research (project No. 16–48–700367 p_a).

References

1. D.M. Terent'ev, Energy-Savings, **3**, 18–21 (2015) (in Russian)
2. I.A. Bashmakov, Problems of Economic Transition, **3**, 75–98 (2016) (in Russian)
3. V.G. Gagarin, Vestnik MGSU, **3**, 152–157 (2011) (in Russian)
4. Z. Wenxiao., Information of China Construction, 11, 135–138 (2007)
5. I.K. Savin, Kommunal'nyj kompleks Rossii [The utilities complex of Russia], **10**, 12–14 (2008) (in Russian)
6. L. Peng, Manual of Energy Efficiency in Buildings of China (Building Technology Publ., Beijing, 2007)
7. A.E. Kiryudcheva, V.V. Shishkina, D.V. Nemova, Construction of Unique Buildings and Structures, 5, 19–30 (2016) (in Russian)
8. M.I. Nizovcev, V.T. Belyj, A.N. Sterlyagov, Construction: new technologies – new equipment, 7, 38–42 (2016) (in Russian)
9. M.I. Nizovcev, A.N. Sterlyagov, Stroitel'stvo i rekonstrukcija [Construction and Reconstruction], **6**, 112–119 (2016) (in Russian)

10. F. Ascione, N. Bianco, F. de' Rossi, G.P. Vanoli, Proceedings of the 2012 Winter Simulation Conference, 567–581 (2012)
11. A.N. Kozlobrodov, E.A. Ivanova, Vestnik of TSUAB, **1,** 133–139 (2016) (in Russian)
12. T.T. Zhang, Y.F. Tan, Y.Z. Li, Journal of Hunan University Natural Sciences, **42**(7), 114–120 (2015)
13. A.N. Khutornoi, A.Ya. Kuzin, A.V. Kolesnikova, N.A. Tsvetkov, Bulletin of the Tomsk Polytechnic University, **309,** 2, 188–192 (2006) (in Russian)
14. A.N. Khutornoi, A.V. Kolesnikova, N.A. Tsvetkov, A.Ya. Kuzin, RU Patent 49067 (2005) (in Russian)
15. N.D. Danilov, P.A. Fedotov, N.S. Akimova, D.F. Petrov, Eurasian Union of Sciences, **7,** 160–162 (2015) (in Russian)

Stress-strain state analysis and optimization of rod system under periodic pulse load

Grigory Grebenyuk[1,*], *Maxim* Veshkin[1], *Vladislav* Maksak[2], and *Olga* Nosireva[1]

[1]Novosibirsk State University of Architecture and Civil Engineering (Sibstrin), 630008 Novosibirsk, Russia
[2]Tomsk State University of Architecture and Building, 634003 Tomsk, Russia

Abstract. The paper considers the problem of analysis and optimization of rod systems subjected to combined static and periodic pulse load. As a result of the study the analysis method was developed based on traditional approach to solving homogeneous matrix equations of state and a special algorithm for developing a particular solution. The influence of pulse parameters variations on stress-strain state of a rod system was analyzed. Algorithms for rod systems optimization were developed basing on strength recalculation and statement and solution of optimization problem as a problem of nonlinear mathematical programming. Recommendations are developed for efficient organization of process for optimization of rod systems under static and periodic pulse load.

1 Introduction

To date, multiple investigations have been performed in the field of design and optimization of rod systems under static and dynamic load [1, 2, 8, 10-17]. In the majority of them, harmonic loads serve as dynamic ones. In a number of studies [2, 3, 5, 6] the emphasis is put on determination of optimal systems properties in optimization problems with constraints of strength, stiffness and lowest frequency of natural vibrations.

Previous research [7, 9-13, 15] were focused on decrease of labor intensity for optimization algorithms, first of all, due to more efficient organization (effective analysis and optimization methods, effective methods of structural state parameters approximation in iteration optimization processes, division of optimization process into levels and stages, etc.). In particular, the work [11] suggests combining the numerical procedure of dynamic analysis and the algorithm of finite difference approximation of structural state parameters to improve efficiency of optimization process for systems dynamic loading.

In recent years, a tendency to build processes of structural optimization without using approximation of structural state parameters appeared [16, 17]. However, development of efficient methods for analysis and optimization of structures under static and periodic pulse load is a relevant task. The purpose of this research is development of convenient and effective methods for analysis and optimization of rod systems under combined static and periodic pulse load.

* Corresponding author: greb@sibstrin.ru

2 Analysis procedure and optimization

2.1 Brief description of dynamic analysis procedure

Equation of equilibrium of a discrete dynamic system (without the account of damping and with regard to the method of finite elements displacement) is written as follows:

$$r_m \cdot \ddot{Z}(t) + r_e \cdot Z(t) + R_F(t) = 0, \tag{1}$$

where r_m , r_e – matrices of mass and stiffness, $R_F(t)$ – vector of dynamic load.

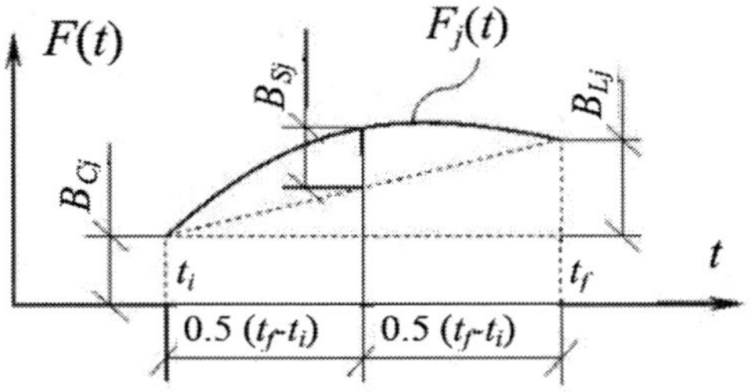

Fig.1. Approximation of dynamic action in a time interval.

Dependence of load value on time of interpulse interval is approximated by invariable, linear and sinusoidal components (Fig. 1).
In case when the interpulse interval exceeds the pulse duration, there are time intervals in which the system executes forced and free vibrations (Fig. 2).
In the first case (interval 1) the solution is as follows [1]:

$$Z(t) = \bar{Z}(t) + \tilde{Z}(t), \tag{2}$$

where – $\bar{Z}(t)$ – general solution of the homogeneous equation.

$$r_m \cdot \ddot{Z}(t) + r_e \cdot Z(t) = 0 \tag{3}$$

$\tilde{Z}(t)$ – particular solution (1), sought in the form of

$$\tilde{Z}(t) = Q \cdot L(t), \tag{4}$$

$L(t)$ – vector of approximating functions, Q – numerical matrix,
constants of integration are found from the initial conditions.

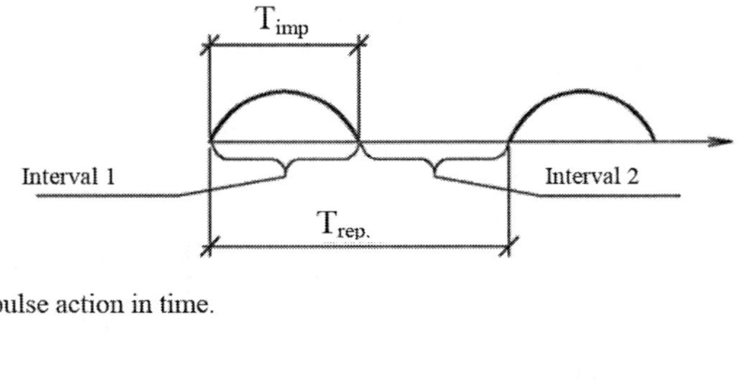

Fig. 2. Periodic pulse action in time.

In the second case (interval 2), the general solution is the solution of the homogeneous equation (3), whereas constants of integration are found from the conditions of the solutions interface in the intervals 1 and 2. Then, the recalculation procedure repeats, in which the initial conditions for defining constants of integration on interval 1 for the next period are the calculation results on interval 2 for the previous period.

A single-span pin-supported beam with its own mass ($\tilde{m} = 0.13659$ t/m) evenly distributed along the longitudinal axis and with three concentrated masses ($m = 0.5$ t) fixed on the longitudinal axis was used as an experimental structure subjected to periodic pulse load. Length of the beam is $l = 4$ m, its stiffness is $EI = 838400$ kN·m^2. Pulse loading was applied to the beam in the form of concentrated force. The size of pulse stayed the same while its shape and duration were varied. Two variants of pulse load application are considered: force at quarter-span (Fig. 3a); and force at mid-span (Fig. 3b).

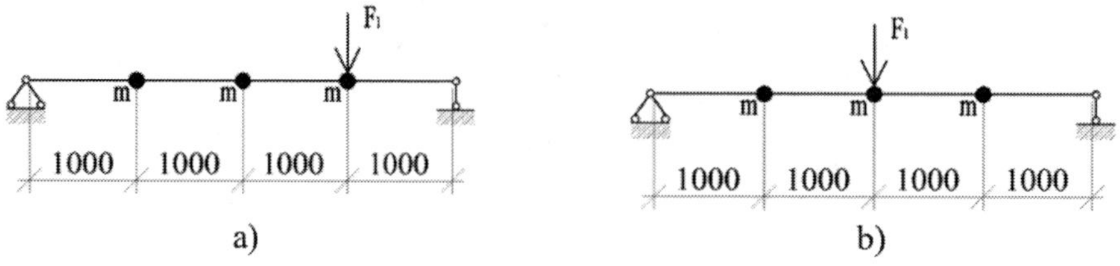

Fig. 3. Design schemes for beam: a) pulse action at quarter-span; b) pulse action at mid-span.

2.2 Analysis of effect of pulse frequency

Influence of pulse repetition frequency n_{rep} (with repetition period T_{rep}) on the maximum value of bending moment in the beam cross-section was analyzed at the initial stage of research. Sinusoidal shape of pulse was accepted for analysis (Fig. 2). First 40 pulses were studied at determined intervals. The analysis was performed with the help of DINAM software based on the method of finite elements displacement and developed at the Structural Mechanics Department of Novosibirsk State University of Architecture and Civil Engineering (Sibstrin). The software allows analyzing rod systems for arbitrary dynamic loads with calculation of natural vibration frequencies.

In the process of the beam vibration under pulse action the maximum values of bending moments were selected in the cross-sections taken every 0.2 m along the beam.

Analysis was performed at the constant duration and amplitude of force action ($T_{imp} = 0.008$ s, $F_{max} = 10$ kN). Frequency of pulse load varied from 3 Hz to 125 Hz, which corresponds to the pulse interval (T_{rep} from 0.33 s up to 0.008 s). The value of the force pulse was 0.05055 kN·s.

The considered frequency range of periodic pulse load contained 2 natural frequencies of the beam: f_1= 15.12 Hz, f_2= 60.13 Hz.

Figure 4 demonstrates a fragment of the dependence graph for bending moment amplitude on pulse frequency. The green line corresponds to the pulse applied to quarter-span, the pink one – to the pulse applied to mid-span.

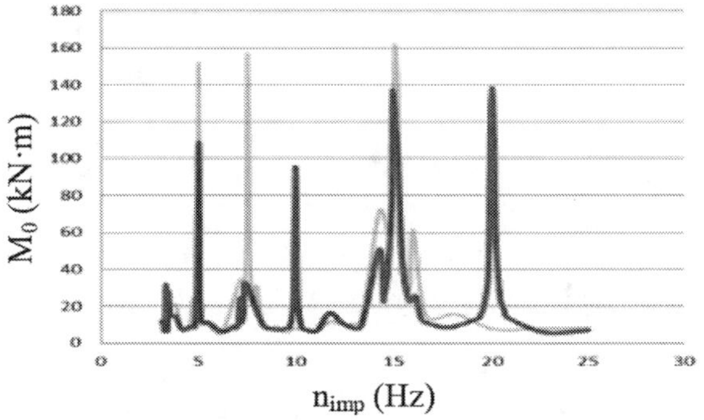

Fig. 4. Dependence of bending moments amplitude on pulse frequency.

As it was assumed, the resonance effect occurred near the natural frequency regions, but the peak values were slightly displaced with regard to natural frequencies. Beside, resonance phenomena were observed at the pulse frequencies divisible by frequencies of natural vibrations. These peaks of amplitudes are obviously caused by harmonic resonance.

2.3 Analysis of the effect of pulse shape and duration

The effect of pulse shape was analyzed on the example of the beam under load at quarter-span. Two pulse shapes were considered (Fig. 5.):

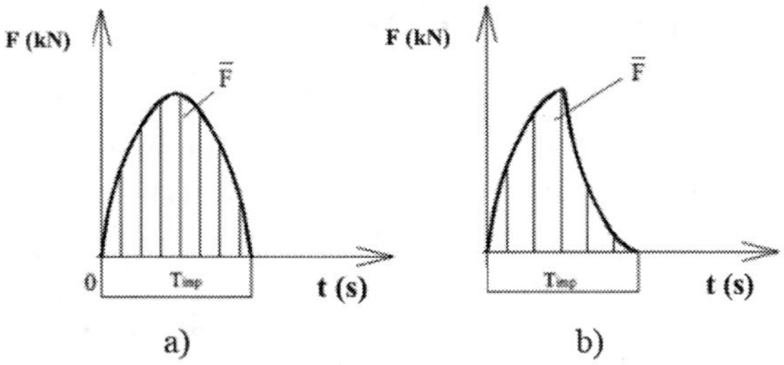

Fig. 5. a) sinusoidal pulse; b) wave pulse.

The pulse size is constant. Figure 6 demonstrates the graph with the analysis results under action of wave pulse (gray line) and sinusoidal pulse (red line). The graph clearly shows that wave pulse is the most dangerous. The maximum value of force under wave pulse action appears to be 30% higher than that under sinusoidal pulse action.

Then, the effect of pulse duration was investigated on the example of a beam under load at quarter-span. Figure 7 presents dependences of the maximum bending moment amplitude for various relations $\boldsymbol{T_{imp}} / \boldsymbol{T_{rep}}$, with

T_{rep} = 0.01663s corresponding to the second own frequency $n_2 = 60.13$;

T_{rep} = 0.044s – non-resonance case.

When increasing the pulse duration T_{imp} the maximum values of bending moment amplitude decrease. In some near-resonance points the graph possesses considerable nonlinearity, which is related to resonance effects.

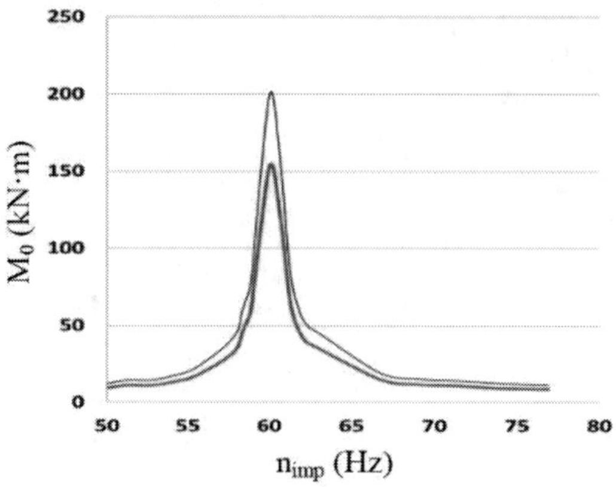

Fig. 6. Dependence of amplitude of bending moments vibrations on pulse frequency.

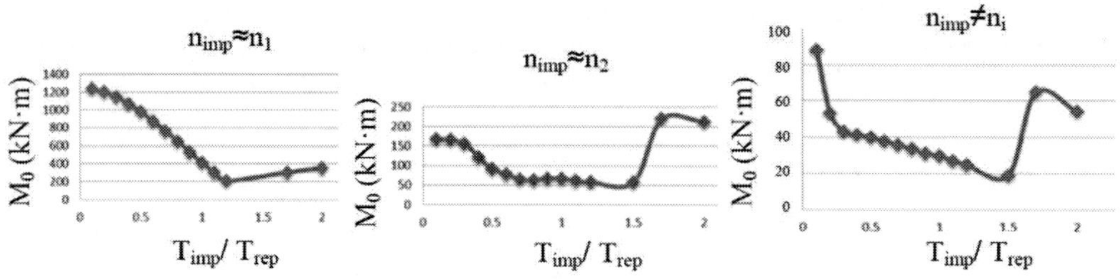

Fig.7. Dependence of M_0 on pulse duration for various values of n_{imp}.

As the result of the investigation wave pulse was found more dangerous than sinusoidal one. Drop of maximum bending moment amplitude is observed in the weakest cross-section when duration of pulse with set frequency was increased; however, in case of two pulses overlap the amplitude shows a sharp rise. Variation of pulse frequency also has a remarkable effect on the values of bending moments amplitudes.

These research outcomes were further used in statement of the frame structure optimization problem with the account of the most negative cases of periodic pulse loading.

2.4 Solution of optimization problem for a steel frame subjected to static and periodic pulse load

A five-floor double-span steel frame made of welded I-shaped elements was taken as an example for optimization problem statement for a structure subjected to combined static and periodic pulse load $q(t)$ (Fig.8). The floor height is 3 m, section span is 6 m. Wave pulse is considered.

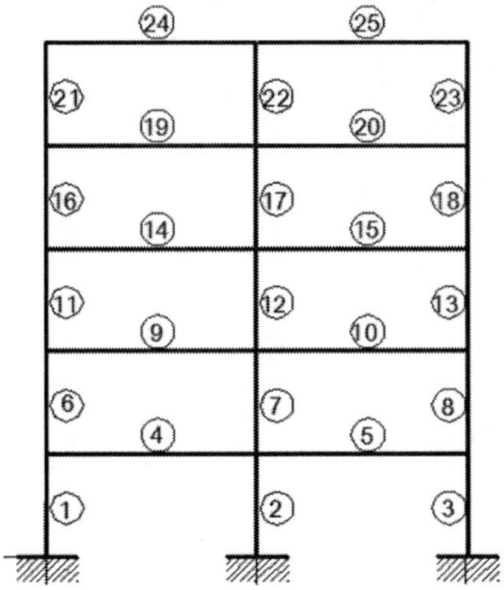

a)

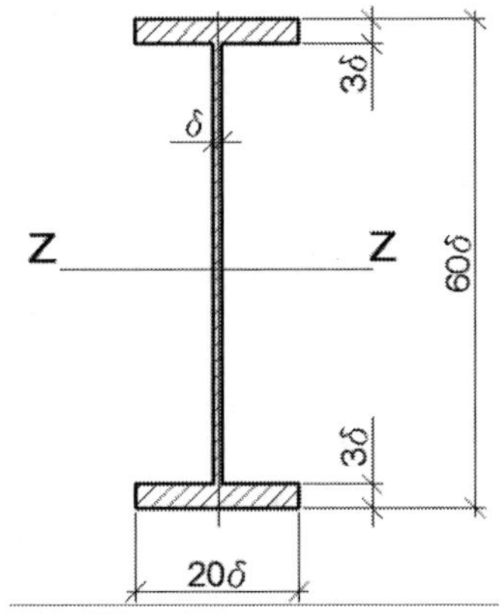

b)

Fig. 8. a) frame analysis scheme; b) shape of elements' cross-sections.

The frame consists of 25 elements (15 columns and 10 cross-beams). Frame loading scheme is showed in Figure 9. Characteristics of cross-sections of columns and cross-beams depend on the δ parameter: cross-section area $\boldsymbol{A} = 174 \cdot \delta^2$; cross-section's moment of inertia $\boldsymbol{I}_z = 110682 \cdot \delta^4$; cross-section's moment of resistance $\boldsymbol{W}_z = 3689.4 \cdot \delta^3$. The following parameters of the frame elements material were accepted: material – steel; material density $\rho = 7850\left(\text{kg/m}^3\right)$; elastic modulus $\boldsymbol{E} = 2.1 \cdot 10^5 (\text{MPa})$; design resistance $\boldsymbol{R} = 200\left(\text{MPa}\right)$.

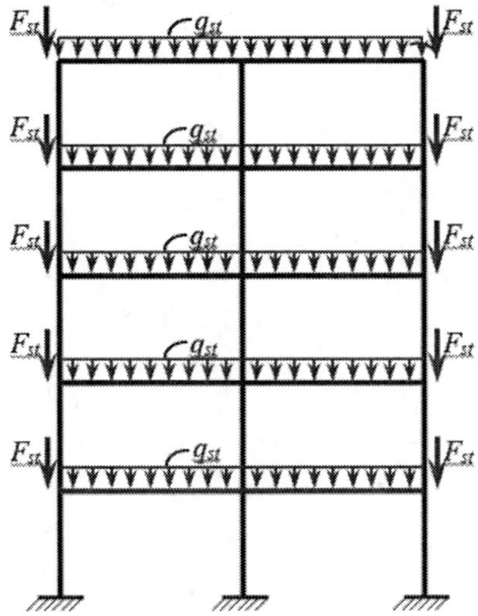

a)

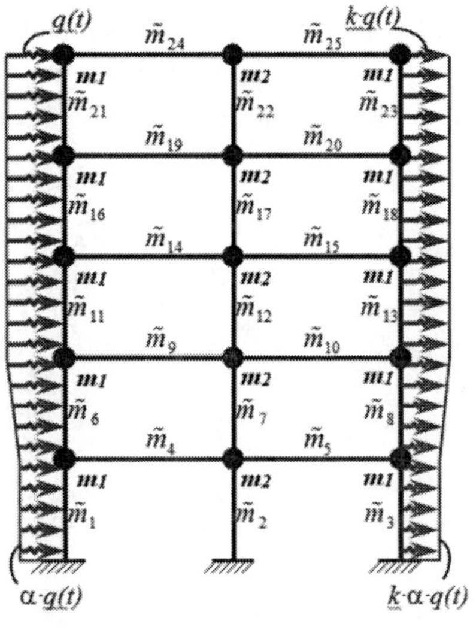

b)

Fig. 9. Frame loading scheme: a) static; b) dynamic.

2.4.1 Optimum design development based on iteration algorithms of strength recalculation

Admitted design with identical cross-sections of frame elements

The function of frame material volume was accepted as an objective function. With identical cross-sections of frame elements the function has the following form:

$$f(X)=(15\cdot l_{col}+10\cdot l_{cb})\cdot 174\cdot X_1^2, \tag{5}$$

where l_{col} – length of columns; l_{cb} – length of cross-beams.

The value of $X_1=\delta$ was recalculated based on fulfillment of the following condition:

$$\max_{i=1,\dots,25}\frac{\dfrac{|M_i|_{\max}}{3689.4\cdot X_1^3}+\dfrac{|N_i|_{cor.}}{174\cdot X_1^2}}{2\cdot 10^5}-1=0, \tag{6}$$

where $|M_i|_{\max}$, $|N_i|_{cor.}$ – maximum modulo value of bending moment and the modulo value of longitudinal force in i frame element.

The problem was solved in the Digital Visual Fortran environment. As the result of the performed analysis the admitted solution $X_1^*=0.00818$ m and the value of objective function $f(X^*)=1.22$ m^3 were obtained. Active strength constraint is in the second element. Constraints on frame joints displacements were disregarded.

Optimum design for two groups of frame elements (columns, cross-beams)

The function of frame material volume was accepted as an objective function.

$$f(X)=15\cdot l_{col}\cdot 174\cdot X_1^2+10\cdot l_{cb}\cdot 174\cdot X_2^2, \tag{7}$$

where $X_1=\delta_1$, $X_2=\delta_2$ – size parameters of columns and cross-beams. Strength constraints similar to (6) were obtained for each group of elements:

– for columns

$$\max_{\substack{i=NEG1(j)\\ j=1,\dots,15}}\frac{\dfrac{|M_i|_{\max}}{3689.4\cdot X_1^3}+\dfrac{|N_i|_{cor.}}{174\cdot X_1^2}}{2\cdot 10^5}-1=0, \tag{8}$$

– for cross-beams

$$\max_{\substack{i=NEG2(j)\\ j=1,\dots,10}}\frac{\dfrac{|M_i|_{\max}}{3689.4\cdot X_1^3}+\dfrac{|N_i|_{cor.}}{174\cdot X_1^2}}{2\cdot 10^5}-1=0, \tag{9}$$

where $NEG1(j)$, $NEG2(j)$ – number of groups of elements in the common list of frame elements.

The problem was solved in the Digital Visual Fortran software environment with the use of the developed IMPULSR software module. The following results were obtained while analyzing:

– the optimal values of the varied parameters $X_1^*=0.00819$ m, $X_2^*=0.00674$ m;

– the value of the objective function in the optimum design $f\left(X^*\right)=1.00\ \text{m}^3$.

<u>*Optimum design for ten groups of frame elements (columns and cross-beams on each floor)*</u>
While performing analysis the elements were divided into 10 groups with respect to the floors (5 groups of columns, *i*=3,5,7,9; 5 groups of cross-beams, *i*=2,4,6,8,10; – number of groups).

As before, the function of the frame material volume was accepted as objective function f(X).

$$f(X)=V(X)=\sum_{i=1,3,5,7,9} 3\cdot l_{col}\cdot 174\cdot X_i^2+\sum_{i=2,4,6,8,10} 2\cdot l_{cb}\cdot 174\cdot X_i^2 \tag{10}$$

where $X_i=\delta_i$ – varied parameter of i group of frame elements.

Strength constraints similar to (8) and (9) were obtained for each group of elements:
– for columns

$$\max_{\substack{j=NGEi(k),\\ k=1,2,3}} \frac{\dfrac{\left|M_{ij}\right|_{\max}}{3689.4\cdot X_i^3}+\dfrac{\left|N_{ij}\right|_{\text{cor.}}}{174\cdot X_i^2}}{2\cdot 10^5}-1\le 0,\ i=1,3,5,7,9; \tag{11}$$

– for cross-beams

$$\max_{\substack{j=NGEi(k),\\ k=1,2}} \frac{\dfrac{\left|M_{ij}\right|_{\max}}{3689.4\cdot X_i^3}+\dfrac{\left|N_{ij}\right|_{cor.}}{174\cdot X_i^2}}{2\cdot 10^5}-1\le 0,\ i=2,4,6,8,10. \tag{12}$$

where i – elements group number; $j=NGEi(k)$ – the number of group i in the common list of frame elements.

The values of varied parameters in the optimum design are given in Tables 1 and 2.

Table 1. Values of varied parameters in columns (mm).

X_1^*	X_3^*	X_5^*	X_7^*	X_9^*
8.03	7.07	6.34	5.94	6.18

Table 2. Values of varied parameters in cross-beams (mm).

X_2^*	X_4^*	X_6^*	X_8^*	X_{10}^*
6.61	6.58	6.55	6.59	6.49

The value of the objective function $f\left(X^*\right)=0.808\ \text{m}^3$.

The following conclusions can be made after the data analysis:

1. Parameters of column cross-sections decrease from first to fifth floor. Cross-sections of columns of fourth and fifth floors are slightly off the dependence, the cross-section of fifth floor column being slightly bigger than that of fourth floor column;
2. The parameters of cross-beam sections are quite stable and fluctuate between 6.61 mm and 6.44 mm. Due to this fact, cross-beams can be integrated in one group for all floors and unified.

Analysis performed with the use of the developed algorithm for strength recalculation with 25 independent varied parameters of frame elements allowed improving the optimum design with the minimum volume $f(X^*) = 0.765\ \text{m}^3$. However, such design leads to the increase of labor intensity for frame production and installation because of considerably higher number of unified groups. In all studied cases maximum horizontal displacement of top frame joints was higher than the set tolerance $[V] = 0.015\ \text{m}=15\ \text{mm}$. It also shall be noted that the use of strength recalculation algorithm does not account constraints on stiffness, stability and vibrations frequency. In this case, the optimization problem shall have the form of a problem for nonlinear mathematical programming [10, 12, 14, 17].

2.4.2 Solution of optimization problem in the form of mathematical programming problem with constraints on strength and stiffness.

Statement and solution of such problems is considered on the example of frame optimization with elements divided into two and ten groups.

<u>*Optimum design for two groups of frame elements (columns, cross-beams)*</u>
The frame optimization process is organized in iterations. The objective function and strength constraints are represented in relations (7), (8), (9); forces in elements are accepted as constant and are found by analyzing the frame in the initial search point. Stiffness constraint has the following form:

$$\frac{V_{max}(X,t)}{[V]} - 1 \le 0, \qquad (13)$$

where V_{max} – horizontal displacement of frame joints; $[V]$ – allowable displacement admitted to be equal to $0.001 \cdot H$, H – frame height. Linear local approximations $V_{max}^k(X,t)$ were built using Taylor series with retention of fixed and linear terms. Elements of gradient of the function $V_{max}(X,t)$ were defined in finite differences:

$$\frac{\partial V_{max}(X_0^k, t_{0P}^k)}{\partial X_i} \approx \frac{V_{max}(X_0^k + \Delta X_i, t_{0P}^k) - V_{max}(X_0^k - \Delta X_i, t_{0P}^k)}{2\Delta X_i}, \quad i=1,2, \qquad (14)$$

where X_0^k – initial point of search and the point of danger at iteration k; ΔX_i – step size at the variable X_i.

The following frame optimization results were obtained:
$X_1^* = 0.0087\ \text{m}$, $X_2^* = 0.00717\ \text{m}$; the value of the objective function in the optimum design is $f(X^*) = 1.14\ \text{m}^3$. The 14% increase in volume if compared to the optimum design which was found using the algorithm of strength recalculation can be explained by active stiffness constraint.

The following results were obtained when using the suggested algorithm and dividing the frame elements into 10 groups:

Table 3. Values of varied parameters in columns (mm).

X_1^*	X_3^*	X_5^*	X_7^*	X_9^*
9.1	8.8	8.2	8.1	7.9

Table 4. Values of varied parameters in cross-beams (mm).

X_2^*	X_4^*	X_6^*	X_8^*	X_{10}^*
7.2	7.1	6.9	6.8	6.7

The value of the objective function in the optimum design equals $f\left(X^*\right) = 0.92\ \text{m}^3$. The frame material volume increased by 13.8% compared to the optimum design previously developed using the strength recalculation algorithm without the account of stiffness constraints, which practically coincides with the case when the frame elements were divided into 2 groups.

3 Conclusion

The results of the conducted research demonstrate that the developed analysis method for rod systems subjected to combined static and periodic pulse load requires less labor intensity while providing highly accurate results. Due to analyzing the effects of various characteristics of periodic pulse loading the most negative factors were determined; they shall be accounted in statement and solution of optimization problems for structures subjected to combined static and periodic pulse loads.

An alternative to the iteration algorithm for development of optimization solutions based on approximation of state parameters is the optimization algorithm without approximations within the selected method of conditional extremum problem solution. Every time when it is required to find the values of state parameters included in expression of constraints one refers to the programs for solving system state equations corresponding to the point of search. This makes the optimization process much more labor intensive and emphasizes the need of simpler methods for recalculation of systems subjected to dynamic loads.

References

1. R. Klaf, D. Zh. Penzien, Dynamics of Structures, (Stroyizdat, Moscow, 1979) (in Russian)
2. L.S. Lyakhovich, A.V. Izhendeev, News of higher educational institutions. Construction, **7**, 11-14 (1998) (in Russian)
3. L.S. Lyakhovich, *Osobye svojstva optimal'nyh sistem i osnovnye napravlenija ih realizacii v metodah rascheta sooruzhenij* [Special properties of optimal systems and the main directions of their implementation in the calculating methods of structures] (TSUAB Publ.House, Tomsk, 2009) (in Russian)
4. L.S. Lyakhovich, Vestnik of TSUAB, **4(45),** 70-80 (2014) (in Russian)
5. R.P. Moiseenko, Vestnik of TSUAB, **2(43)**, 72-79 (2014) (in Russian)
6. R.P. Moiseenko, Stroitel'stvo [Civil Engineering], **6,** 123-126 (2008) (in Russian)
7. L.A. Schmit, Structural Optimization. Some Key Ideas and Insights, in: *New Directions in Optimum Structural Design,* 1-45 (Wiley, 1984)
8. I.B. Lazarev, *Osnovy optimal'nogo proektirovanija konstrukcij. Zadachi i metody* [Fundamentals of optimal structural design. Tasks and methods] (Siberian Transport University, Novosibirsk, 1995) (in Russian)
9. I.M. Rabinovich, *Raschet konstrukcij minimal'nogo vesa pod dejstviem kratkovremennyh dinamicheskih nagruzok* [Calculation of the structures with minimum weight under the effect of short-term dynamic loads] (VIA, Moscow, 1965) (in Russian)

10. T.T. Feng, J.S. Arora, E.J. Haug, Int. J. Numerical Methods in Eng., **1**, 39-52 (1977)
11. Kh. Jakimova, *Proektirovanie optimal'nyh dinamicheski nagruzhennyh konstrukcij. Novye napravlenija v stroitel'nom proektirovanii* [Designing of optimal dynamically loaded structures. New directions in construction design], 245-262 (Strojizdat, Moscow, 1989) (in Russian)
12. G.I. Grebenyuk, V.I. Roev, M.A. Dimitrov, Proceedings of the 5th All Russian Seminar *"Problems of optimal structural design"*, 119-131 (Sibstrin, Novosibirsk, 2005) (in Russian)
13. E. Hogg, K. Choi, V. Komkov, *Structural Sensitivity Analysis* (Mir, Moscow, 1998) (in Russian)
14. G.I. Grebenyuk, V.V., Bezdelev, *Voprosy dinamiki i prochnosti v mashinostroenii* [dynamics and strength issues in engineering], 34-40 (OmSTU, Omsk, 1983) (in Russian)
15. W.C. Mills-Curran, L.A. Schmit, AIAA Journal, **23(1),** 132–138 (1985)
16. G.I. Grebenyuk, M.S. Veshkin, Izvestiya of Altai State University, **1,** 36-39, (AltGU, Barnaul, 2012) (in Russian)
17. T.L. Dmitrieva, Le Chan Min Dat, Nguen Van Ty, *Realizacija algoritmov chislennoj optimizacii v sovremennyh programmnyh kompleksah* [Implementation of numerical optimization algorithms in modern software] (IRNTU, Irkutsk, 2015) (in Russian)

Calculation of road reliability based on mathematical theory of reliability

Vladimir Efimenko[1,*], *Rostislav* Moisejenko[1]

[1]Tomsk State University of Architecture and Building, 634003 Tomsk, Russia

Abstract. The paper presents the mathematical theory of reliability as related to the theory of road reliability. The concepts of "initial reliability" and "durability" are combined in a single algorithm. The presented theoretical method is implemented in two stages: calculation of initial reliability and calculation of durability. This sequence enables analytical calculation at given levels of initial reliability and durability of roads and their elements. It is shown that initial reliability and durability should be calculated separately. This method has proved to be effective in applied sciences dealing with the theory of structural reliability.

Introduction

Russian and foreign scientists pay great attention to calculating the reliability of flexible pavements. The key findings are presented in works by N. N. Ivanov [1], I. A. Zolotar [2, 3], S. Yu. Rokas [4], V. A. Semyonov [5], V. V. Stolyarov [6, 7]. Some findings of foreign scientists can be presented in papers [8, 9].

In works by Russian authors, some terms adopted in the mathematical theory of reliability tend to be interpreted in a different way. These differences are reflected in regulatory documents. For example, mathematicians define the concept of reliability as follows [10]: "Reliability means the ability of a technical facility to maintain its characteristics (parameters) within certain limits under certain operating conditions". Experts A. M. Polovko and S. V. Gurov draw attention to a common mistake in papers of Russian researchers: the concept of reliability is associated with operating time of a facility [10]. Operating time of a facility is characterized by durability. This concept is defined as follows [10]: "Durability is the ability of a technical facility to remain serviceable before the onset of the critical condition, subject to the established maintenance and repair system".

In most publications dealing with road reliability, the term "durability" is not used at all, since the terms "reliability" and "durability" are combined into a single concept [2]: "The road reliability as an integrated transport facility is its ability to provide safe design traffic at an average speed close to the optimal one during its standard or specified service life".

V. V. Stolyarov [6, 7] even replaces the theory of reliability by the theory of risk. Using the "risk" as a synonym of "failure", his theory does not propose anything new. A. M. Polovko and S. V. Gurov emphasize [10]: "At present, there is no scientific theory of risk:

* Corresponding author: svefimenko_80@mail.ru

fundamental concepts and definitions have not been formulated; mathematical models have not been developed; there is no engineering calculation method".

Incorrect formulations can lead to errors in calculating the reliability of structures. Therefore, it is important to further improve the theory of road safety. The solution of this problem will contribute, in particular, to formulation of the scientific language for the theory of road reliability. This language should correspond to other applied sciences dealing with the reliability of facilities and rely on the existing terms of the mathematical theory of reliability.

The aim of this study is to show that the road reliability can be calculated according to the general algorithm based on the existing mathematical theory of reliability.

Methodology

To achieve this aim, the following calculations were performed: (1) calculation of initial reliability (to determine the probability of failure-free operation (P_0) at initial numerical characteristics of random variables); (2) calculation of durability (to determine the probability of failure-free operation during time t). As an example, parameters of flexible pavement are calculated according to permissible elastic deflection.

The first problem is solved in the following sequence. The strength condition of the road structure according to the permissible elastic deflection is written as [3]:

$$E_{gen} \geq E_{req} . \quad (1)$$

As recommended by A. R. Rzhanitsyn [11], the condition (1) can be represented as:

$$g = E_{gen} - E_{req} \geq 0 , \quad (2)$$

where g is the bearing capacity reserve of a roadbed and pavement.

It is shown in [13] that the probability distribution function of random variables, such as elastic modulus, corresponds to the normal law.

The probability of failure-free operation (reliability index) under normally distributed random variables is calculated by the formula:

$$P(g > 0) = 0{,}5 + \Phi(\beta) , \quad (3)$$

where Φ is the Laplace integral; β is the reliability index.

$$\beta = \frac{\overline{E}_{gen} - \overline{E}_{req}}{s_g} , \quad (4)$$

where s_g is the standard deviation of the bearing capacity reserve; $\overline{E}_{gen}$ is the average value of the elastic modulus of the road structure; $\overline{E}_{req}$ is the average value of the required elastic modulus.

The second problem is solved in the following sequence. Durability function has the form [14]:

$$P(t) = P_0 e^{-\int_0^t \lambda(t)dt}, \quad (5)$$

where $\lambda(t)$ is the failure rate at time t.

The study [2] proves that the exponential law of failure rate is not applicable to calculating the reliability of roads. A number of studies [4, 5] note that the Weibull law is most appropriate for calculating parameters of roads [15]:

$$\lambda(t) = \alpha\lambda t^{(\alpha-1)}, \quad (6)$$

where λ is the parameter that has a dimension of failure rate (year^{-1}); α is the dimensionless parameter.

According to the expressions (5, 6), the durability function takes the form:

$$P(t) = P_0 e^{-\lambda t^{\alpha}}. \quad (7)$$

The mean time of failure-free operation is determined by the formula:

$$T_0 = P_0 \int_0^{\infty} e^{-\lambda t^{\alpha}} dt. \quad (8)$$

Results

Let us consider an example of calculation according to the proposed approach. For comparative calculation, we consider the example presented in [12]. The initial data are presented in Table 1.

Table 1. Parameters of pavement layers and materials.

Material	Layer thickness h, cm.	Elastic modulus E, MPa, calculated according to elastic deflection
Dense bituminous asphalt concrete, grade BND 90/130	5	2,400
Open-graded asphalt concrete of the same grade	6	1,400
Graded crushed stone, 40–80 mm	18	450
Crushed stone mix C1	16	300
Sand-and-gravel mix C4	22	275
Sandy loam, $W_p = 0.68W_T$	–	44.6

By reducing multilayer pavement to two-layer one, we can calculate the parameters using the appropriate nomogram: $E_{gen} = 312$ MPa; $E_{min} = 202$ MPa; $K_{str} = 1.54$.

To calculate the reliability by formulas (3, 4), it is necessary to determine the standard deviation s_g [7]:

$$s_g = \sqrt{(C_{E_{gen}} \cdot E_{gen})^2 + (C_{E_{req}} \cdot E_{req})^2}\,, \tag{9}$$

where $C_{E_{gen}}$ is the coefficient of variation of the general elastic modulus of the pavement; $C_{E_{req}}$ is the coefficient of variation of the required elastic modulus of the structure.

The study [5] shows that if the probability of the random pavement parameter changes according to the normal law, then the coefficient of variation of this parameter is limited to $C_g < 0.2$. Assume that $C_{E_{gen}} = 0{,}19$; $C_{E_{req}} = 0{,}15$. Then the standard deviation is

$$s_g = \sqrt{(C_{E_{gen}} \cdot E_{gen})^2 + (C_{E_{req}} \cdot E_{req})^2} = \sqrt{(0.19 \cdot 312)^2 + (0.15 \cdot 202)^2} = 66.57 \text{ MPa.}$$

According to the expression (4), the average value of the bearing capacity reserve of the road structure is equal to $\bar{g} = \bar{E}_{gen} - \bar{E}_{req} = 312 - 202 = 110$ MPa. The reliability index is equal to:

$$\beta = \frac{\bar{E}_{gen} - \bar{E}_{req}}{s_g} = \frac{\bar{g}}{s_g} = \frac{110}{66.57} = 1.65\,.$$

The probability of failure-free operation of the road structure is calculated by the formula (3):

$$P(g > 0) = 0{,}5 + \Phi(\beta) = 0{,}5 + \Phi(1.65) = 0.5 + 0.4505 = 0.9505\,.$$

At the coefficient K_{str} = 1.54, P = 0.9505. According to Table 8.1 [12], at K_{str} = 1.29, the probability of failure-free operation of the structure is equal to P = 0.98. The discrepancy between the results is explained by different values of the coefficient of variation. The coefficients of variation can be calculated so that the calculated probability in the example will be greater than the tabulated probability.

However, the result in the example is not flawless, since the E_{min} module is determined at the total load for the entire service life of the road, while the E_{gen} module is determined for the initial period of operation. Therefore, the difference ($E_{gen} - E_{min}$), i.e. the bearing capacity reserve (g), is overestimated, which leads to an increase in the reliability index (β) and an increase in the probability of failure-free operation (P).

For comparison, below is the calculation made using the Barber's formula, which approximates the nomogram for layer-by-layer calculation of pavement with sufficient accuracy [2]:

$$E_{gen} = \frac{E_1 E_2 \left[1 + \left(\frac{2h}{D}\right)^2 \left(\frac{E_1}{E_2}\right)^{2/3}\right]^{1/2}}{E_1 - E_2 \left\{1 - \left[1 + \left(\frac{2h}{D}\right)^2 \left(\frac{E_1}{E_2}\right)^{2/3}\right]^{1/2}\right\}}\,. \tag{10}$$

The calculation from bottom to top gives the following results: 5th layer – soil ⟶ E_{gen} = 87.2 MPa; 4th–5th layers ⟶ E_{gen} = 120.77 MPa; 3rd–4th layers ⟶ E_{gen} = 179.54 MPa; 2nd–3rd layers ⟶ E_{gen} = 208.412 MPa; 1st–2nd layers ⟶ E_{gen} = 240.52 MPa;

Recalculation of the probability gives the following result.

$$\bar{g} = \bar{E}_{gen} - \bar{E}_{req} = 240{,}52 - 202 = 38{,}52\,\text{MPa} \rightarrow$$

$$s_g = \sqrt{(C_{E_{gen}} \cdot E_{gen})^2 + (C_{E_{req}} \cdot E_{req})^2} = \sqrt{(0{,}19 \cdot 240{,}52)^2 + (0{,}15 \cdot 202)^2} = 54{,}83\,\text{MPa} \rightarrow$$

$$\rightarrow \beta = \frac{\bar{g}}{s_g} = \frac{38{,}52}{54{,}83} = 0{,}7 \rightarrow P(g > 0) = 0{,}5 + \hat{O}(\beta) = 0{,}5 + \hat{O}(0{,}7) = 0{,}5 + 0{,}258 = 0{,}758.$$

The results of calculations according to the nomogram and Barber's formula (10) differ significantly. Therefore, the conclusion about the sufficient accuracy of the Barber's formula [2] needs to be corrected.

Durability can be calculated by formulas (6, 7, 8) at known values of the parameters (α, λ) in the Weibull function (6). These parameters can be determined only by experiment [5, 16]. Despite numerous experimental studies of pavement strength published in the literature, the function of road failure rate has not been studied yet. This is due to the fact that the durability function (7) is not used in modern calculations of road safety at all.

To complete the example, approximate values are assigned to the parameters (α, λ). Assume that values of the failure rate function for the first and tenth years of operation are known: $\lambda(1) = 0.001$ (year^{-1}); $\lambda(10) = 0.1$ (year^{-1}). Then, using the function (6), two equations are set up:

$$\alpha \cdot \lambda \cdot 1^{(\alpha-1)} = 0.001.$$

$$\alpha \cdot \lambda \cdot 10^{(\alpha-1)} = 0.1.$$

From these equations, the roots can be determined: $\alpha = 3$, $\lambda = 0.000\,(3)$. The durability function (7) takes the form:

$$P(t) = P_0 e^{-\lambda t^{\alpha}} = 0.9505 \cdot e^{-0{,}000(3)t^3}.$$

Subject to formula (8), the mean time of failure-free operation is calculated by the formula

$$T_0 = P_0 \int_0^{\infty} e^{-\lambda t^{\alpha}}\, dt = 0{,}9505 \int_0^{\infty} e^{-0{,}000(3)t^3}\, dt.$$

Numerical integration by trapezoidal method gives the result: $T_0 = 12.05$ (year). The obtained result agrees with the data in Table 4.6 [12], from which it follows that the service life of a projected road is $T_{sl} = 12$–16 years. Therefore, the assumed values $\lambda(1) = 0.001$ (year^{-1}); $\lambda(10) = 0.1$ (year^{-1}) correspond to regulatory recommendations.

Conclusion

The paper shows that the mathematical theory of reliability is effective for calculating the reliability of roads. A large amount of statistical data accumulated in regions of Russia makes it possible to develop theoretical methods for calculating the reliability of pavements [12]. The theoretical method is implemented in two stages: calculation of initial reliability and calculation of durability. Only this sequence enables analytical calculation at given levels of initial reliability and durability of roads and their elements.

The materials of the paper can be useful in developing new regulatory documents and conducting experimental studies to determine the parameters included in the failure rate function.

References

1. N.N. Ivanov, *Konstruirovanie i raschjot nezhjostkih dorozhnyh odezhd* [Design and calculation of flexible pavement] (Transport, Moscow, 1973) (in Russian)
2. I.A. Zolotar, V.K. Nekrasov, S.V. Konovalov, Yu.M. Yakovlev, M.S. Koganzon, *Povyshenie nadjozhnosti avtomobil'nyh dorog* [Improving of the road reliability] *Повышение надёжности автомобильных дорог* (Транспорт, Москва, 1977) (in Russian)
3. I.A. Zolotar, *Obespechenie nadjozhnosti avtomobil'nyh dorog po prochnosti pri ih proektirovanii, stroitel'stve i ekspluatacii* [Ensuring the roads reliability for strength by their design, construction and operation] (Military Academy of Domestic Front and Transport, St. Petersburg, 1996) (in Russian)
4. S.Yu. Rokas, *Statisticheskij kontrol' kachestva v dorozhnom stroitel'stve* [Statistical quality control in road construction] (Transport, Moscow, 1977) (in Russian)
5. V.A. Semyonov, *Kachestvo i odnorodnost' avtomobil'nyh dorog* [Quality and roads homogeneity] (Transport, Moscow, 1989) (in Russian)
6. V.V. Stolyarov, *Proektirovanie avtomobil'nyh dorog s uchjotom teorii riska* [Design of highways with regard to the risk theory], part I (SSTU, Saratov, 1994) (in Russian)
7. V.V. Stolyarov, *Proektirovanie avtomobil'nyh dorog s uchjotom teorii riska* [Design of highways with regard to the risk theory], part II (SSTU, Saratov, 1994) (in Russian)
8. M.I. Dater, W.R. Hudson, R.C.G. Haas, Selection of optimal pavement designs considering reliability, performance and costs, in Transp. Res. Rec., 67–79 (1974)
9. M. Matsuo, K. Kuroda, Soils and Foundations, **14**, 2 (1974)
10. A.M. Polovko, S.V. Gurov, *Osnovy teorii nadjozhnosti* [Fundamentals of the reliability theory] (BHV-Peterburg, St. Peterburg, 2006) (in Russian)
11. A.R. Rzhanitcyn, *Teorija raschjota stroitel'nyh konstrukcij na nadjozhnost'* [Calculating theory of structures for reliability] (Strojizdat, Moscow, 1978) (in Russian)
12. O.P. Afinogenov, V.N. Efimenko, S.V. Efimenko, *Konstruirovanie i raschjot dorozhnyh odezhd* [Designing and calculation of road pavement] (Kuzbassvuzizdat, Kemerovo, 2008) (in Russian)
13. V.K. Apestin, A.M. Shak, Yu.M. Yakovlev, *Ispytanie i ocenka prochnosti nezhjostkih dorozhnyh odezhd* [Tests and evaluation of the strength of non-rigid pavements] (Transport, Moscow, 1977) (in Russian)
14. Rekomendacii po ocenke nadjozhnosti stroitel'nyh konstrukcij [Recommendations for evaluation the reliability of building structures] (Uralskiy Promstrojniiproekt, Sverdlovsk, 1974) (in Russian)
15. B.V. Gnedenko, Yu.K. Belyaev, A.D. Solovyov, *Matematicheskie metody v teorii nadjozhnost* [Mathematical methods in the theory of reliability] (Nauka, Moscow, 1965) (in Russian)
16. S.I. Malafeev, *Nadjozhnost' tehnicheskih sistem. Primery i zadachi* [Reliability of technical systems. Examples and tasks] (Lan Publ., St. Petersburg, 2012) (in Russian)

Influence of metal frame on heat protection properties of a polystyrene concrete wall

Nikolay Tsvetkov[1,*], *Andrei* Khutornoi[1], *Alexandr* Kozlobrodov[1], *Sergei* Romanenko[2], *Yuri* Shefer[2], and *Artem* Golovko[1]

[1]Tomsk State University of Architecture and Building, 634003 Tomsk, Russia
[2]National Research Tomsk Polytechnic University, 634050 Tomsk, Russia

Abstract. The use of novel thermal-efficient building materials and technologies that allow increasing the level of thermal protection of external envelope structures and reducing the time for construction are of practical interest and represent a relevant task in the conditions of rapidly changing and increasing requirements to energy efficiency of buildings. This research aims at simulating the process of spatial heat transfer in a multilayer non-uniform structure of an external cast-in-place framed wall produced from polystyrene concrete with a stay-in-place formwork. Based on the physico-mathematical model developed with the use of ANSYS and COMSOL software complexes, parametric analysis of the impact of various factors on thermal behavior of the external wall was performed with the account of heat-stressed frame elements. The nature of temperature fields distribution in a polystyrene concrete structure was defined, and its thermal protection properties were investigated. The impact of a metal frame on thermal protection properties of a wall was found to be insignificant.

Introduction

The issue of energy efficiency improvement in buildings under construction with regard to the price-quality ratio and reduction of the negative effect of energy-related technologies on the environment is the high-priority task all over the world. Low-rise buildings in Russia have high energy saving potential, for its share in the country exceeds 50% of the total number of newly built constructions [1]. Since heat loss in buildings through external walls can reach up to 45% [2], search for novel thermal-efficient building materials and technologies that provide high level of thermal protection of external envelope structures is a relevant task.

One of the techniques of constructing main building frame that is gaining popularity is construction of cast-in-place polystyrene concrete walls. This composite material consists of cement, mineral and porous filler, water and several admixtures to provide such properties as strength, decay resistance, water resistance, etc. [1].

* Corresponding author: nac.tsuab@yandex.ru

Due to low thermal conductivity of polystyrene concrete a conclusion can be made that thermal protection of buildings with such walls will be high enough and will allow constructing buildings in Siberia without additional insulation. However, metal frame put inside the wall structure may have negative impact on thermal protection properties of external walls [3, 4-8].

In this regard, the purpose of this paper is evaluation of metal frame elements' impact on thermal protection properties of cast-in-place external walls, which is a relevant task with high practical importance.

Numerical experiment

Polystyrene concrete is one of prospective building materials, [9, 10] so is the cast-in-place construction technology with a stay-in-place formwork [11, 12]. In [3], the technique is suggested for construction of quickly erected frame buildings from polystyrene concrete, structural scheme of which is presented in Figure 1. External wall of a building contains internal load-bearing frame consisting of columns 2 joint by cross-beams 3, with fixed formed studs 5, and a foundation 1 (Fig.1). For low-rise buildings (1-3 floors) load-bearing structures, columns, cross-beams and flooring beams can be produced, for example, from steel tubes with the cross-section of 50×100 mm and the wall thickness of 4 mm joint by welding. If necessary, in order to increase the bearing capacity, fire resistance and seismic resistance steel tubes can be filled with heavyweight and/or lightweight concrete. The complete technology of the building construction is presented in [3].

The subject of the study described in this paper is a cast-in-place wall produced from D200 polystyrene concrete with the thickness of $\delta 1$ (Fig. 2), operating in extreme conditions. The wall structure includes a stay-in-place formwork from magnesium oxide boards with the thickness of $\delta 2 = 10$ mm, fixed to U-shaped studs with the size of 50×50 mm and thickness of $\delta 3 = 0.4$ mm. The load-bearing frame is produced from tubes with rectangular cross-section, the size of 50×100 mm, and thickness of $\delta 4 = 4$ mm.

Numerical experiments were performed with the use of ANSYS Mechanical APDL 17.0. and COMSOL Multiphysics 5.2 software complexes and were aimed at defining the impact of studs and location of internal metal frame inside the wall on the common thermal state of the structure.

Thermal state is investigated for three types of wall thickness: variant 1 – $\delta 1 = 0.6$ m; variant 2 – $\delta 2 = 0.45$ m; variant 3 – $\delta 1 = 0.3$ m. The location of metal frame inside the wall is different for each variant: variant A – at the distance of 100 mm from the internal surface of the wall; variant B – in the center of the wall; variant C – 100 mm from the external surface of the wall.

Physico-mathematical statement of a problem

For mathematical problem statement, geometrical parameters of envelope structure elements and their thermophysical characteristics presented in Table 1 are considered given, generally they may depend on the temperature.

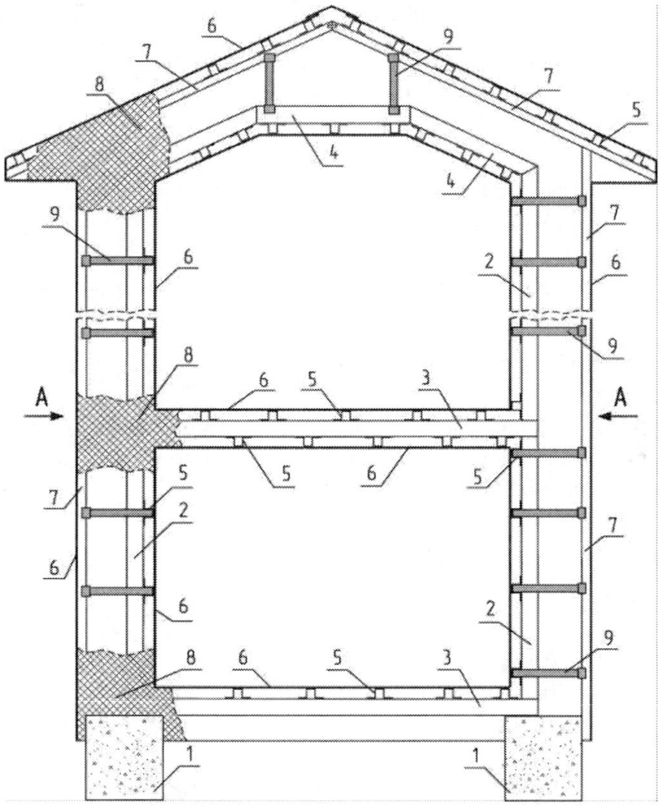

Fig. 1. Cross-sectional view of the building: 1 – foundation; 2 – column of internal frame; 3 – cross-beam of internal frame; 4 – flooring beam; 5 – formed stud; 6 – formwork board; 7 – external frame; 8 – lightweight concrete; 9 – temporary spacer.

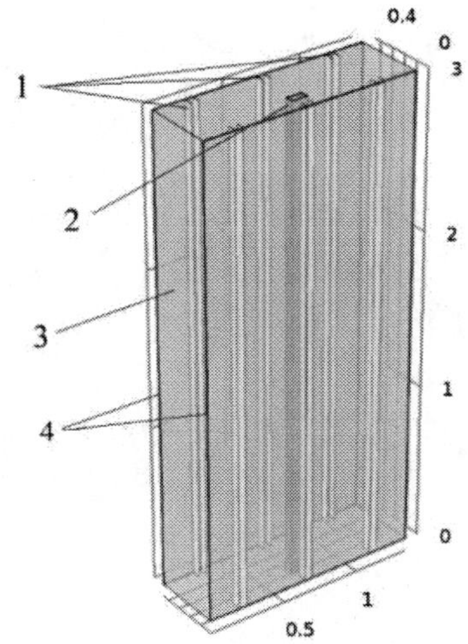

Fig. 2. Geometrical parameters of the structure: 1 – U-shaped stud; 2 – steel tube with rectangular cross-section; 3 – polystyrene concrete; 4 – magnesium oxide formwork.

Table 1. Values of thermal conductivity coefficients of wall materials.

No.	Material	Thermal conductivity coefficient λ, W/(m·°C),
1	Metal profile	57
2	Steel tube	50
3	Polystyrene concrete	0.07
4	Magnesium oxide	0.24

Established parameters are temperatures of outdoor t_{otd} and indoor t_{ind} air, as well as thermal conductivity coefficients for external and internal surfaces. The indoor and outdoor temperature values are +20 and -39 °C, and heat transfer coefficients for external and internal surfaces, respectively, are α_{otd}= 23 W/(m^2·°C) and α_{ind} = 8.7 W/(m^2·°C).

Stable spatial heat transfer in the studied wall fragment is described in Cartesian coordinate system as a system of nonlinear three-dimensional equations for thermal conductivity with corresponding boundary conditions for all structural elements' surfaces [8].

Calculation results

As the result of numerical study, the influence of studs and location of metal frame inside the wall on the common thermal state of the studied structure was determined.

Figure 3 illustrates temperature fields in the wall section for variant 1A in the raster image without (a) and with the account of studs (b).

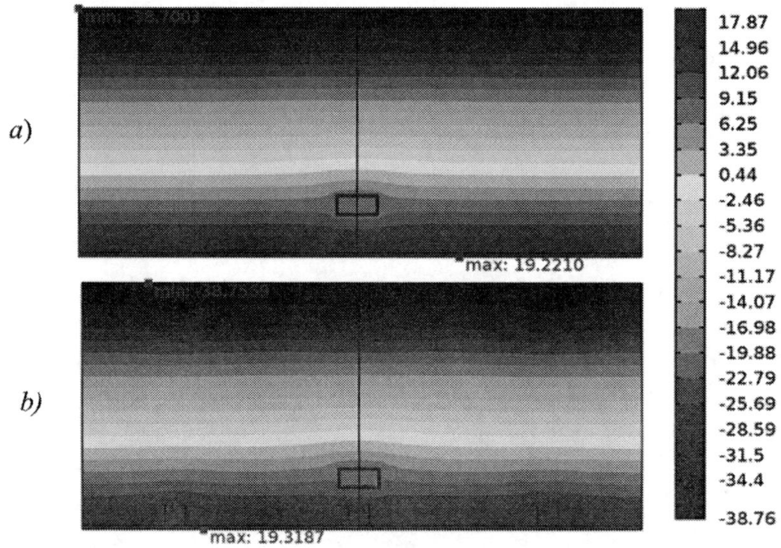

Fig. 3. Temperature field in the analyzed wall fragment: *a* – without studs; *b* – with studs

Comparative analysis of temperature fields in the considered variants of the wall structure demonstrates that presence of studs does not cause any remarkable variation in temperature field and their impact can be referred to as inconsiderable.

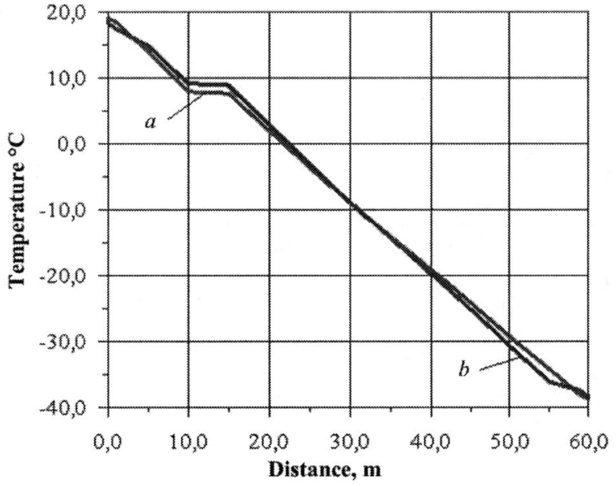

Fig. 4. Temperature distribution in the wall cross-section: *a* – without studs; *b* – with studs

This is supported by the calculation results given in Figure 4. It is visible from the Figure that the temperature difference of about 1 °C is observed in regions located around 10-15 cm away from the internal and external wall surfaces.

Based on such result a conclusion can be made that the impact of studs can be neglected when analyzing more geometrically complex spatial structures. This will enable to build a more uniform grid and simplify the task solution without considerable increase of error while dividing the analyzed area into finite elements.

Figures 5 and 6 demonstrate distribution of temperature in the wall thickness with different bearing frame locations for variants 1 and 2. According to the analysis of the presented results, the visible temperature variation is observed around the frame tubes. The maximum temperature difference in the studied cases is 5 °C. Such behavior of temperature distribution in the wall indicates the fact that location of the load-bearing frame also has no significant effect on thermal state of the structural fragment under study. However, if we consider the obtained data on temperature distribution from the point of creation of favorable microclimate in premises, location of the bearing frame in proximity to the internal wall is preferable, since half of the wall is in a higher temperature.

Figures 7 and 8 illustrate respectively the temperature field in a raster image and a vector image of the heat flow rate for a wall with the thickness of 0.6 m and a bearing frame located in the center.

The values of temperature and heat flow rates showed in these Figures visualize the thermal state of an envelope structure in different points of the studied wall fragment and enables to evaluate the impact of size and location of bearing frame elements.

The temperature of the internal surface of the wall for the considered case is higher than 19°C; it remarkably increases the dew point temperature which equals 15.2 °C even with quite high relative humidity $\varphi = 75$ %.

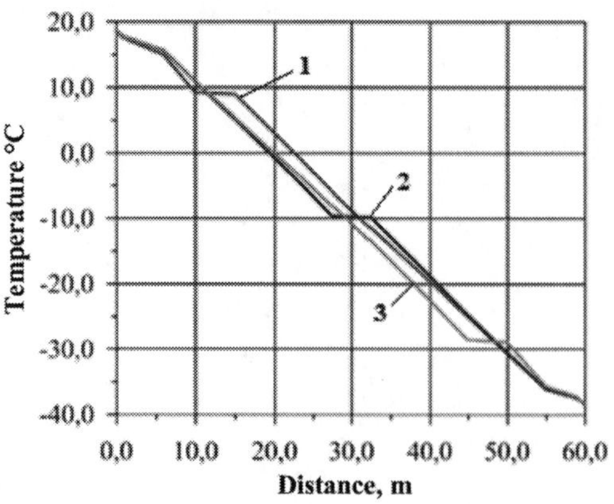

Fig. 5. Temperature distribution in the wall cross-section with different frame locations for variant 1:
1 – internal part of the wall; 2 – center; 3 – external part of the wall

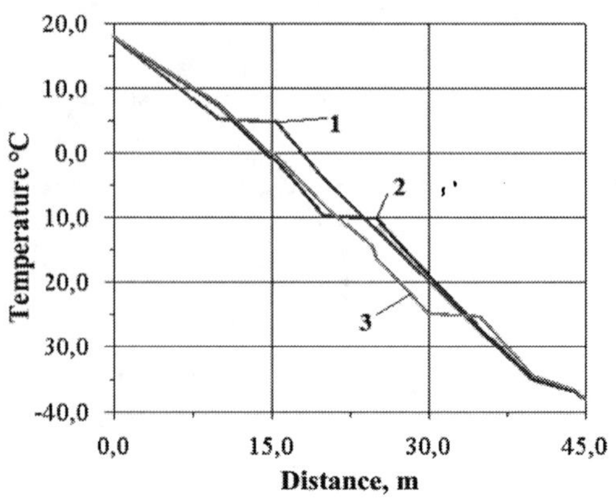

Fig. 6. Temperature distribution in the wall cross-section with different frame locations for variant 2:
1 – internal part of the wall; 2 – center; 3 – external part of the wall

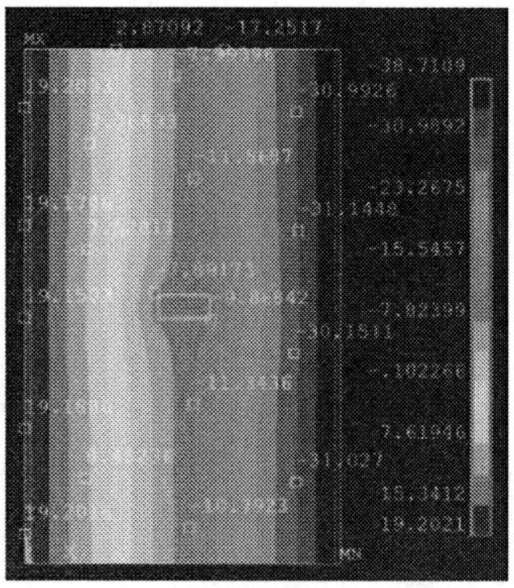

Fig. 7. Temperature field in the wall with the thickness of δ = 0.6 m

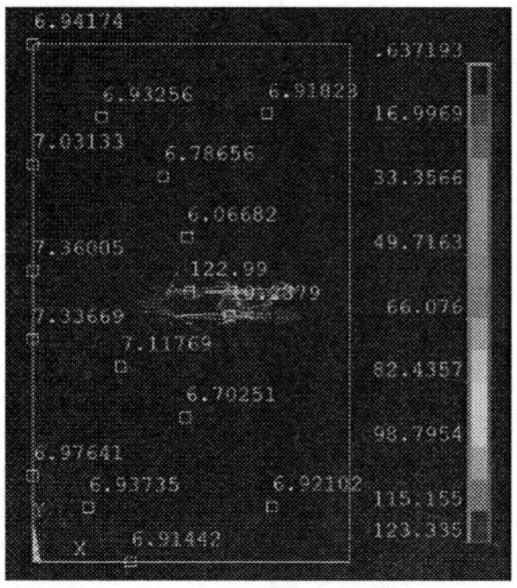

Fig. 8. Vector image of heat flow rate in the wall with the thickness of δ = 0.6 m

Since relative humidity of indoor air does not normally exceed 60%, this confirms operation efficiency of the analyzed types of envelope structures from the point of condensate formation on the internal surface of the studied types of walls in the given climate conditions. Similar situation is observed for other studied structures of walls with the thickness of 0.45 and 0.3 m. Due to performed analysis it was found that for a cast-in-place wall made of polystyrene concrete even with the thickness of δ = 0.3 m in the most heat-stressed region of a structure the temperature on the internal surface of the wall is higher than 16°C.

For the climate conditions of Tomsk city the standard value of thermal resistance of external walls currently is 3.75 m^2·°C/W, whereas for the studied structured from polystyrene concrete such values are 4.08, 6.23 and 8.37 m^2·°C/W for walls with the thickness of δ = 0.3, 0.45 and 0.6 m respectively.

Thus, cast-in-place structure of an external wall produced from polystyrene concrete meets modern standard thermal protection requirements.

Conclusion

The obtained research results show that metal profile studs sized 50×50 mm have no significant impact on thermal state of structures for all considered types of walls; this enables to simplify the task solution without considerable increase of error and neglect such impact in further research.

It was found that location of load-bearing frame inside the wall also has insignificant effect on common thermal state of a structure. However, it is reasonable to install the bearing frame closer to the internal surface of the envelope structure for creation of a favorable microclimate in premises.

Based on the obtained numerical study results a conclusion can be made that the impact of metal studs can be neglected while performing further research on unstable thermal processes and while analyzing more geometrically complex spatial structures, including various heat-stressed elements. This will enable to build a more uniform grid while dividing the analyzed area into finite elements, reduce the analysis time and simplify the task solution without considerable increase of error.

The research was performed with the financial support of the Russian Foundation for Basic Research (project No. 16–48–700367 p_a).

References

1. T. Fadeeva, *Proceedings of International Scientific Conference "Innovative Development of Modern Science"*, **4**, 286-288 (2014) (in Russian)
2. E. Malyavina, Teplopoteri zdanija. Spravochnoe posobie [Heat losses in buildings. Reference book], **144** (2011) (in Russian)
3. Yu.V. Shefer, Bystrovozvodimoe jenergojeffektivnoe karkasnoe zdanie [Quickly erected energy efficient frame buildings patent], RF Patent 2503781 (2014) (in Russian)
4. A. Kozlobrodov, D. Tsvetkov, Vestnik of TSUAB, **3**, 298–307 (2013) (in Russian)
5. AISI–S100. North American Specification for the Design of Cold-Formed Steel Structural Members, 149 (American Iron and Steel Institute, Washington, 2001)
6. EN 1993–1–3:2006. Eurocode 3: Design of steel structures. Part 1-3: General rules. Supplementary rules for coldformed members and sheeting (Brussels, 2006)
7. M.V. Leshhenko, V.A. Semko, Magazine of Civil Engineering, **8**, 44-55 (2015) (in Russian)
8. Y. Bazhenov, E. Korol, V. Erofeeva, E. Mitina, *Ograzhdajushhie konstrukcii s ispol'zovaniem betonov nizkoj teploprovodnosti. Osnovy teorii, metody rascheta i tehnologicheskoe proektirovanie* [Envelope structures produced from low thermal conductivity concrete. Fundamental theory, analysis methods and technological design] (2008) (in Russian)
9. V. Sokolov, *Konstruirovanie kompleksnyh paro-, teplo- i gidroizoljacionnyh polistirolbetonov* [Design of complex insulating polystyrene concretes resistant to steam, heat and water], 200 (2015) (in Russian)
10. A. Tsydypova, L. Urkhanova, *Proceedings of the Conference "Knowledge intensive technologies and innovations"*, 410-417 (2014) (in Russian)
11. V. Yakubson, Magazine of Civil Engineering, **2**, 7-8 (2009) (in Russian)
12. A. Zolotkov, Magazine of Civil Engineering, **1**, 14-21 (2012) (in Russian)

Prevention of uncontrolled progressive collapse of a high-rise brick building

Oleg Pakhmurin[1,*]

[1]Tomsk State University of Architecture and Building, 634003 Tomsk, Russia

Abstract. In recent years, there is a trend towards increasing the failure rate of buildings and structures as a result of unforeseen situations. This study is aimed at assessing the operational reliability of the structure to prevent emergencies and progressive collapse. The behavior of structures affected by various factors was analyzed. The operational condition of a building with stone walls and a rigid structural frame constructed in the middle of the 20th century was considered. By means of a detailed instrumental examination, effects of long-term operation, climatic factors and violations during construction on the serviceability of the building were analyzed. Geotechnical conditions of the site were investigated. Necessary re-calculations for load-bearing structural elements and foundations were carried out. An expected failure diagram for vertical load-bearing structures of the building is presented. Proposals for preventing progressive collapse were developed on the basis of the author's technique.

Introduction

Due to various factors, the operational condition of buildings aged 50 years old or above does not always meet the regulatory requirements. These factors include climatic and industrial factors, period of operation, violations of operating rules, as well as deviations from the design and building regulations. This is especially true for buildings from such building materials as brick, mortar and wood.

Operational reliability assessment of buildings and their structural elements are addressed by a number of scientists and specialists such as V.S. Abrashitov, M.D. Boyko, A.N. Dobromyslov, O.V. Kabantsev, V.G., Kazachek, M. Kusainov, A.I. Malganov, V.S. Plevkov, G.P. Tonkikh, M. Rimshin, A.G. Roytman, I.A. Fizdel, and others. [1-13].

The subject of this study is an instrumental examination of actual physical and mechanical parameters of load-bearing structural elements to prevent uncontrolled progressive collapse of the building in case of failure of its load-bearing structures. For this purpose, author's design solutions were used in the form of crib piers installed in the area of potential collapse.

[*] Corresponding author: orpahmurin@mail.ru

Research Methodology and Results

As a typical example for solving this problem, we will consider a high-rise building with reinforced concrete foundations, brick bearing walls, reinforced concrete and wooden ceilings, and a wooden rafter system.

The building under consideration is located in Tomsk (Russia). It is a five-storey building with a U-shaped plan constructed in the 1950s (Fig. 1). The overall dimensions of the building in plan view along layout axes are 73.618 × 32.600 m. The building was designed and erected on the basis of a rigid structural diagram with an incomplete load-bearing frame. Vertical bearing structures along the internal axes are represented by brick piers with a cross section of 640×510 mm, and brick walls with a thickness of 380 mm. There are longitudinal load-bearing walls along the outer axes. The walls are made of ordinary clay brick on cement-sand mortar. The thickness of the external walls is 640 mm. The thickness of the basement walls is 770 mm. The external walls are made of clay bricks on a cement-sand mortar.

The building is corridor-type. Rigidity and stability of the structure is ensured by a system of longitudinal and transverse walls monolithically connected with each other, rigidity of the incomplete load-bearing frame and floor disks that are rigid in their plane. Frame girders are made of reinforced concrete. They are laid in the longitudinal direction. Metal beams made of rolled section No. 25 are laid in the transverse direction of the building.

a)

b)

Fig. 1. General view of the building: northern (*a*), central and southern blocks (*b*).

The building has a hip roof designed according to the general building plan. The main load-bearing structures are represented by jack rafters with substructures of two types: posts and purlins. The rafter system is made of wood. Intermediate and attic floors are made of wood according to the beam pattern. The main load-bearing elements of the attic floor include metal, reinforced concrete and wooden beams laid in the longitudinal and transverse directions of the building.

According to the results of the examination performed in accordance with [14, 15], the current condition of foundations, vertical load-bearing structures, floors and roof was determined. Geometric parameters were measured with laser and metal measuring tapes. Strength parameters of reinforced concrete structures were estimated by the shock-pulse method using an ONIKS 2.5 electronic strength tester for non-destructive strength testing. A destructive test was performed with press equipment for testing the strength of materials (brick and mortar) extracted from the load-bearing walls of the building. Moisture content of wooden structures was measured with a Delta 200L moisture meter. All facilities that were necessary to assess the operational reliability of the building were constructed in compliance with the applicable regulatory requirements.

Hydrogeological conditions of the site were studied. The results show that the soils under the foundations of the building are moistened and functional only to a limited extent. Load-bearing stone structures of the external walls were analyzed. The results show that they are water saturated at the level of the ground floor and the plinth almost along the entire perimeter, both outside and inside the building. Due to high hygroscopic properties of bricks, moisture in load-bearing walls rises to a height of two–three meters. The masonry of the walls is watered not only from below, through the plinth, but also from above, due to damaged integrity of interfacing between the roofing and cornice sections of the walls. It should be noted that the bricks and mortar in the masonry of the plinth part of the walls are under destruction.

Due to saturation of the masonry with ground water, atmospheric precipitation and household water accompanied with cyclic exposure to negative temperatures, the strength of the masonry has significantly deteriorated. Crushing and disintegration of bricks are observed in exposed sections of the masonry of the ground floor walls (Fig. 2).

a) *b)*

Fig. 2. Destruction pattern of the masonry of ground-floor wall partitions.

Samples of bricks and mortar taken from the walls of the building were tested. The results show that the strength of the ground-floor and basement walls changes from "zero" to M 10. According to the compression test, the strength of bricks corresponds to grade M 25 ... M 50.

Re-calculations for the masonry of the building walls at actual strength characteristics of their materials and operating permanent and temporary loads [16] established by building codes and regulations [17] showed that the load-bearing capacity of the masonry in the external basement walls, the ground and first floors does not meet the regulatory requirements [18]. The masonry of the walls and piers in the areas where concentrated loads are transferred from reinforced concrete and metal beams does not meet the requirements for crushing strength of walls and masonry piers of the basement, ground, first, second and third floors of the building.

The condition of the end wall in the northern block is of particular concern. The load-bearing walls are seen to be outside the plane and buckled. At the same time, the masonry is segregated into separate vertical columns along the thickness due to loss of frost resistance and deteriorated strength parameters of bricks and mortar. Cracks in the walls result from out-of-tolerance deformations of soils under the foundations due to watering, as well as from the compliance of the masonry in the plinth part of the wall due to loss of compressive strength and frost resistance. Due to the combined effect of these factors, the end wall and adjacent sections of the longitudinal external walls do not meet the regulatory requirements [18], are in a state of disrepair [13, 14], and may result in collapse (Fig. 3) [19, 20].

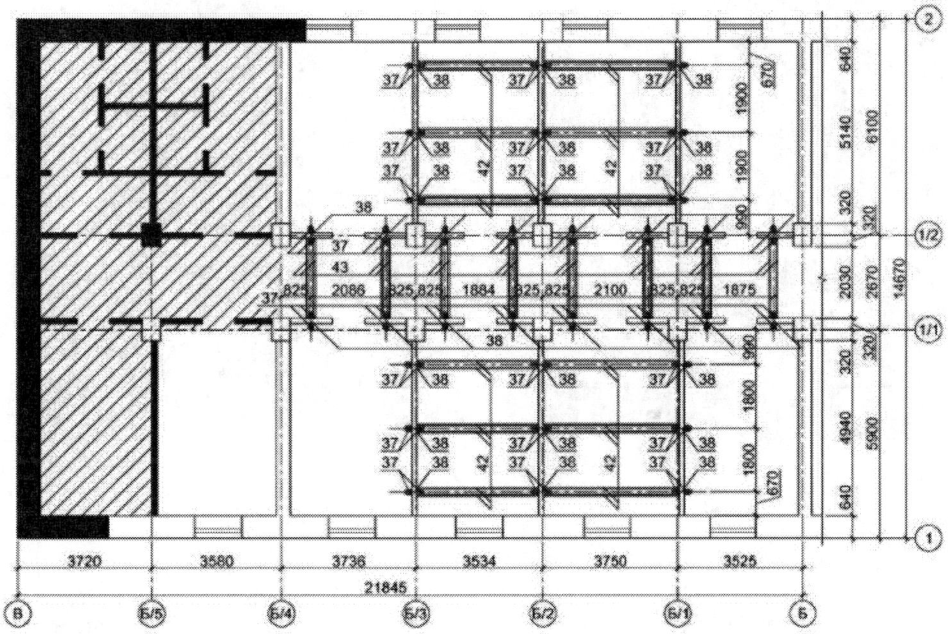

Legends:

- Floor sections to be dismantled;

- Brick piers, partitions and sections of walls to be dismantled.

Fig. 3. Expected collapse diagram for the northern block of the building and proposals for prevention of uncontrolled progressive collapse.

Developed deformations and movements are so significant that these areas cannot be restored or strengthened and should be dismantled. The end wall of the building should be dismantled. To prevent uncontrolled collapse along the end wall of the building, relieving crib piers [21] along the entire height of the building are proposed to be installed.

The basic idea behind installing crib piers in the expected collapse area is that loads from the floors will be transferred to these piers during collapse of the load-bearing walls in the end part of the building. The overall dimensions of the piers and the number of beams in the lower row were selected so that the stresses in soils under them did not exceed the calculated resistance of watered foundation soils. The number of piers along the load-bearing wall and the distance between them were calculated in such a way that they took active loads from the floors and all overlying structures. Cross sections of individual beams for the crib were selected so that their bending and shearing strength was ensured. Frame piers were used as emergency prevention measures for the rest of the building.

It should be noted that the end wall collapsed as expected, in full accordance with the prediction (Fig. 4).

Fig. 4. The real collapse of the northern block of the building.

Conclusion

The operational reliability of a building should be assessed on the basis of an integrated approach, including diagnostics of its foundation and load-bearing frame, the strength and deformability of intermediate floors and a roof. The proposed measures to prevent uncontrolled progressive collapse of a building in a state of disrepair proved to be effective and prevented the uncontrolled progressive collapse of the building.

References

1. V.S. Abrashitov, *Tehnicheskaja jekspluatacija i obsledovanie stroitel'nyh konstrukcij* [Technical operation and inspection of buildings and structures] (ASB Publ., Moscow, 2005) (in Russian)
2. M.D. Boyko, *Tehnicheskoe obsledovanie i remont zdanij i sooruzhenij* [Technical inspection and repair of buildings and structures] (Strojizdat, Leningrad, 1986) (in Russian)
3. A.N. Dobromyslov, *Diagnostika povrezhdenij zdanij i inzhenernyh sooruzhenij* [Damage diagnostics of buildings and building structures] (ACV, Moscow, 2006) (in Russian)
4. A.N. Dobromyslov, *Defekty v konstrukcijah pri stroitel'stve* [Defects in structures during construction] (ASV, Moscow, 2009) (in Russian)
5. Yu.V. Ivanov, *Rekonstrukcija zdanij i sooruzhenij: usilenie, vosstanovlenie, remont* [Reconstruction of buildings and structures: reinforcement, restoration, repair] (ASB Publ., Moscow, 2009) (in Russian)
6. O.V. Luzhin, A.B. Zlochevskij, I.A. Gorbunov, V.A. Volohov, *Obsledovanie i ispytanija sooruzhenij* [Inspection and testing of structures] (Strojizdat, Moscow, 1987) (in Russian)
7. M. Rimshin, *Obsledovanie i ispytanie zdanij i sooruzhenij* [Inspection and testing of buildings and structures] (Vysshaja shkola, Moscow, 2007) (in Russian)
8. A.A. Kalinin, S.D. Sokova, A.N. Topilin, Obsledovanie i ispytanie konstrukcij zdanij i sooruzhenij [Inspection and testing of buildings and structures] (Infra, Moscow, 2010)

9. M. Kusainov, Miny zamedlennogo dejstvija v stroitel'nom proizvodstve [Time bombs in construction industry] (Saryarka, Astana, 2009) (in Russian)
10. G.P. Tonkikh, V.S. Plevkov, A.I. Malganov, O.V. Kabancev, *Ocenka tehnicheskogo sostojanija stroitel'nyh konstrukcij zdanij i sooruzhenij* [Assessment of the technical condition of buildings and structures] (Pechatnaja manufaktura, Tomsk, 2009) (in Russian)
11. V.A. Rogonskij, A.I. Kostric, V.F. Shcheryakov, *Ekspluatacionnaja nadjozhnost' zdanij* [Operational reliability of buildings] (Strojizdat, Leningrad, 1983) (in Russian)
12. A.G. Roytman, *Preduprezhdenie avarij zhilyh zdanij* [Prevention of accidents in residential buildings] (Strojizdat, Leningrad, 1990) (in Russian)
13. I.A. Fizdel, *Defekty v konstrukcijah, sooruzhenijah i metody ih ustranenija* [Defects in buildings and structures and methods for their elimination] (Strojizdat, Москва, 1979) (in Russian)
14. GOST 31937 – 2011. Buildings and constructions. Rules of inspection and monitoring of the technical condition (Standartinform, Moscow, 2014) (in Russian)
15. SP 13–102–2003. Regulations. Rules for inspection of bearing structures in buildings and construction (GUP CPP, Moscow, 2003) (in Russian)
16. O.G. Kumpyak, Z.R. Galyautdinov, O.R. Pakhmurin, V.S. Samsonov, *Zhelezobetonnye i kamennye konstrukcii* [Reinforced and masonry structures] (ASV, Moscow, 2014) (in Russian)
17. SP 20.13330.2011. Loads and actions. Revised edition of SNiP 2.01.07–85* (GUP CPP, Moscow, 2011) (in Russian)
18. SP 15.13330.2012. Masonry/reinforced masonry structures. Revised edition of SNiP II–22–81* (FAU «FCS», Moscow, 2012) (in Russian)
19. O.G. Kumpyak, Z.R. Galyautdinov, O.R. Pakhmurin, A.A. Petukhof, *News of science*, Proceedings of the international scientific conference, 73-82 (3-4 August 2015, Czech Republic, Karlovy Vary – Russia, Moscow)
20. O.R. Pakhmurin, V.V. Scherbenkova, Proceedings of the I International Scientific Conference of Students and Young Researchers, 65-66 (TSUAB Publ., Tomsk, 2014) (in Russian)
21. O.G. Kumpyak, Z.R. Galyautdinov, G.I. Odnokopylov, O.R. Pakhmurin, A.V. Kudyakov, Y.A. Mun, Sposob obespechenija zhivuchesti avarijnogo zdanija ili sooruzhenija i bezopasnosti provedenija remontno-vosstanovitel'nyh rabot [Technique for providing durability of dangerous buildings or structures and safety of repair and restoration work], RF Patent 2441967 (2012) (in Russian)

MATEC Web of Conferences **143**, 01007 (2018) https://doi.org/10.1051/matecconf/201814301007
YSSIP-2017

Control of stress-strain state in double-span reinforced concrete beams

Nataliya Statsenko[1,*], and *Valery* Mitasov[1]

[1]Novosibirsk State University of Architecture and Civil Engineering (Sibstrin), 630008 Novosibirsk, Russia

Abstract Considering the tendency of reinforced concrete to redistribute load depending on the amount and location of effective reinforcement a series of experiments were performed using a non-traditional approach with set strain behavior of reinforced concrete beams during crack formation. The influence of location and number of cracks on load redistribution in double-span reinforced concrete beams was assessed. During the experiments comparative analysis of moments at support and at span was performed using uniform method of experiment. The paper presents mechanisms of load redistribution caused by non-linear deformations of a cracked element and plastic properties of compressed concrete.

Introduction

The peculiarity of RC structures strain in statically indeterminate systems is the ability of self-regulation of stress-strain state depending on the actual load. This is well demonstrated in design of flat RC cast-in-place ribbed floorings with beam elements, when reinforcement of multispan beam elements leads to formation of plastic hinges and allows using schemes with equal moments. In fact, we control stressed state of a RC structure by installation of reinforcement in multispan beam or plate structural systems. Such techniques are widely used in design and civil engineering. However, reinforcement is not the only stress state regulator .

Krasnoyarsk scientific school founded by N. Abovsky and including such scientists as L. Endzhievsky, V. Savchenkov, A. Deruga, N. Marchuk, G. Sterekhova, V. Palagushkin, N. Andreev, P. Svetashkov, O. Maksimova, etc., have been studying the influence of various regulation parameters on stress-strain state of structures for many years. Sources [1-4] formulate the principles for creation of automatic control system for stress-strain state based on system approach to the controlled structure design with consideration of internal actions and energy required for managing.

Vast experimental and theoretical material has been accumulated on investigation of load redistribution and bearing capacity of statically indeterminate structures with consideration of plastic properties of reinforcement and concrete [5-13]. It is obvious that

* Corresponding author: stacnat@yandex.ru

both theory and practice of RC construction impose the necessity of such approaches for designing RC structural systems.

While working on the energy theory of reinforced concrete and a new end-to-end method of reinforced concrete analysis based on energy diagram method [14-16], practicality of cracks generation while producing RC structures was proved both theoretically and experimentally.

Experimental results for pre-cracked beams showed that such cracks serve for internal control of stress-strain state. This finding opens new perspectives for further extension of reinforced concrete theory and practice which are based on energy theory of reinforced concrete resistance.

The purpose of the research, thus, is to assess the effect of number and location of pre-cracks on load redistribution in double-span reinforced concrete beams.

Experiment

In recent years, the orientation for providing seismic resistance of buildings and structures has evolved in the CIS and other countries. It is characterized by the development of active seismic protection and isolation systems, the use of which allow managing within certain limits the regulation of stresses in buildings and structures during earthquakes [17-19].

Regulation (redistribution) of load in reinforced concrete structures ensures their normal functioning in varied operating conditions. The task of regulation consists in prior redistribution of load to provide favorable state for certain operating conditions.

Program of experiments was developed for studying the peculiarities of strain processes and crack formation processes in beams with solid cross-sections and in pre-cracked beams. During the experiments comparative analysis of stress-strain state of the above-mentioned beams was performed using uniform method of experiment.

Specimens preparation. Four series of double-span RC beams with rectangular cross-section (b=100mm, h=200mm) were prepared for the experiments (Table 1), three out of the four series consisted of pre-cracked beams of different types.

Beams were reinforced with a flat steel frame. The frame consists of longitudinal principal rebars of A-400 type 6 mm in diameter and transverse rebars of A-240 type 6mm in diameter placed with the pitch of 80 mm. Reinforcement ratio is μ =0.15%.

All specimens were produced of fine grained concrete with the components ration of 1:2.5 (cement of type M400:sand) with W/C=0.45 and filler fineness modulus of 2.5-5 mm, filler strength corresponding to the type B20.

Reference prisms 100×100×400 mm in size and cubes 100×100×100 mm in size were prepared and tested along with experimental specimens using the П-250M hydraulic press to define the strength of concrete. Prism strength values for fine grained concrete in experimental specimens during testing are given in Table 1.

Table 1. Program of experiment.

Beam reference number	Scheme of pre-cracks	Longitudinal reinforcement, d_s,mm / μ,%	Prism strength / Cross-section diameter, mm
Б-1-1а	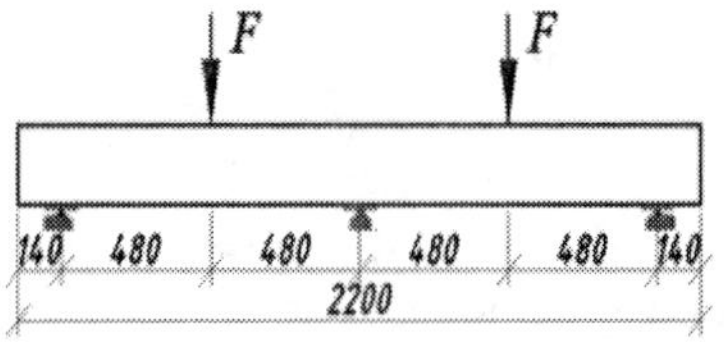	Ø6 / 0,15	16.22MPa / 100x200
Б-1-1б			
Б-1-1в			

Б-2-2а	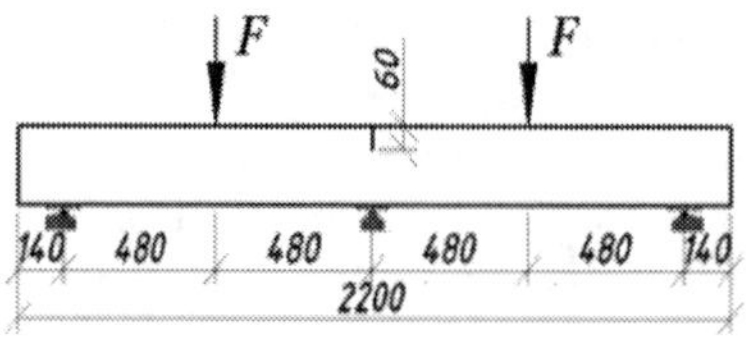		
Б-2-2б			14.5MPa 100x200
Б-3-3в			
Б-3-3а	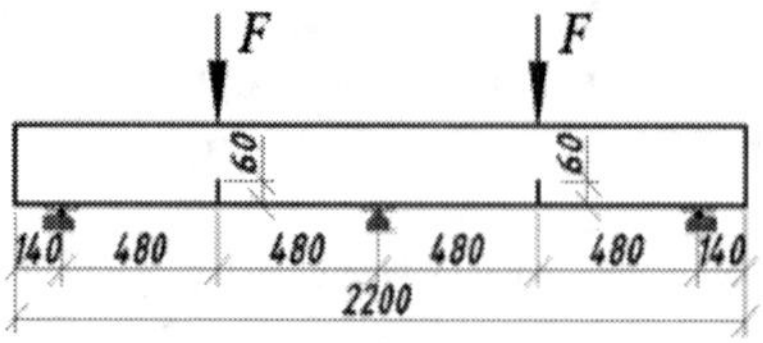		
Б-3-3б			17.72MPa 100x200
Б-3-3в			
Б-5-5а	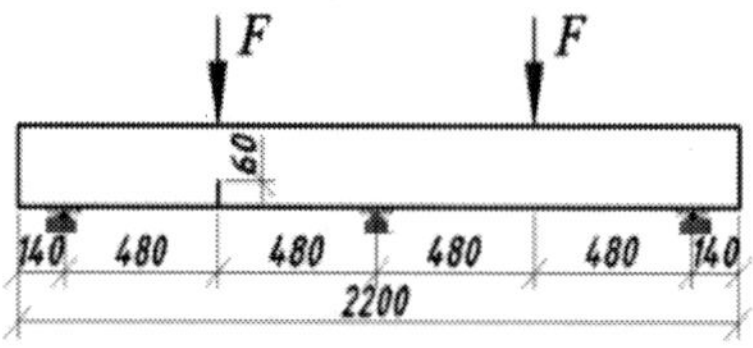		
Б-5-5б			17.23 MPa 100x200
Б-5-5в			

Beam identification code: first letter «Б» stands for "beam". First digit marks the series, the second digit marks the variant of pre-cracks location: 1 – beams with solid cross-sections, 2 – beams with pre-cracks in mid support, 3 – beams with two pre-cracks in the middle of each span, 5 – beam with one pre-crack in the middle of one span (Table 1). Pre-crack was organized by installing a steel zink-coated plate 100x60mm in size and 0.3mm wide.

Loading was performed by applying two concentrated forces, one force for each span at the distance of half of the span. Considering this loading scenario, the ratio of support moment to the span moment defined from the elastic system is $M_{sup}/M_{sp}=1.2$. For the purpose of load redistribution due to pre-cracks formation and opening, reinforcement of support and span cross-section was installed according to elastic system calculation. Practically it was achieved by setting the same amount of reinforcement in support and in span.

The load was transmitted to the beam through a pin-supported crossbar. The load was sustained for 15 minutes at each stage, during that time the readings of sensors were recorded and the pre-cracks opening width was measured.

The moment of cracks formation, their propagation pattern in height and opening width were recorded visually and using the МПБ-2 microscope with the scale of 0.05 mm. The width of regular cracks was measured at the level of tensile reinforcement. During pauses between loading sessions, the height, width and propagation pattern of cracks at a certain loading state were marked on the bleached side surface of the beam.

Concrete strain and beam deflection were measured up to bearing capacity exhaustion, which corresponded to the yield stress in reinforcement and further crush of beam concrete in the compression area.

Results

All beams were tested using a uniform technique. As a result of the experiments, comparative data were obtained on deformation of beams with different variants of pre-cracks location.

Moment values for support reactions in beams with pre-cracks were found different from those obtained for elastic deformation scheme. This difference was equally observed for all beams series almost from the initial loading stages, which corresponds to formation of stochastic cracks or to opening of pre-cracks in span or support beam cross-sections. Figure 1 presents graphs of support moments and span moments variation for 4 beam series. The graphs also include the values of moments obtained from both experiments and elastic deformation calculation.

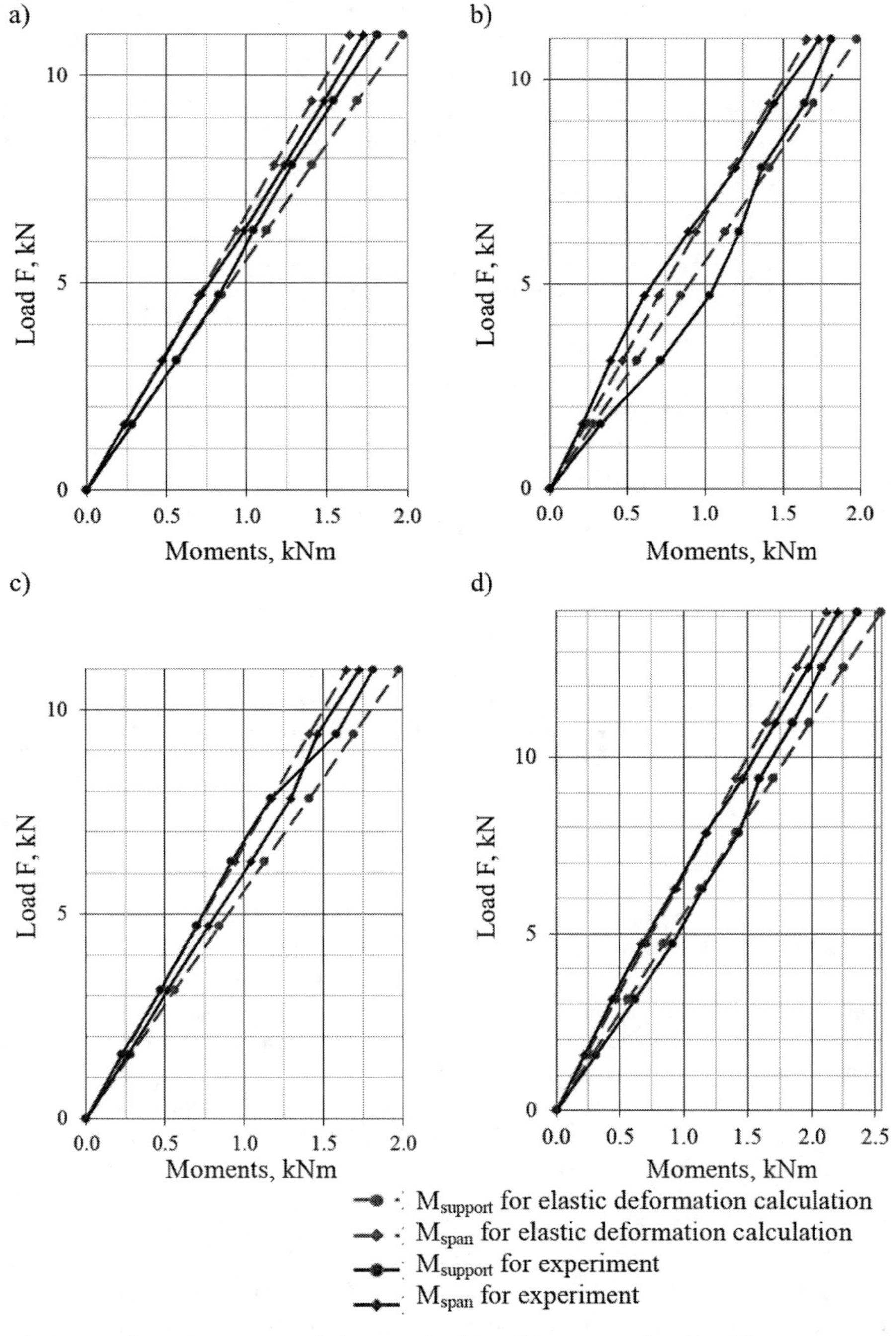

Fig. 1. Span and support moments in beams: a) with solid cross-section; b) with two pre-cracks in each mid-span; c) with a pre-crack in mid-support; d) with one pre-crack in the middle of the first span.

Having analyzed the obtained graphs, one can notice that under low loads (loading stage 1-3) the experimental values for beams series 1, 2 and 5 correspond closely to the calculated ones. With stochastic cracks occurring or pre-cracks opening, experimentally obtained moment values differ a lot from those calculated for elastic deformation scenario.

Considering redistribution of load after stochastic cracks occurrence or pre-cracks opening, the value of support moment under load F=14kN in each span was M=2.36 kNm for beam series 1, 2, 5 and M=1.878 kNm for beam series 3, which is significantly lower than the limit value M=2.54 kNm. The moment at support reached the value of 2.54 kNm under the load in span roughly equal to F=16 kNm. As a result of formation and opening of cracks in beam series 1, 2, 5 due to redistribution of loads compared to elastic deformation scenario the load at which support cross-section reaches its limit state increased by 14%.

The results of RC beams testing were used to obtain graphs of midspan deflection (Fig. 2).

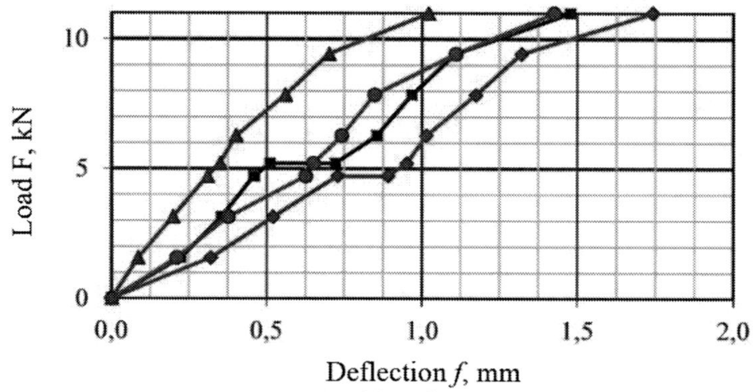

Fig. 2. Graph of dependence of deflection f on load F.

The load-deflection curve for beams with solid cross-section in span has a clearly visible region of deflection growth under permanent load, while in pre-cracked beams such regions are not observed. The maximum deflection in span is 1.48 mm with the load of 11 kN in each mid-span.

The following was found for pre-cracked beams:

– In beams with pre-cracks in each mid-span the deflection at initial stage is higher that that in beams with solid cross-section. Slight variation of the curve and transition to the region of less growth correspond to opening and propagation of pre-cracks. The maximum deflection is 1.43 mm with the load of 11 kN in each mid-span.

– In beams with a pre-crack in the middle of span No.1 the values of deflection in each mid-span differ considerably , approx. by 2.7 times. Deflections in the second span are higher than in beams with solid cross-section. However, the deflection values in span No.1 compared to those in beams series 1 is 45% lower. Maximum deflection in the first span is 1.02 mm, in the second span - 1.74 mm under the load of 11 kN in the mid-span.

Generation and opening of pre-cracks considerably varies load distribution compared to that within elastic system calculation, which indicates the regulatory function of pre-cracks.

The conducted research serves as transition to regulation of deformation in reinforced concrete beams, and in prospect it will allow managing crack resistance by pre-cracks generation.

Conclusion

General approach for pre-cracks generation in frame systems is developed. The number and location of such cracks are proved to have direct effect on load redistribution and stress-strained state.

At the moment, full-scale experiments are being prepared with structures (reinforced concrete hollow-core flooring slabs) damaged up to serviceability limit state with pre-cracks in the maximum moment area.

The suggested method of "plastic hinges" creation by cracks generation will allow using them as energy sinks.

In the beginning of 2018, dynamic experiments with RC frames prepared in scal 1:2 are planned in Kazakh State Research and Experimental Design Institute on Earthquake Engineering and Architecture. Such experiments will allow developing recommendation on the use of new energy sinks in buildings and structures located in seismic regions.

References

1. N.P. Abovsky, *Upravljaemye konstrukcii* [Controllable structures] (KrasGASA Publ., Krasnoyarsk, 1998) (in Russian)
2. A.A. Krasovsky, *Spravochnik po teorii avtomaticheskogo upravlenija* [Handbook on the automatic control theory] (Science, Moscow, 1987) (in Russian)
3. L.V. Endzhievskiy, *Upravljaemye konstrukcii.* [Controllable structures. Scientific educational complex] (KrasGASA, Krasnoyarsk, 2001) (in Russian)
4. N.P. Abovsky, *Sovremennye aspekty aktivnogo obuchenija. Stroitel'naja mehanika. Teorija uprugosti. Upravlenie stroitel'nymi konstrukcijami* [Modern aspects of active learning. Structural mechanics. Elasticity theory. Structural control] (Sib. feder. un-t, Krasnoyarsk, 2008) (in Russian)
5. S.M. Krylov, *Pererasspredelenie usilij v staticheski neopredelimyh zhelezobetonnyh konstrukcijah* [Redistribution of forces in statically indeterminate reinforced concrete structures] (Stroiizdat, Moscow, 1964) (in Russian)
6. *Rukovodstvo po raschetu staticheski neopredelimyh zhelezobetonnyh konstrukcij.* ed. by L.G. Bal'yan [Guide on the calculation of statically indeterminate reinforced concrete structures] (Strojizdat, Moscow, 1975) (in Russian)
7. N.I. Karpenko, *Obshhie modeli mehaniki zhelezobetona* [General models of reinforced concrete mechanics] (Stroiizdat, Moscow, 1996) (in Russian)
8. S. Kumaraswamy, IOSR-JMCE, **6**, 38-43 (2013)
9. M. Leps, M. Sejnoha, Computers & Structures, **81,** 18-19 (2003)
10. M.N.S. Hadi, L.C. Schmidt, International Journal of Materials & Product Technology, **19**, 3-4 (2003)
11. S.-S. Lu, W.-Z. Zhen, Journal of Harbin Institute of Technology (New Series), **17**, 4 (2010)
12. Z.P. Bazant, Journal of Structural Engineering, **110**, 2015-2036 (1984)
13. H. Akbarzadeh, A.A Maghsoudi, Mater Des, **31**(3), 1130-1147 (2010)
14. V.M. Mitasov, *Osnovnye polozhenija teorii soprotivlenija zhelezobetona* [The basic provisions of the theory of reinforced concrete resistance] (NGASU (Sibstrin), Novosibirsk, 2010)

15. V.M. Mitasov, V. Adishev, M.A. Logunova, Proceedings of the International Scientific Conference "Modern Construction Science, State and Development Prospects", 50-53 (Alma-Ata, October 27-29, 2016)
16. V.M. Mitasov, M.A. Logunova, Proceedings of the International Scientific Conference "Science, Technical Management and Engineering in Construction: State and Prospects", 271-273 (Karaganda, April 29-30, 2016) (in Russian)
17. N.V. Martynov, *Aktivnaja sejsmozashhita: varianty razvitija i kriticheskij analiz prakticheskih vozmozhnostej* [Active earthquake protection: options for development and critical analysis of practical possibilities] (monograph, Simferopol, 2013) (in Russian)
18. V.M. Mitasov, V. Adishev, N.V. Statsenko, Proceedings of the International Scientific Conference "New building trends in the XXI century", 18-21 (Almaty, June 27, 2017) (in Russian)
19. E.K. Nurmagambet, *Sooruzhenija v ekstremal'nyh uslovijah* [Constructions under extreme conditions] (KazNTU, Almaty, 2005) (in Russian)

Modeling of storage tank settlement based on the United States standards

Alesya Gruchenkova[1,*], *Aleksandr* Tarasenko[2], and *Petr* Chepur[2]

[1]Surgut Institute of Oil and Gas, Surgut Branch of the Tyumen Industrial University, 628404 Surgut, Russia
[2]Tyumen Industrial University, 625000 Tyumen, Russia

Abstract. Up to 60% of storage tanks in operation have uneven settlement of the outer bottom contour, which often leads to accidents. Russian and foreign regulatory documents have different requirements for strain limits of metal structures. There is an increasing need for harmonizing regulatory documents. The aim of this study is to theoretically justify and to assess the possibility of applying the U.S. standards for specifying the allowable settlement of storage tanks used in Russia. The allowable uneven settlement was calculated for a vertical steel tank (VST-20000) according to API-653, a standard of the American Petroleum Institute. The calculated allowable settlement levels were compared with those established by Russian standards. Based on the finite element method, the uneven settlement development process of a storage tank was modeled. Stress-strain state parameters of tank structures were obtained at the critical levels established in API-653. Relationships of maximum equivalent stresses in VST metal structures to the vertical settlement component for settlement zones of 6 to 72 m in length were determined. When the uneven settlement zone is 6 m in length, the limit state is found to be caused by 30-mm vertical settlement, while stresses in the wall exceed 330 MPa. When the uneven settlement zone is 36 m in length, stresses reach the yield point only at 100-mm vertical settlement.

Introduction

Uneven settlement is one of the causes that give rise to limit states in metal structures of vertical steel tanks (VST), which often lead to accidents [1–3]. Effects of uneven foundation settlement on the stress-strain state of storage tanks were studied by Russian scientists, such as V. B. Galeev, V. V. Lyubushkin, V. A. Burenin, P. P. Borodavkin and others, as well as by foreign scientists — J. S. Clarke [4], E. E. DeBeer [5], D. A. Greenwood [6] and others. However, despite accumulated research findings and long-term experience in tank construction, researcher have failed to resolve a number of issues related to assessing the stress-strain state of VST, which is subject to uneven foundation settlement [7]. For non-axisymmetric strains of a tank shell, it is not always possible to determine high-stress zones in metal structures [8].

* Corresponding author: alesya2010-11@yandex.ru

To prevent unacceptable stresses in the VST structure, Russian and foreign regulatory documents specify the criteria that determine allowable uneven settlement levels. For example, regulatory documents of the Russian Federation (GOST, RD, SA, RB) and the USA (API) contain the specific formulas, relationships and tabular values, which allow the maximum settlement level to be determined for VST of various designs and sizes.

Internationalization, globalization of energy markets, and, therefore, the involvement of foreign contractors for constructing storage tanks in the Russian Federation make it necessary to harmonize the Russian regulatory framework in the field of storage tank construction. An important problem is that the storage tanks constructed according to Russian and western standards have some differences in terms of design and materials [11]. In turn, this affects the overall stiffness of the structure, which determines the deformation behavior of the VST shell in case of uneven settlement. Thus, the authors aim to theoretically justify and to assess the possibility of applying the U.S. standards for determining the stress-strain state of storage tanks of Russian standard sizes.

Materials and Methods

The authors suggest considering the possibility of using the requirements of API-653 standard (Appendix B) [12] for specifying allowable settlement levels in Russian storage tanks. In order to justify the possibility of using the American API standard, it is necessary to perform a number of verifications of the stress-strain state for VST structure at given uneven settlement parameters, to determine effective stresses in structural elements, and to compare them with allowable values for a certain steel grade (09G2S). The authors will consider the most common storage tank of the Russian Federation — VST-20000 with a fixed roof and reinforced concrete foundation ring.

To perform the tasks set in the paper, the authors applied the classical principles of structural mechanics, the theory of elasticity, deformable body mechanics, and the finite element method. Being a numerical method, the finite element method can be used to solve a wide range of physical problems, which are mathematically formulated through systems of differential equations. This method is suitable for a deep analysis of the stress-strain state in metal structures. Theoretical studies were carried out using methods of mathematical analysis and numerical modeling with the ANSYS software certified by the Federal Nuclear and Radiation Safety Authority of Russia.

The reliability of the obtained results is confirmed by satisfactory correlation with widely known analytical solutions. Moreover, it is supported by the research methodology proposed by Russian and foreign scientists, using regulatory documents and statistical data, minimal errors of calculated and experimental data, and the compliance of the data with scientific concepts.

The methods described above made it possible to obtain new relationships of the VST stress-strain state to dimensions of the uneven settlement zone. These relationships reflect the real deformation behavior of metal structures, taking into account the physical and geometric nonlinearity of the applicable model.

Experimental Part

According to API-653 [12], the maximum allowable settlement level is calculated on an individual case basis for each storage tank with uneven settlement components. For this purpose, it is necessary to level the outer bottom contour and to plot a distortion curve. Then uneven settlement zones are identified, and their length S_{arc}, the so called "settlement arc", is determined. The maximum allowable vertical component, S_{max}, is calculated for each settlement arc according to the formula (1). This formula (API-653, Appendix B.3.2.2) takes VST diameter, height, and steel grade into account. Table 1 presents additional dimensionless coefficients and type of VST design selected according to the tank diameter.

$$S_{\max} = \min\left[\frac{D}{H} \cdot K \cdot S_{arc} \cdot \frac{Y}{E}\right], \tag{1}$$

where $S_{\max}$ – maximum allowable value of the vertical settlement component, inches

S_{arc} – effective length of the uneven settlement arc, feet;

D – tank diameter, feet

Y – yield strength of tank metal, lbf/in^2;

E – Young's modulus, feet;

H – tank height, feet;

K – dimensionless parameter depending on the type and diameter of the tank, selected according to the data in Table 1 and Table 2.

Table 1. Selection of the Coefficient K Depending on VST Diameter, according to API-653 [12].

VST diameter, feet	**D ≤ 50**	**(50… 80]**	**(80…120]**	**(120…180]**	**(180…240]**	**(240…300]**	**300 < D**
K, VST with a fixed roof	10.5	5.8	3.9	2.3	-	-	-

Table 2. Parameters of VST-20000 for Calculating Allowable Settlement, according to (1).

Parameter	**S_{arc}, m**	**D, m**	**Y(09G2S), MPa**	**E, MPa**	**H, m**	**Coefficient K**
Value	6.72	45.6	325	210,000	14	2.3

In order to verify the calculated parameters of VST-20000, length of the settlement arc S_{arc} was assumed to be from 6 to 72 meters. Other parameters of the tank are presented in Table 2. Using the formula (1), we obtained maximum settlement values for points of the outer bottom contour depending on the size of VST-20000 uneven settlement zone. In Figure 1, these relationships are presented graphically.

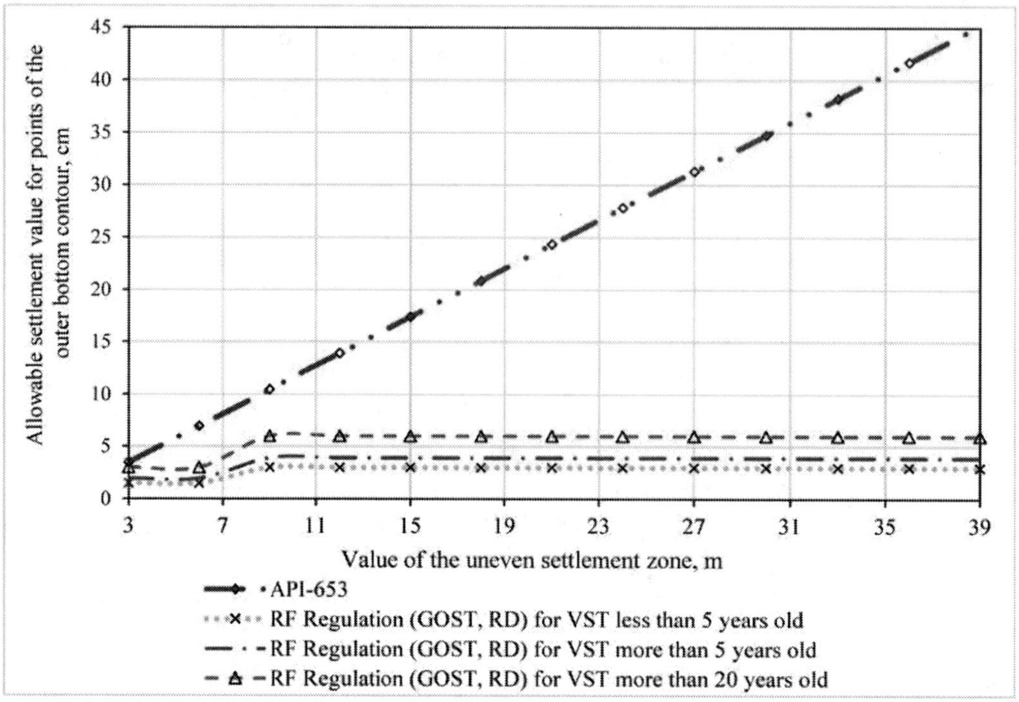

Fig. 1. Requirements of Russian and U.S. technical regulations that determine maximum settlement values for points of the outer bottom contour depending on the size of the VST uneven settlement zone.

The graph also shows maximum settlement values according to the industry standard RD-23.020.00-KTN-283-09 of OAO AK Transneft [13]. It can be seen that the maximum tolerances for uneven settlement in API-653 [12] are much broader than in the Russian standard, which is especially evident when the uneven settlement zone exceeds 10 m. This difference is explained by greater stiffness of tanks constructed according to API.

To calculate the stress-strain state of the VST-20000 tank at the maximum values determined in [14], the authors suggest using the finite element model for a storage tank developed in [15]. Based on the model developed with the ANSYS Workbench 16.1 software, it is possible to determine equivalent stresses and strains of VST metal structures when specifying settlement parameters — length of the uneven settlement zone and vertical component [16–18]. The calculation algorithm, boundary and contact conditions, as well as loading parameters for the model are described in detail in [15].

Results and Discussion

Figures 2–4 show the results of numerical simulation of the VST-20000 settlement process for uneven settlement zones with a length of 12 and 72 m and vertical components with a size of 10, 30, 100 mm.

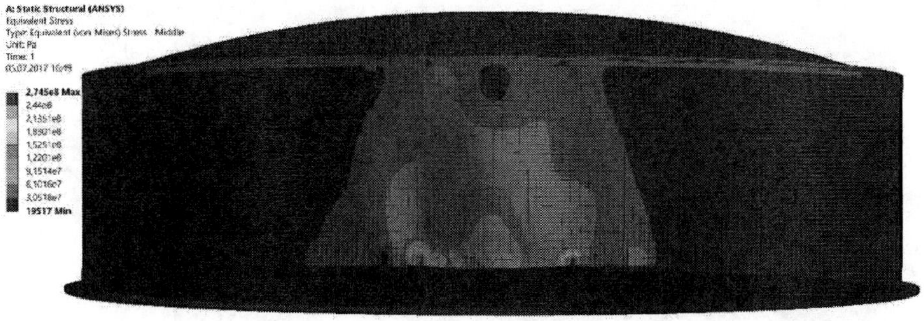

Fig. 2. Effective stresses at 10-mm settlement with 12-meter-long uneven settlement zone.

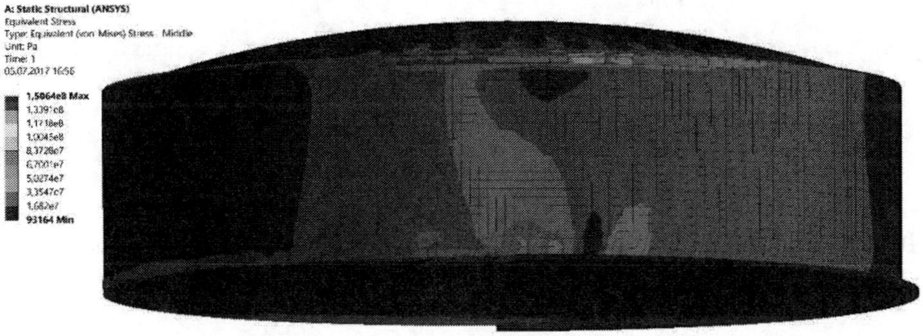

Fig. 3. Effective stresses at 30-mm settlement with 72-meter-long uneven settlement zone.

Fig. 4. Effective stresses at 100-mm settlement with 72-meter-long uneven settlement zone.

As a result of post-processing treatment, we obtained relationships of maximum equivalent stresses in VST metal structures to the vertical settlement component for zones with a length of 6 to 72 m. The graph below (Fig. 5) indicates three limiting zones. In zone I, stresses in metal structures do not exceed the allowable values specified in Russian technical regulations (188 MPa). In the second zone, stresses exceed the allowable values. In the third zone, inadmissible plastic strains occur.

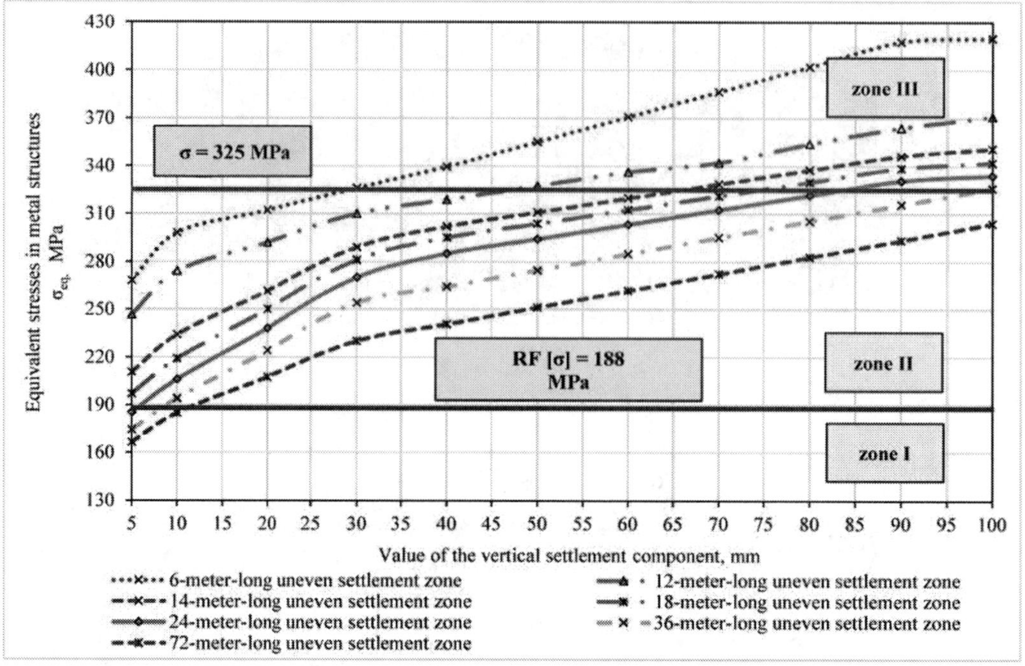

Fig. 5. Relationships of maximum equivalent stresses in VST metal structures to the vertical settlement component for uneven settlement zones with a length of 6 to 72 m.

The obtained relationships indicate that the limit state in the VST-20000 design is reached at a 100-mm vertical settlement and uneven settlement zone with a length up to 72 m. According to [12], the allowable settlement for tanks based on the API standard is distortion with a 500-mm vertical component and uneven settlement zone with a length of 45 m. This clearly suggests that the U.S. standards are not directly applicable to assessing uneven settlement of Russian vertical steel tanks. In order to use approaches of the standard [12] and to harmonize the Russian standards as related to the assessment of uneven foundation settlements, additional calculations are required.

Conclusion

The comparison of the allowable settlement levels according to [12] with those specified in Russian standards [13] shows that the settlement tolerances established in Russian documents are much more stringent than the U.S. ones (Fig. 1). Finite-element modeling of settlement allowed us to determine stresses and strains caused by non-axisymmetric deformation of the structure. The results based on the numerical model reflect the real deformation behavior of VST metal structures, which confirms the scientific and practical value of this study. Based on the calculations performed, it can be concluded that the criteria presented in [12] cannot be used to determine the allowable uneven settlement of VST tanks constructed according to the Russian standards. For this purpose, it is necessary to conduct additional studies of the stress-strain state for determining the coefficients in formula (1) that would reflect stiffness characteristics of the VST-20000 tank. Then the proposed approach [12] could be extended to the Russian tank construction industry.

This paper was prepared as the basic part of the State Assignment for project No. 7.7858.2017/BCh: Scientific Justification of the Technique for Determining the Stress-Strain State of Large-Size Oil Storage Tanks in Case of Uneven Bottom and Foundation Settlements.

References

1. G.G. Vasilyev, A.A. Salnikov, Neftyanoe khozyaystvo [Oil Industry] **2**, 106-108 (2015) (in Russian)
2. I.M. Rozenshtein, *Avarii i nadezhnost' stal'nyh rezervuarov* [Accidents and reliability of steel tanks] (Nedra, Moscow, 1995) (in Russian)
3. A. Akhavan-Zanjani, A. Fakher, S.R. Maddah-Sadatieh, A numerical study on the effect of uneven settlements of oil storage tank, 2nd International Conference on New Developments in Soil Mechanics and Geotechnical Engineering (Nicosia, North Cyprus, May 28–30, 2009)
4. J.S. Clarke, *Recent tank bottom and foundation problems* (Division of Refuting, 1971)
5. E.E. De Beer, Annales de L'Institut Belge du Petrole **6**, 25-40, (1969)
6. D.A. Greenwood, Proceedings of the Conference on Settlement of Structures, 35-97 (British Geotechnical Society, Cambridge, UK, 1974)
7. A.S. Gorelov, *Neodnorodnye gruntovye osnovanija i ih vlijanie na rabotu vertikal'nyh stal'nyh rezervuarov* [Inhomogeneous ground base and their influence on the operation of vertical steel tanks] (Nedra, SPb, 2009) (in Russian)
8. I.V. Slepnev, *Naprjazhenno-deformirovannoe uprugoplasticheskoe sostojanie stal'nyh vertikal'nyh cilindricheskih rezervuarov pri neravnomernyh osadkah osnovanij* [Stress-strain elastic-plastic state of steel vertical cylindrical tanks with irregular immersion of bases] (MISI, Moscow, 2014) (in Russian)
9. EN 14015-2004. Specification for the design and manufacture of site built, vertical, cylindrical, flat-bottomed, above ground, welded, steel tanks for the storage of liquids at ambient temperature and above (2004)
10. BS 2654. Manufacture of vertical steel welded non-refrigerated storage tanks with butt-welded shells for the petroleum industry (1989)
11. A. Gruchenkova, A. Tarasenko, P. Chepur, D. Tarasenko, AIP Conference Proceedings **1800**, 040019 (2017)
12. API 653. Tank Inspection, Repair, Alteration, and Reconstruction, Fourth Edition (USA, Washington, 2009)
13. RD-23.020.00-KTN-283-09. Rules of repair and reconstruction of oil storage tanks 1000-50000 cubic meters in capacity (2009)
14. A.A. Tarasenko, P.A. Konovalov, F.F. Zekhniev, P.V. Chepur, D.A. Tarasenko, Soil Mechanics and Foundation Engineering **53**(6), 405-411 (2017)
15. A. Tarasenko, P. Chepur, A. Gruchenkova, MATEC Web of Conferences **73**, 01025 (2016)
16. G.E. Korobkov, R.M. Zaripov, I.A. Shammazov, *Chislennoe modelirovanie naprjazhenno-deformirovannogo sostojanija i ustojchivosti truboprovodov i rezervuarov v oslozhnennyh uslovijah jekspluatacii* [Numerical modelling of the stress-strain state and stability of pipelines and tanks in complicated operating conditions] (Nedra, Saint Petersburg, 2009)
17. V.A. Bruyaka, V.G. Fokin, E.A. Soldusova, N.A. Glazunova, I.E. Adeynov, *Inzhenernyj analiz v ANSYS Workbench* [Engineering analysis in ANSYS Workbench] (SamSTU, Samara, 2010) (in Russian)
18. A.V. Beloborodov, *Ocenka kachestva postroenija konechno-jelementnoj modeli v ANSYS* [Evaluation of the finite element model construction quality in ANSYS] (Ural State Technical University, Ekaterinburg, 2005)

Sound insulation of lightweight partition walls with regard to structural sound transmission

Olga Lelyuga[1,*], and *Sergey* Ovsyannikov[1]

[1]Tomsk State University of Architecture and Building, 634003 Tomsk, Russia

Abstract. The known methods of acoustical calculation in buildings disregard the phenomenon of structural sound transmission, whereas its effect can reach from 2 to 12 dB. The purpose of this paper is to develop the calculation method for sound transmission and vibrations in connected vibroacoustic systems. Theoretical research methods were used based on the theory of statistical energy analysis (SEA) and the theory of self-consistent sound fields with regard to dual nature of sound formation - resonance and inertia. Based on M. Sedov's method of sound fields consistency, a calculation method for sound insulation was developed with integration in SEA methodology. Use of the developed method allows predicting sound transmission through a double-panel partition with the account of adjacent structures.

1 Introduction

One of the topical tasks of civil and architectural engineering is design of partitions with relatively small mass. This allows decreasing both load on load-bearing structures and material consumption. The main requirement to partitions consists in standard sound insulation performance. The use of multilayer structures provides high sound insulation with considerable reduction of area weight [1, 2]. For effective sound insulation in partitions structural parameters of these partitions shall be accounted along with conditions of sound vibration transmission in adjacent structures. Experimental studies show [3] that structural sound transmission effect on sound insulation of partitions can reach from 2 to 12 dB.

Acoustic and vibration sound sources cause sound vibration in the building frame in the form of elastic waves: longitudinal, transverse or bending [4]. A combination of factors cause disturbance of acoustic comfort in rooms both adjacent to and remote from the source of sound or vibration. Mechanical vibrations of walls, floorings or pipelines occurring in one room are transmitted to a considerable distance with little vibration loss and are emanated in the air as audible sound. Lightweight structures with big surface area emanate sounds particularly well. Vibrations can be caused by direct action (footsteps, door slam, vibrations of pipelines and technical appliances) or can be the consequence of airborne sound action.

* Corresponding author: olga.startseva27@gmail.com

The methods for buildings and structures calculation known to date [5-10] disregard the phenomenon of structural sound transmission. In this relation, theoretical and experimental research is required to investigate transmission of sound and vibrations in connected vibroacoustic systems of building models.

Sound is transmitted through joints of double panels, for instance, lightweight partitions adjoined to load-bearing walls and floorings, joints between interior and exterior structures, in case the latter consist of two panels, etc. (Fig. 1) Joints of double-panel structures may have elastic plugs or gaskets under flooring plates. Thickness of adjoined panels may vary [4].

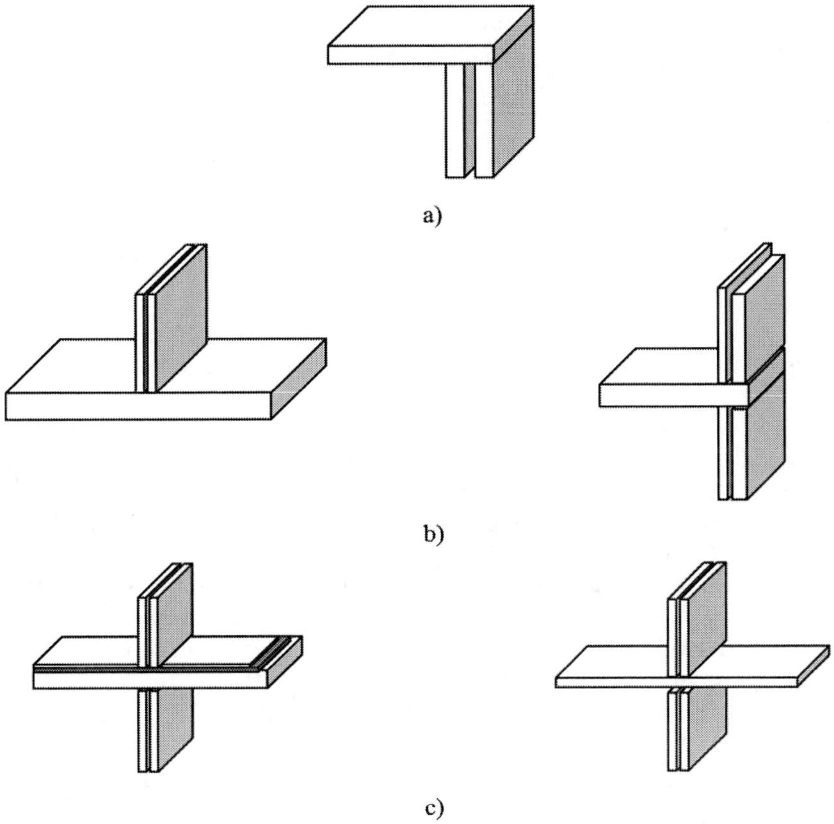

Fig. 1. Joints of double-panel partitions: a) angle joint; b) T-shaped joint; c) X-shaped joint.

Stiffness of joints with elastic plugs is characterized by the elastic modulus of the plug material and its geometrical parameters. Stiff joints can be obtained by equaling the elastic modulus of the plug material to that of the main structures material.

The problem of sound transmission through the joint of a two-panel partition can be solved using the SEA theory by writing a system of equations which describes energy relation of separate structural elements [4]. Energy balance equations account resonance properties and dissipative characteristics of such elements which are considered as subsystems in the common system of the studied building fragment. The SEA method in general statement used for defining energy coupling coefficients only accounts resonant sound transmission while disregarding the non-resonant one. Such omission can lead to considerable error in forecasting sound transmission.

The most complete analysis of sound transmission through a two-panel partition joint can be performed using the M. Sedov's theory of sound fields consistency [10].

The method of two-panel partitions calculation [10] developed by the research group of M. Sedov is based on the theory of sound fields consistency in the partition plane and the plate wave field formed by natural waves. The method accounts the dual nature of sound formation - resonance and inertia. The main paths of sound transmission through two-panel structures herein include not only those through plates and air space, but also paths of sound vibrations propagated along the adjoined structures. This method enables to account all forms of sound transmission while performing SEA.

The paper presents the calculation method for sound insulation based on M. Sedov's method of sound fields consistency, with integration in SEA methodology.

The alternative approach to calculation of resonant sound transmission is suggested by F. G. Leppington [14]. However, this approach has a number of limitations like the need of ideal boundary conditions which cannot be easily adopted in the structure of SEA used in this paper.

2 Research method

2.1 SEA Model

The first stage of using SEA includes building a physical model in which the whole studied system is divided into elements and subsystems. To represent energy coupling of subsystems in the SEA model, let us consider the system of three panels (structural subsystems) and two air volumes of a room (acoustic subsystems).

The scheme of energy coupling of room partitions is presented in Figure 2.

The subscript indices of *I*, *II* are used to mark the room numbers, and the subscript indices of *1*, *2* and *3* denote the panels which exchange bending wave energy. The sequence of the indices shows the direction of energy flow. The superscript characters indicate the type of wave field in the model elements, *A* – stands for acoustic field in the room, and *b* – for bending wave field in a panel. W_1, W_2, W_2, W_I и W_{II} – energies in subsystems, P_1, P_2 and P_3 – external power supplied into the subsystems, $P_{1,diss}$, $P_{2,diss}$, and $P_{3,diss}$ – dissipated power in the subsystems, P_{12}, P_{21}, P_{23}, P_{32} and P_{13}, P_{31} – power of energy exchange between subsystems.

Powers that are lost in subsystems can be expressed in the following way:

$$P_{1,diss}^{b} = \omega \cdot \eta_1^b \cdot W_1^b \quad (1)$$

$$P_{2,diss}^{b} = \omega \cdot \eta_2^b \cdot W_2^b \quad (2)$$

$$P_{3,diss}^{b} = \omega \cdot \eta_3^b \cdot W_3^b, \quad (3)$$

where η_1^b, η_2^b, η_3^b – internal loss factors in subsystems in the process of bending waves propagation; $\omega = 2\pi f$ – angular frequency taken as a geometric mean frequency of one-third octave and octave band.

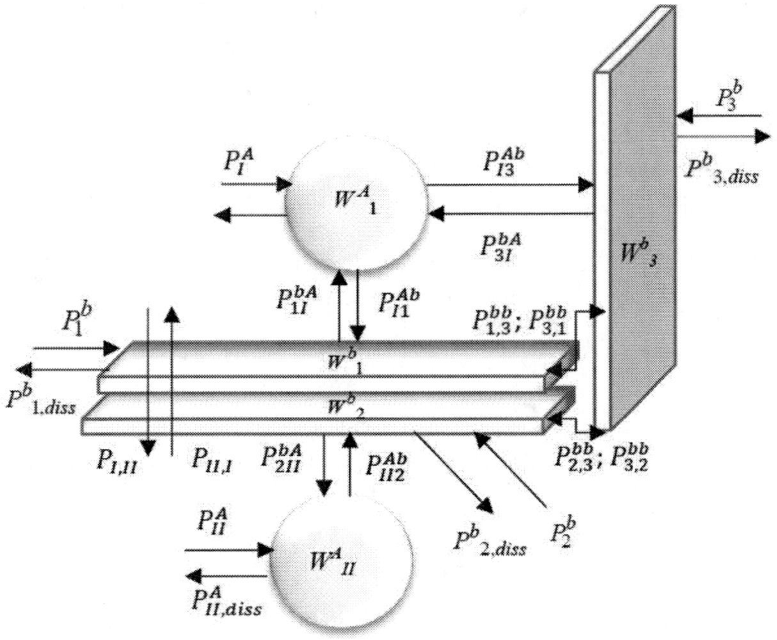

Fig. 2. Scheme of energy exchange between acoustic and structural subsystems in the process of bending wave propagation in panels.

Now, let us write an energy balance equation for each panel and for two acoustic subsystems:

$$P_I^A = \omega \cdot \eta_{I,I}^A \cdot W_I^A - \omega(\eta_{1I}^{bA} \cdot W_1^b + \eta_{3I}^{bA} \cdot W_3^b), \tag{4}$$

where $\eta_{I,I}^A = \eta_I^A + \eta_{I1}^{Ab} + \eta_{I3}^{Ab} + \eta_{I,II}^{AA}$

$$P_{II}^A = \omega \cdot \eta_{II,II}^A \cdot W_{II}^A - \omega \cdot \eta_{2II}^{bA} \cdot W_2^b, \tag{5}$$

where $\eta_{II,II}^A = \eta_{II}^A + \eta_{II2}^{Ab} + \eta_{II,I}^{AA}$

$$P_1^b = \omega \cdot \eta_{1,1}^b \cdot W_1^b - \omega \cdot (\eta_{I1}^{Ab} \cdot W_I^A + \eta_{3I}^{bb} \cdot W_3^b)\ , \tag{6}$$

where $\eta_{1,1}^b = \eta_1^b + \eta_{1I}^{bA} + \eta_{13}^{bb}$

$$P_2^b = \omega \cdot \eta_{2,2}^b \cdot W_2^b - \omega \cdot (\eta_{II2}^{Ab} \cdot W_{II}^A + \eta_{32}^{bb} \cdot W_3^b), \tag{7}$$

where $\eta_{2,2}^b = \eta_2^b + \eta_{2II}^{bA} + \eta_{23}^{bb}$

$$P_3^b = \omega \cdot \eta_{3,3}^b \cdot W_3^b - \omega \cdot (\eta_{I3}^{Ab} \cdot W_I^A + \eta_{13}^{bb} \cdot W_1^b + \eta_{23}^{bb} \cdot W_2^b), \tag{8}$$

where $\eta_{33}^b = \eta_3^b + \eta_{3I}^{bA} + \eta_{31}^b + \eta_{32}^{bb}$

In a number of cases, including full-scale measurement of propagation of noise and sound vibration in a building, the power of source can be evaluated using the measured or calculated sound pressure level Lpk in a room where the sound source is located [11]:

$$P_k^A = \frac{A_k}{4 \cdot 10^{12}} \cdot 10^{0.1 \cdot L_{pk}}\ , \tag{9}$$

where L_{pk} – sound pressure level measured at a number of different points in a room and averaged; A_k – equivalent absorption in a room (k – acoustic subsystem).

$$A_k = 0{,}163\frac{V}{T}, \quad (10)$$

where T – reverberation time, s; V – room volume, m^3.

Other powers are equal to 0, for no sound or vibration source affects the panel and the second room.

The solution consists in composing linear power equations to calculate the unknown energies $W_1^b, W_2^b, W_3^b, W_I^A, W_{II}^A$ in each subsystem. Sound level is found for a room to define its sound insulation:

$$L_k = 10lg\frac{W_k^A \cdot c_o}{P_o \cdot V_k}, \quad (11)$$

where c_0= 344 m/s – speed of airborne sound; P_0 = 10^{-12} W – power threshold value.

2.2 Energy coupling factors

2.2.1 Energy coupling factors for panel and air volume of a room

Energy coupling factors for a panel and air volume of a room can be found using the Cramer's formula [12]

$$\eta_{ik}^{bA} = \frac{p \cdot c_o \cdot \sigma_i}{2\pi \cdot f \cdot p_i \cdot h_i}, \quad (12)$$

where $\rho \cdot c_0$= 410 H s/m^3 – acoustic impedance of air (ρ – air density, c_0 – speed of airborne sound); ρ_i and h_i – volumetric weight and thickness of i structure; σ_i – sound radiation coefficient.

$$\sigma = \frac{U \cdot \lambda_c}{\pi^2 \cdot S} \cdot \sqrt{\frac{f}{f_c}}, \text{ with } f << f_c \quad (13)$$

$$\sigma = 0.45 \cdot \sqrt{\frac{U}{\lambda_c}}, \text{ with } f = f_c \quad (14)$$

$$\sigma = 1, \text{ with } f << f_c, \quad (15)$$

where S – panel area; U – panel perimeter; λ_c – wave length at cutoff frequency; f_c – cutoff frequency which can be found in the following formula:

$$f_c = \frac{\sqrt{3} \cdot c_0^2}{\pi h \cdot c_l}, \quad (16)$$

where c_l – longitudinal wave velocity in a panel.

The energy coupling factor for air volume and a panel η_{ki}^{Ab} can be found using inverse factor η_{ik}^{bA} defined through the formula (12) as follows:

$$\eta_{ki}^{Ab} \cdot n_k^A = \eta_{ik}^{bA} \cdot n_i^b, \quad (17)$$

where n_k^A – density of natural vibrations (modal density) in a room; k; n_i^b – density of natural bending vibrations in a panel.

2.2.2 Internal sound energy loss in a room

In the SAE model, internal sound energy loss in a room is described by internal loss factor η_k^A:

$$\eta_k^A = 13.5 \cdot \frac{A_k}{f \cdot V_k} \ , \tag{18}$$

where f – geometric mean frequency of a band; A_k – equivalent sound absorption in a room; V – room volume, m^3.

3 Results

Coefficients of energy coupling between panels

Resonant transmission is considered by the SAE model, whereas non-resonant room-to-room transmission through a wall is considered by the method of mutual coupling of sound fields.

It is suggested to determine coefficients of energy coupling between rooms using inertial transmission coefficient in a formula of R.J.M. Craik [13] which can be presented in as follows:

$$\eta_{I,II} = \frac{c_0 \cdot S \cdot \tau_{ij}}{8 \cdot \pi \cdot f \cdot V} \, , \tag{19}$$

where c_0 – wave velocity in the air, S – surface area, τ_{ij} – inertial transmission coefficient for i and j structures, V – room volume, m^3.

The inertial sound transmission coefficient is calculated using M. Sedov's formula [10]:

$$\tau = \tau_{nu} + \tau_{1u} \cdot \tau_{2u} \ , \tag{20}$$

where n indices denote sound transmission through plates with elastic coupling. At the same time,

$$\tau_{nu} = 1 / \left\{ \left[\frac{\pi^2}{p_o^2 c_o^2} \cdot \frac{m'^2 f^2}{F_{1u.cp}^2} \left(\frac{f^2}{f_o^2} - 1 \right)^2 \right] + 1 \right\} . \tag{21}$$

Here, $m' = m_1 + m_2$, where $m_1 m_2$ – surface of the first and the second structures, F_{1u} – response function of the first plate affected by sound [10], f_0 – resonant frequency of the system "mass-elasticity-mass":

$$f_o = 60 \sqrt{\frac{m_1 + m_2}{L m_1 m_2}} \tag{22}$$

L – distance between the plates, m.

Inertial sound successive transmission through the first and the second plates divided by air space is determined using the following coefficients:

$$\tau_{1u} = 1 / \left(\frac{\pi^2}{p_o^2 c_o^2} \cdot \frac{m_1^2 f^2 cos^2 \theta_1}{F_{1u.cp}^2} + 1 \right) \tag{23}$$

$$\tau_{2u} = 1 / \left(\frac{\pi^2}{p_o^2 c_o^2} \cdot \frac{m_2^2 f^2 cos^2 \theta_2}{F_{2u.cp}^2} + 1 \right) \ , \tag{24}$$

where θ_2 – angle of incidence between the second plate and the sound waves forming natural vibrations in the air space along the planes of the plates. In the majority of actually used structural solutions for two-panel partitions $\cos\theta_2 = L/\sqrt{a^2 + L^2}$ [10].

4 Conclusion

This paper presents a physical model of a two-panel partition with an adjoined panel. Systems of equations describing energy coupling between separate structural elements are produced.

The suggested calculation method accounts both resonant and non-resonant sound transmission forms while defining energy coupling coefficients. The obtained outcomes are based on complete analysis of sound transmission through the joint of a two-panel partition.

Further extension of the performed research require experimental study of the developed calculation method. The experiment shall solve both direct and inverse problems of SEA with calculation of energy coupling coefficients between panels and between air volumes.

References

1. A.A. Kochkin, I.L. Shubin, *Zvukoizoljacija legkih ograzhdajushhih konstrukcij zdanij iz jelementov s vibrodempfirujushhimi slojami* [Sound insulation of lightweight envelope structures using elements with vibration damping layers], Doctoral thesis, Vologda State University, 2013 (in Russian)
2. V.N. Bobylev, V.A. Tishkov, P.A. Grebnev, D.V. Monich, Vestnik Volzhskogo regional'nogo otdeleniya RAASN [Proceedings of Volga Regional Department of Russian Academy of Architecture and Construction Sciences], **18**, 96-99 (2015) (in Russian)
3. B.G. Kreitan, *Zashhita ot vnutrennih shumov v zhilyh domah* [Protection against internal noises in residential houses], (Strojizdat, Moscow, 1990) (in Russian)
4. S.N. Ovsyannikov, *Rasprostranenie zvukovoj vibracii v grazhdanskih zdanijah* [Propagation of sound vibration in civil buildings], (TSUAB Publishing house, Tomsk, 2000) (in Russian)
5. G.L. Osipov, *Shumy i zvukoizoljacija* [Noises and sound insulation] (Strojizdat, Moscow, 1967) (in Russian)
6. V.I. Zaborov, E.M. Lalaev, V.N. Nikolski, *Zvukoizoljacija v zhilyh i obshhestvennyh zdanijah* [Sound insulation in residential and public houses] (Strojizdat, Moscow, 1979) (in Russian)
7. V.I. Zaborov, M.I. Mogilevskiy, V.N. Myakshin, E.P. Samoilyuk, *Spravochnik po zashhite ot shuma i vibracii zhilyh i obshhestvennyh zdanij* [Handbook on residential and public buildings protection against noise and vibration] (Budivjel'nik, Kiev, 1989) (in Russian)
8. I.I. Bogolepov, *Promyshlennaja zvukoizoljacija* [Industrial sound insulation], (Sudostroenie, Leningrad, 1986) (in Russian)
9. V. Blazi, *Spravochnik proektirovshhika* [Handbook of a civil engineer. Building physics], (Tehnosfera, Moscow, 2005)
10. M.S. Sedov, News of Higher Educational Institutions. Construction and Architecture, **2**, 37-42 (1990) (in Russian)
11. V. Anert, V. Reihard, *Osnovy tehniki zvukousilenija* [Fundamentals of sound reinforcement equipment], (Radio i svjaz', Moscow, 1984) (in Russian)

12. L. Cremer, M. Heckl, E. Ungar, *Structure-borne sound*, (Springer Verlag, Berlin, 1973)
13. R.J.M. Craik, R.S. Smith, Applied Acoustics, **61,** 223-245 (1999)
14. F.G. Leppington, K.H. Heron, E.G. Broadbent, S.M. Mead, Proceedings of the Royal Society of London, **412**, 309-337 (1987)

Criteria of choosing building structures for rooftop boiler rooms

Artyom Plotnikov[1,*], and *Tatiana* Zhilina[1]

[1]Tyumen Industrial University, 625000 Tyumen, Russia

Abstract. The paper investigates parameters of noise and vibration distribution in the territory of residential area depending on the structural materials and power of independent heat supply systems. Rooftop boiler rooms are decentralized heat supply systems in buildings.Today, residential areas are strongly affected by noise and vibrations. Adverse effects are isolated by buildings materials, protective shields and floating floors. Rooftop boiler rooms located in Tyumen city were investigated within this research. Structures of rooftop boiler rooms were analyzed. Acoustic analysis results and the parameters of equivalent continuous sound level are presented. An option for improvement of rooftop boiler rooms structures is suggested. Comparison of capital investments in construction and installation activities is carried out. Conclusion on capital investments required for noise protection is made.

Introduction

Development of cities and residential areas leads to growth of engineering networks in order to provide comfortable living conditions in new districts. In particular, the issue of providing heating to the Russian cities is connected with certain climate conditions and long heating period [1] and is one of the top priorities of the government [2].

The existing techniques of providing heating are represented by two standard solutions [3]:

- centralized heating supply systems, i.e. heating supply of a big residential area of an industrial complex with a common heat network of one or several heating energy sources;
- decentralized heating supply systems (independent), i.e. heating supply of one consumer from one heating energy source.

This paper investigated the parameters of noise and vibration distribution [4] in the territory of residential area [5, 6] with regard to structural building materials and power of decentralized (independent) heating supply systems. The purpose of the study is to substantiate the minimum number of reasonable criteria for choosing structures for construction of rooftop boiler rooms.

Rooftop boiler rooms are decentralized heating supply systems in buildings. One of the major issues of independent heating sources organization is the proximity of technical equipment which uses natural gas [7] in devices that do not belong to the category of

*Corresponding author: stereo-twist@yandex.ru

hazardous production facilities of high hazard [8] to the residential premises. Side effects of such proximity are noise and vibrations, which inherently go along with each other and affect the residential area. The issue of proximity of people, technical equipment and engineering networks is highly relevant for urban planning of new residential districts and infill development [9-13].

Materials and Methods

The methods of noise and vibration control that are used in construction and energy production industries shall account both cost and efficiency. People are isolated from adverse effects by means of building materials, protective shields and floating floors, i.e. by everything that absorbs and reflects noise and vibration.

Experiment

Rooftop boiler rooms located in the city of Tyumen (Russia), in "Mys" residential district under construction (planning area No. 14), were selected for experimental research within the study. Full-scale measurements [14] were carried out by using ASSISTENT TOTAL sound level meter.

Structures of several rooftop oiler rooms were analyzed. Then, the analysis of structural properties of sound absorption [15, 16] and full-scale measurements were performed. The following conclusion was made based on the results: structures are selected with the account of minimum cost and, consequently, have minimum protective effect. The used structural materials of walls, foundations and roof are presented in Figure 1.

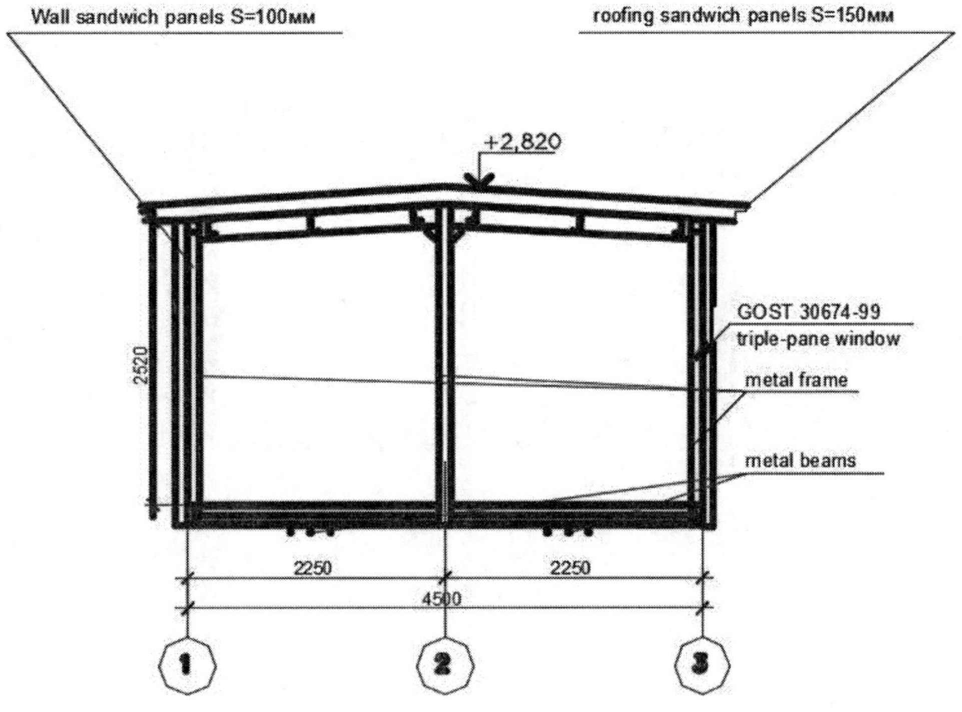

Fig. 1. Structural elements of rooftop boiler rooms.

Boiler rooms with the power of 1.0; 1.5; 2.0 MW were accepted for analysis.

After conducting acoustic analysis of boiler rooms [15, 16, 17] and full-scale measurements of sound level outside the boiler rooms the following results were obtained (Table 1).

Table 1. Parameters of equivalent continuous sound level in the existing boiler rooms.

Capacity, MW	Values of the analyzed parameters for octave band center frequencies in a boiler room [18, 19], Hz								Equivalent continuous sound level, dB
	63	125	250	500	1000	2000	4000	8000	
1.0	31.5	37.7	45.1	41.4	38.0	37.0	33.0	34.2	47
1.5	34.3	41.1	53.1	45.4	43.0	32.0	38.0	38.2	52
2.0	37.3	43.2	53.1	45.4	43.0	42.0	38.0	38.2	56
Standard parameters [20]	63	52	45	39	35	32	30	28	40

Data in Table 1 demonstrate the actual parameters in a boiler room for the values of 250-8000 Hz are higher than octave band center frequencies. This interval includes spectra of human hearing.

The following improvements of the boiler rooms structures are suggested: increase thickness of roof elements and wall panels and replace the type of glazing by installing triple-pane glass.

Modified boiler room structures are presented in Figure 2.

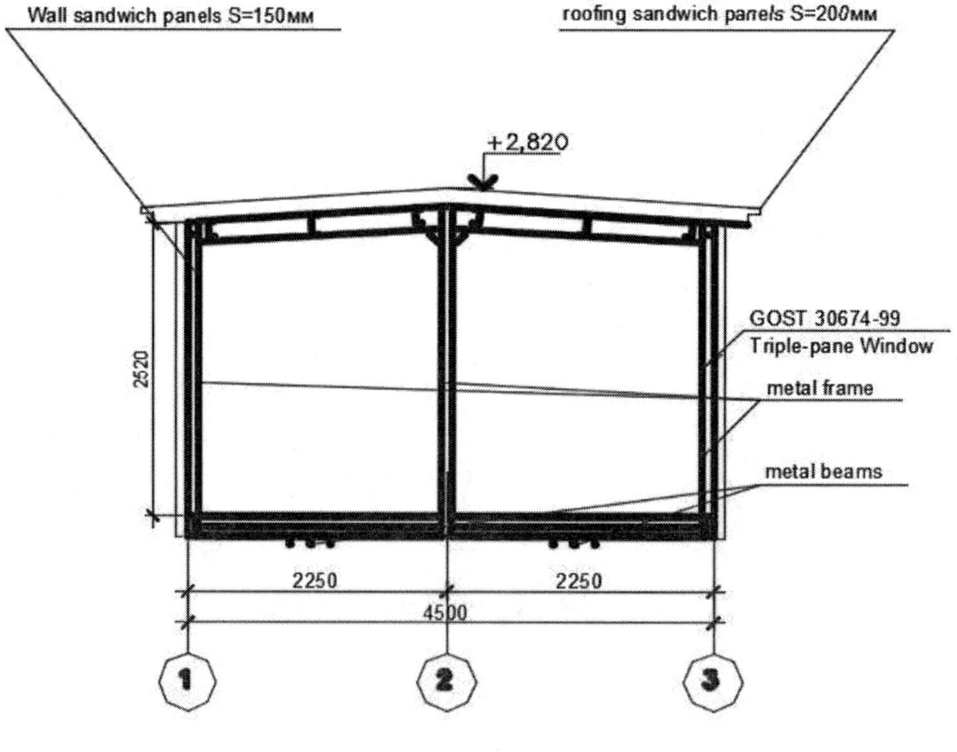

Fig. 2. Modified structural elements of rooftop boiler rooms.

Repeated acoustic analysis and full-scale measurements were performed after modifying structures and increasing the level of noise protection. The result sound level parameters are given in Table 2.

Table 2. Parameters of equivalent continuous sound level in the improved boiler rooms.

Capacity, MW	Values of the analyzed parameters for octave band center frequenciesin a boiler room [18, 19], Hz								Equivalent continuous sound level, dB
	63	125	250	500	1000	2000	4000	8000	
1.0	31.5	36.7	41.1	40.4	36.0	29.0	31.0	27.2	37
1.5	34.3	39.1	43.1	41.4	39.0	28.0	33.0	29.2	44
2.0	37.3	40.2	43.1	41.4	39.0	32.0	33.0	29.2	47
Standard parameters [20]	63	52	45	39	35	32	30	28	40
Sound level variation before and after improvement									
1.0	0	-1	-4	-1	-2	-8	-2	-5	-10
1.5	0	-2	-10	-4	-4	-4	-5	-9	-8
2.0	0	-3	-10	-4	-4	-10	-5	-9	-9

Improvement of boiler room structures results in decrease of acoustic load parameters, including spectra of human hearing. The parameters of acoustic load variations are given in the table.

Results

Envelope structures of a boiler room were modified to improve noise protection parameters. Repeated analysis of a boiler room sound absorption was performed with the account of all construction requirements [3].

Figure 3 demonstrates the comparison of capital investments in construction and installation of an independent rooftop heating supply source.

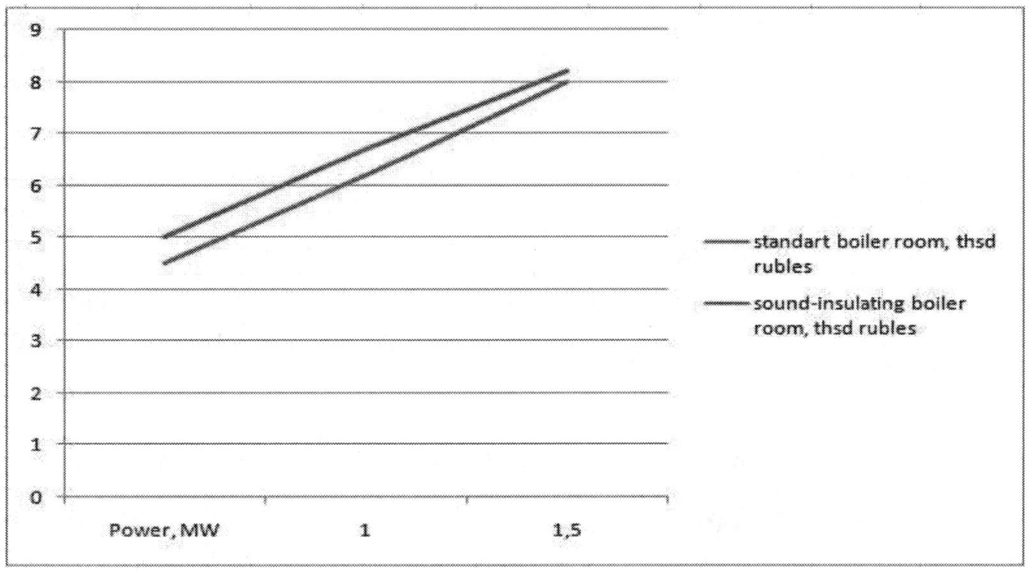

Fig. 3. Comparison of capital investment.

Production plant prices were used to perform economic comparison of the boiler room cubicle cost. The results of calculations are presented in Table 3.

Table 3. Cubicle cost before and after improvement

Power capacity, MW	Standard boiler room cubicle (Fig.1)	Sound-insulating boiler room cubicle (Fig.1)
	price, thousand rubles	price, thousand rubles
1.0	4.5	5.0
1.5	6.2	6.7
2.0	8	8.3

Data from Table 3 demonstrate that the increase of cubicle cost is between 4 and 11 % depending on the power capacity of the boiler room.

Conclusion

According to the full-scale measurement and analysis results, one can conclude that rooftop boiler rooms are a bigger source of noise and vibration. The reduced noise level of rooftop boiler rooms increases the level of living comfort and the improved structures provide higher level of noise protection of boiler rooms without considerable increase of capital investments. Further research in the field will be focused on development of recommendations to production plants on improvement of structures of boiler rooms cubicles.

References

1. SP 131.13330.2012. Building climatology. Revised edition of SNIP 23-01-99* (2012) (in Russian)
2. A.S. Plotnikov, T.S. Zhilina, Nauchnoe obozrenie. Tehnicheskie nauki [Scientific review. Engineering sciences] **6**, 84 (2016) (in Russian)
3. SP 89.13330.2016. Boiler installations. Revised edition of SNIP II-35-76 (2016) (in Russian)
4. W.J. Bottega, *Engineering Vibrations* (Taylor & Francis Group, 2015)
5. V.B. Tupov, Proceedings of the V All-Russian scientific conference "Protection against Increased Noise and Vibration", 57 (2015) (in Russian)
6. L. David; G. Benoît, INTER-NOISE and NOISE-CON Congress and Conference Proceedings, InterNoise16, 1149-1158 (2016)
7. N.K. Kiryushina, V.N. Stepanov, Proceedings of the V All-Russian scientific conference "Protection against Increased Noise and Vibration", 25 (2015) (in Russian)
8. GOST 5542-2014. Natural gas for industrial and household purpose (2015) (in Russian)
9. A.Preis, H.Honorata, S.Malina, K.Jedrzej, Noise Control Engineering Journal **64**, 1 (2016)
10. M.V.M.Oliveira Filho, P.H.T.Zannin, Performance of Sound Insulation in Buildings **21**, 3 (2016)
11. B. Fernando, Z.P.H. Trombetta, Noise Control Engineering Journal **63**, 1 (2015)
12. M.Ermann, J.Wiley, *Architectural Acoustics* (John Wiley & Sons Inc, 2017)
13. Brown, Alan Lex, INTER-NOISE and NOISE-CON Congress and Conference Proceedings, InterNoise16, 550-556 (2016)

14. GOST 23337-2014. Noise. Methods of noise measurement in residential areas and in the rooms of residential, public and community buildings (2015) (in Russian)
15. Recommendations for the acoustic calculation of boiler houses. Series ЖЗ-172 (Moscow, 1984) (in Russian)
16. GOST R EN 12354-5-2012. Building acoustics. Calculation methods for building acoustic properties on the base of their element's properties. Part 5. Engineering equipment noise (2014) (in Russian)
17. A.A. Ivanovich, A.V. Bacunova, I.L. Shubin, Vestnik of TSUAB **6,** 71 (2015) (in Russian)
18. G.L. Osipov, V.N. Bobylev, *Zvukoizoljacija i zvukopogloshhenie* [Acoustic isolation and sound absorbing] (Astrel, Moscow, 2004) (in Russian)
19. A.I. Antonov, A.V. Batsunova, I.L. Shubin, Privolzhsky Scientific Journal **3,** 85 (2015) (in Russian)
20. SP 51.13330.2011. Noise protection. Revised edition of SNIP 23-03-2003 (2011) (in Russian)

Numerical simulation of deformation behavior of aluminum alloy sheets under processing by groove pressing method

Evgeny Moskvichev[1,*], *Alexander* Kozulin[1], *Vladimir* Krasnoveikin[1,2], and *Vladimir* Skripnyak[1]

[1]National Research Tomsk State University, 634050 Tomsk, Russia
[2]Institute of Strength Physics and Materials Science, Siberian Branch of Russian Academy of Science, 634055 Tomsk, Russia

Abstract. The results of theoretical estimation of capabilities of the material structure modification of 1560 aluminum alloy sheets under processing by severe plastic deformation are presented in this paper. Severe plastic deformation of flat specimens is effected by the constrained groove pressing method in original dies with trapezoidal teeth. The numerical simulation results of the sheet specimen treatment process by severe plastic deformation were used for dies designing. The stress-strain state of flat aluminum alloy specimens and the steel dies at high processing temperature, support reaction force during pressing and the degrees of plastic strain accumulation at the optimum mode of pressing were estimated. The main numerical result is the value of accumulated plastic strain in the specimen per one pressing cycle which is about 1.14. Large degrees of strain are the reasons of grain structure and material texture changes, which leads to inevitable change of its physical-mechanical properties. Increasing the number of pressing cycles leads to proportional increase of the degree of accumulated plastic strain.

Introduction

Modification of coarse-grained crystalline structure of light aluminum alloys up to ultrafine-grained state with the use of severe plastic deformation is necessary to improve their physical-mechanical properties [1]. Specimens from sheet products, unlike the bulk specimens cannot be subjected to widely spread processing techniques, such as equal channel angular pressing and high pressure torsion [1-3], due to their shape.

New processing techniques by severe plastic deformation, for instance constrained groove pressing (CGP) [4] can be used as an alternative in these cases. During CGP, the specimen undergoes shear deformation between two grooved dies. Large amount of plastic strain can be accumulated in local parts of the plate subjected to shear deformations. One processing cycle consists of four stages: groove pressing; first flattening; sheet shifting and repetitive groove pressing; second flattening. One cycle is expected to provide a uniform

[*] Corresponding author: em_tsu@mail.ru

treatment of the flat billet [5]. Results of groove pressing application to magnesium, copper, titanium, aluminum and nickel alloys are described in details in [4].

Therefore, broadening the list of alloys able to withstand structural changes with simultaneous enhancement of their physical-mechanical properties is of great concern. This refers to aluminum alloy 1560 (Russian grade) which is widely applied in aerospace industry. Another relevant issue is development of physical-mathematical models that describe adequately the treatment process of materials of this class to define the tendencies of stress-strain state in the specimen under treatment [6-8].

The existing theoretical estimates of different authors demonstrate quite contradictory results. This could be the reason to apply various approaches in idealization of properties and boundary conditions during simulation. Moreover, the adequate physical-mathematical model will enable to assess the values of stress-strain state of flat specimen and the dies and support reaction force during pressing. Thus, it will help to eliminate errors while designing the dies and optimize treatment modes.

The main objective of the this study was focused on defining the particular aspects of sheet product treatment while using the set of tools for CGP for structure modification of flat specimens from 1560 aluminum alloy.

Materials and methods

Numerical simulation techniques were used to produce the set of tools. While choosing the optimum geometric dimensions of the dies set based on simulation and analysis of numerical results it is required to estimate the stress-strain state of specimens during treatment and support reaction force within the whole deformation range. Mathematical model for solution of the set objective is given as a differential equation system of continuum mechanics using elasto-plastic behavior model of the material under study. Finite element method for plane strain condition was used for solving the system of equations. Figure 1 illustrates geometrical models of dies with trapezoidal teeth for pressing of flat specimens and their further flattening. The dimensions of teeth height, interface distance between teeth, top horizontal width of teeth, projection of sloping plane on x axis and the thickness of the specimen were equal to 1.5 mm. Boundary conditions for the models are given as follows (1):

$$u_y\Big|_{BC} = -u_0\,,\; u_{x,y,z}\Big|_{AD} = 0\,,\; u_{x,z}\Big|_{AB,CD} = 0\,, \qquad (1)$$

where $u_{x,y,z}$ are components of displacement vector on the corresponding edges of a model, the displacement value $u_0 =$ 1.5 mm. The conditions of frictionless contact are predetermined at the boundaries of contact of the die with the flat specimen.

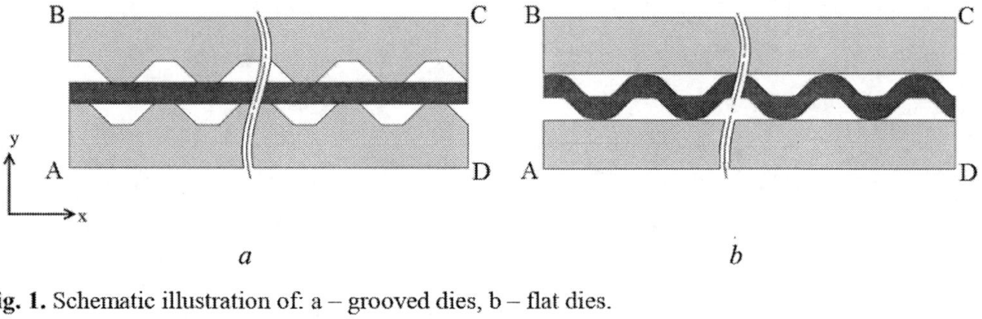

Fig. 1. Schematic illustration of: a – grooved dies, b – flat dies.

Adequacy of the constitutive equation (2) is defined by the ability to describe the effects of elastic behavior, plastic flow and hardening. In actual practice during experimental data processing and their use in calculation modules of popular engineering applications the condition of conducting power approximation is quite often followed. In this case experimental stress-strain dependences are required to be given in true coordinates. Additional condition is the necessity of introducing temperature dependence while conducting the experiment. Thus, elasto-plastic deformation behavior of aluminum alloys can be approximated by the following functions, where elastic behavior is described by the Hooke's law, while plastic flow is described by simplified Johnson-Cook power law equation [9, 10]:

$$\sigma = \begin{cases} E\varepsilon, \ \varepsilon \le \varepsilon_y, \\ (A + B(\varepsilon_{pl})^n)(1 - T_*^m), \ \varepsilon > \varepsilon_y; \end{cases} \tag{2}$$

where ε_y - yield strain; ε_{pl} - effective plastic strain; E – Young's modulus, GPa; A – corresponds to the yield stress ($\sigma_{0.2}$) of the material, MPa; B – strain hardening factor, MPa; n – strain hardening index; m – temperature softening index; T_* - homologous temperature, given as $(T - T_{room})/(T_{melt} - T_{room})$. Approximation coefficients are defined from the experimental data based on the methodology described in [10, 11].

A number of experimental works on uniaxial tension of flat specimens at the strain rate 0.01 1/s was conducted to find the constants of the constitutive equation. Experimental results are obtained as the true diagrams of elasto-plastic behavior of 1560 aluminum alloy (chemical composition: 91 % Al – 6.12 % Mg – 0.59 Mn - ~ 2.29 % other) by the temperature 297 and 473 K. Elasticity modulus (E) of the material is defined while measuring velocity of sound wave transmission using ultrasonic detector HARFANG VEO 128:16.

It is suggested that all components of the dies behave in completely elastic manner during the experiment; the material used in the experiment was quenched steel with the following specification: E (473 K) = 206 GPa; ν = 0.3; $\sigma_{0.2}$= 785 MPa.

Results

Figure 2 illustrates the approximation curves of plastic flow of aluminum alloy 1560 used in numerical simulations, temperature 297 and 497 K.

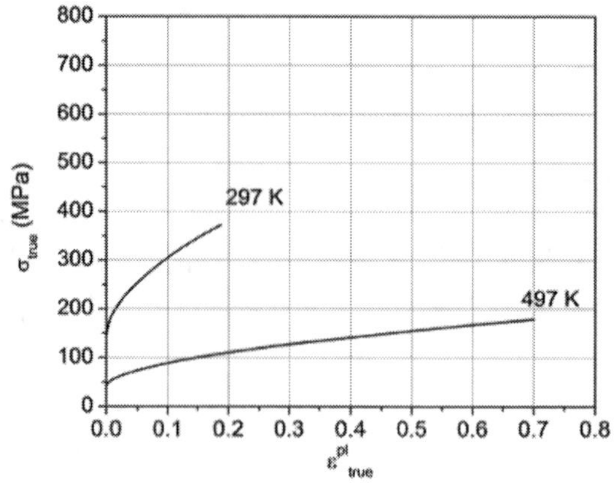

Fig. 2. True stress – true strain approximations curves of the 1560 alloy used in the numerical simulations.

Results of experimental data processing for conducting approximation are presented in Table 1. Approximation results conform well with the experimental data for the alloys AA5180, AA5754 described in [9], which are the closest analogues of the alloy under study. It was shown that during quasistatic loading increase of the ambient temperature leads to decrease of aluminum alloys resistance to plastic strain and increase of the degrees of ultimate strain.

Table 1. Data for approximation of elasto-plastic behavior of aluminum alloy 1560 under tension.

E (GPa)	*A* (MPa)	*B* (MPa)	*n* (MPa)	*m*
78	150	573	0.55	0.308

Estimate of the plastic strain value in the specimen at the stage of pressing can be conducted by analytical methods [4] using the equation (3). The von Mises equivalent strain can be obtained from the equation:

$$\varepsilon_{eff}=\sqrt{\frac{2}{9}\left[(\varepsilon_x-\varepsilon_y)^2+(\varepsilon_y-\varepsilon_z)^2+(\varepsilon_z-\varepsilon_x)^2\right]+\frac{4}{3}\left[(\varepsilon_{xy})^2+(\varepsilon_{yz})^2+(\varepsilon_{zx})^2\right]}. \quad (3)$$

Shear strain in accordance with Figure 3*a* is defined as: $\varepsilon_{xy}=\frac{\gamma_{xy}}{2}$, where $\gamma_{xy}=\frac{\Delta x}{x}=\tan 45°=1$ – engineering shear strain. The groove teeth angle α=45° is optimal from the point of reaching the maximum effect of plastic strain accumulation.

It is accepted that strain is a simple shear strain and it does not have longitudinal and transverse extension, therefore:

$$\varepsilon_x=\varepsilon_y=\varepsilon_z=\varepsilon_{yz}=\varepsilon_{zx}=0, \quad \varepsilon_{xy}=\sqrt{\frac{2\gamma_{xy}}{3}}.$$

And thus theoretical value of equivalent strain equals:

$$\varepsilon_{eff}=\left(\frac{1}{\sqrt{3}}\right)\approx 0.58.$$

Theoretical value of equivalent strain about 0.58 can be called accumulated effective plastic strain, it is maximum for this stage of pressing and as can be seen from the formulas it depends only on the geometry of the die and its sizes.

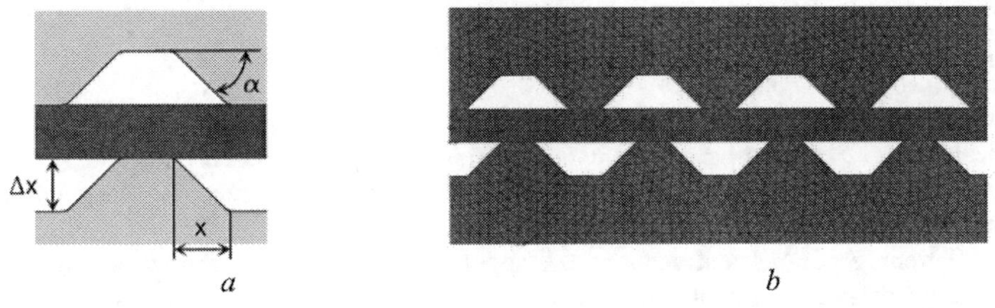

Fig. 3. The scheme of one segment of set of tools for pressing (*a*) and finite element model for calculation (*b*).

Finite element model applied in calculations is shown in Figure 3*b*. The density of the grid and the size of finite elements were selected with regard to convergence condition of numerical results. For development of finite-element grid two-dimensional linear triangular elements were used.

Results of numerical simulation are shown in Figure 4. Two stages of the treatment process are shown, i.e. groove pressing and flattening. During one stage of pressing shear strain in the specimen is accumulated non-uniformly. It can be seen that accumulation of effective plastic strain occurs in the shear zone. In numerical calculation at the given loading conditions the value of equivalent strain equals ~0.57 and differs from the analytical estimate by approximately 2 %.

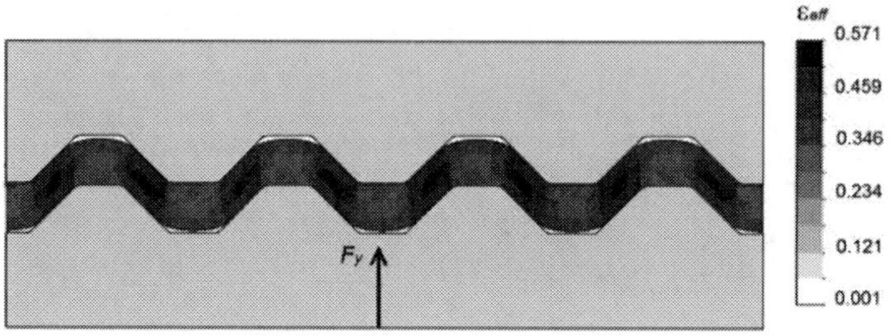

a

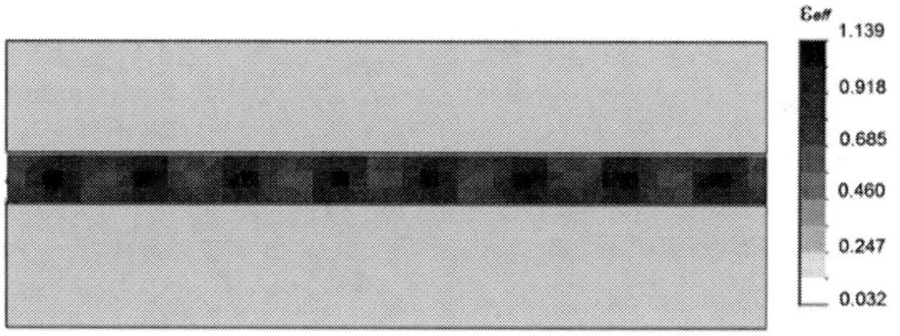

b

Fig. 4. Equivalent strain distributions in groove pressing during half cycle: *a* – first pressing, *b* – first flattening.

The values of equivalent strain obtained numerically in [6, 7] exceed the analytical values by 20% on average. This effect is explained by the authors as excessive idealization of theoretical calculation, namely assumption on absence of friction and all the components of strain tensor except for ε_{xy} being equal to zero. In practice the specimen undergoes combined stress state that brings a significant contribution to accumulation of effective plastic strain. The difference with results of other authors can be explained by assumption on plain strain state during simulation in this work – absence of strain in the direction of *z* axis. Moreover, the strain values by the same sections and sizes of grooved parts of other scientists also do not coincide.

During specimens flattening (Fig. 4*b*) the accumulated plastic strain is summarized in the shear zone, i.e. during half cycle its value in local zones is increased up to ~1.14. After the first stage of groove pressing the strained specimen is significantly elongated and its thickness decreases by 30% in the shear zone. However, due to constrained environment of further flattening geometrical sizes of the specimen recover. This aspect should be taken into account during treatment, the edges of specimens always must have dead stop. The sizes of horizontal grounds on teeth must be equal to projection of the shear zone. This condition is required for the next two stages of treatment with the grounds displacement per value equal to x (Fig. 3*a*) and providing the uniform accumulation of plastic strain along the whole length of the specimen.

The value of support reaction force F_y (Fig. 4*a*) was estimated and equaled 10.5 kN during pressing of material with section area S_{xz}=100 mm^2 and the thickness of 1.5 mm. While preparation of the set of tools its sizes shall be chosen considering the length and width of flat material and capability of the pressing tools. Thus, the force required for treatment increases proportionally to the surface of material.

Conclusion

The estimate of stress-strain state of flat billets in the original dies at the temperature 497 K was conducted based on the results of numerical simulation. The degrees of plastic strain accumulation under the optimal pressing modes were defined. Numerical results correlate with the analytical estimates within admissible errors. The maximum value of accumulated plastic strain along the whole length of the material per one treatment cycle equals 1.14, which is commensurable with one treatment cycle of bulk specimens by equal channel angular pressing. Large degrees of strain are the reason of refinement of the internal material structure [12], thus leading to changing of its physical-mechanical properties. Increase in the number of pressing cycles leads to proportional increase of the degree of accumulated strain.

Simulation results will be used while designing the components of the dies for treatment of sheet aluminum alloy 1560 up to ultrafine-grained state. Considering simulation results and the capacities of pressing tools one can define the maximum surface of the flat billet and the sizes of the dies components.

The work was supported by the RF President Grant MK-5914.2016.1.

References

1. M.V. Markushev, Letters on Mater., **1**(1), 1 (2011)
2. Y. Estrin, M.Yu. Murashkin, R.Z. Valiev, Fundamentals of Aluminium Metallurgy: Production, Processing and Applications, 468-503 (2010)
3. A.A. Kozulyn, V.A. Skripnyak, V.A. Krasnoveikin, V.V. Skripnyak, A.K. Karavatskii, Russ. Phys. J., **57**(9), 1261-1267 (2015)
4. A.K. Gupta, T. S. Maddukuri, S.K. Singh, Progress in Materials Science, **84**, 403–462 (2016)
5. E.N. Moskvichev, V.A. Skripnyak, V.V. Skripnyak, A.A. Kozulin, D.V. Lychagin, Letters on Materials, **6**(2), 141-145 (2016)
6. C. Yoon, A. Krishnaiah, U. Chakkingal, H.S. Kim, Comput. Mater. Sci., **43,** 641–645 (2008)
7. E. Hosseini, M. Kazeminezhad, Comput. Mater. Sci., **48**, 166–173 (2010)
8. K. Peng, X. Mou, J. Zeng, L.L. Shaw, K-W. Qian, Comput. Mater. Sci., **50**, 1526–1532 (2011)
9. R. Smerd, S. Winkler, C. Salisbury, M. Worswick, D. Lloyd, M. Finn, International Journal of Impact Engineering, **32,** 541–560 (2005)
10. R. Bobbili, V. Madhu, A. K. Gogia, J. Mater. Res. Technol., **5**(2), 190–197 (2016)
11. N.S. Brar, V.S. Joshi, B.W. Harris, AIP Conf. Proc., **1195**, 945-948 (2009)
12. Vladimir Krasnoveykin, Vladimir Skripnyak, Alexander Kozulin, Olga Senatova, Advanced Materials Research, **1040**, 107-112 (2014)

Features of road-climatic zoning of territories

Vladimir Efimenko[1], *Sergey* Efimenko[1], and *Alexey* Sukhorukov[1,*]

[1]Tomsk State University of Architecture and Building, 634003 Tomsk, Russia

Abstract. The paper reflects the Russian and foreign experience of road-climatic zoning. It demonstrates the need to refine the applicable taxonomic zoning system using the taxonomic scheme "zone – subzone – road district". When creating a database for road-climatic zoning, the authors of this paper, in contrast to the scheme adopted in Russia and abroad, take into account territorial patterns of water and heat processes that occur in road structures. The paper justifies soil characteristics for designing pavements in the study area.

Introduction

The presented studies focus on a set of connections and patterns that characterize water and heat processes in subgrades of road facilities in Western Siberia. These processes are affected by natural and climatic conditions that are typical for a particular region.

Being a linear engineering structure, a highway can be located in various natural, climatic, soil and hydrological conditions along its entire length. These conditions are elements of the geographic complex. When designing and constructing highways, experts rely on results of road-climatic zoning to take elements of the geographic complex into account. The road-climatic zoning, which is reflected in technical regulations of the Russian road industry, is based on the results of long-term and thorough research obtained mainly for the road network in the European part of the former USSR. The "zone" taxon is taken as a unit of road-climatic zoning.

Road-climatic zoning of territories has formed the basis for developing building codes, directives and guidelines that are in effect not only in Russia [1, 2] but also in China [3, 4], Vietnam [5], the USA [6], Germany [7], the UK, Sweden [8], and other countries, including post-soviet states [9–12].

Regardless of the continent and area occupied by a particular country, road-climatic zoning of territories is based on similar principles and includes three main stages (Fig. 1).

* Publishing author: av_suhar@mail.ru

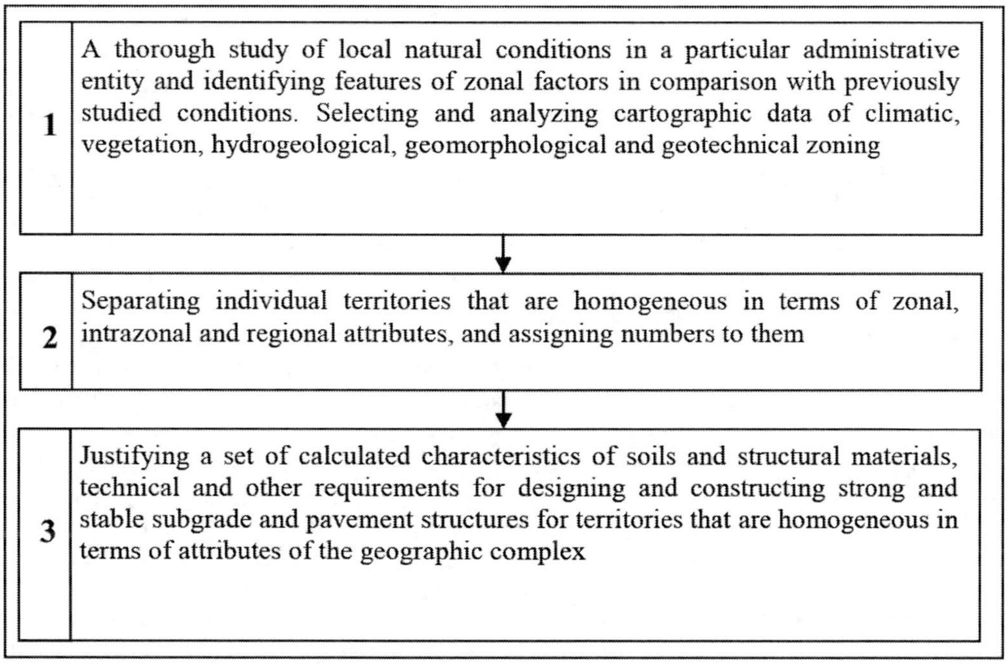

Fig. 1. The main stages of road climatic zoning.

The aim of the study presented in this paper is to ensure the reliable functioning of highways in Western Siberia.

The scientific novelty of the study lies in an effort to identify regional connections and patterns that reflect the influence of the geographic complex on water and heat processes in road structures located in territories that are homogeneous in terms of attributes of the geographic complex.

Problem Statement

According to the maps given in [1, 2], the territory of the Russian Federation has zonal differentiation and is divided into five road-climatic zones, which differ significantly in terms of natural, climatic and geotechnical conditions. In turn, zones are divided into nine subzones according to sectoral road regulations [1] and into 13 subzones according to the code of rules [2].

The US Interagency Guidance for Road Research [6] identifies 97 physical and geographic districts according to the bearing capacity of soils and their changes under the influence of moisture, as well as according to the availability of road building materials.

By analyzing road-climatic zoning in China, it is possible to conclude that it is based on methodological techniques previously used in the former Soviet Union. This is why there are identical road-climatic zones on the borders between the two countries. Depending on moisture degree, depth of soil freezing, and temperature, the territory of China is divided into seven road-climatic zones [13]. With regard to terrain and climate conditions, 33 subzones and 19 road districts are separated within each road-climatic zone [13]. In terms of climatic conditions, the most extreme road districts are located in the northern China. They belong to road-climatic zone I, which is characterized by temperatures below 0°C and the presence of permafrost [13].

However, as previously noted by specialists [14–17], the taxonomic "zone–subzone" system does not allow satisfactory division of the climatically diverse Russian territory for ensuring high-quality design and construction of transport infrastructure facilities. Therefore, we suggest that another unit, "road district", should be introduced to the taxonomic zoning system [18–21].

The road district is a genetically homogeneous territory and is a lower element of the taxonomic system "zone-subzone-district". The road district is characterized by specific climate, geology, topography, and other physical elements. Road structures of the road district (primarily subgrades and pavements) located in similar soil and hydrological conditions should be characterized by similar strength and resistance. In this context, strength means the ability to withstand loads without breaking the continuity, while resistance is the ability to change its state so that the strains associated with these changes do not exceed the permissible ones. The concept of subzone implies a territory with uniform microrelief, and the subsystem zone characterizes the earth's surface with uniform distribution of heat and moisture, determining the development of specific interrelated types of soil and vegetation.

It should be noted that a lower unit, "section" taxon, can be used for designing and constructing unique facilities. In case of mass construction, this unit can be neglected.

For road zoning based on the taxonomic scheme "zone – subzone – road district", a special method was developed [14] and tested in five territorial road administrations in Western Siberia. The proposed algorithms and solutions allowed us to localize the boundaries of road-climatic zones I–II, II–III, III–IV, as well as 112 road districts within 14 administrative entities.

Methodology Development

It should be especially noted that the main task of refining the existing road-climatic zoning practices in relation to individual Russian regions and territories is to separate territories within which the similar road structures are of uniform strength and stability. Target zoning for designing and constructing highways requires the calculation of soil characteristics for introducing subgrade and pavement structures with a specified level of reliability [22].

When calculating required strength of pavements, experts use strength and strain characteristics as the main mechanical soil properties of a subgrade. They include calculated relative moisture content W_{rel}, resilient modulus E_{sl}, angle of internal friction φ_{sl}, and specific cohesion C_{sl}. Soil strength and strain characteristics depend on moisture content, density, structure, genesis, and loading conditions. Therefore, they should be determined in the following sequence. Initially, we calculated moisture content and then, using the established relationships, find values of E_{sl}, φ_{sl} and C_{sl} at the calculated moisture.

The analyzed results of our studies [14, 15, 22] allowed us to approach the problem of forecasting design soil characteristics in a way different from the existing probability calculation methods [23]. In particular, in relation to the study area, spring moisture of soils in the working subgrade layer is found to depend to a large extent on its freezing conditions, which can be indirectly estimated by accumulation pattern of negative air temperatures during October–December period.

Considering that a significant part of the study area is characterized by excessive moisture, while soil and climatic conditions predetermine the redistribution of moisture in the subgrade in liquid form, we used a method proposed by professor I. A. Zolotar [23] for calculating moisture content of subgrade soils.

The reliability assessment of the moisture calculation methods used in the research in relation to clay soils in areas with shallow and deep groundwater level was carried out by comparing results of mathematical modeling with field observations of water and heat conditions of subgrades in Tomsk and Kemerovo regions. At reliability of forecast P = 0.95, the error of moisture calculation was found not to exceed 4%, i.e. to be within the accuracy limits of the measurement by existing methods [10].

For example, based on a set of theoretical and experimental studies, 112 road districts were delineated in Western Siberia. Road-climatic zone II, which has been previously separated according to principle of excessive moisture and normalized in its boundaries, does not include 60% of the Kemerovo Region, about 20% of the Altai Territory and at least 20% of the Novosibirsk Region. We proposed simulation models taking into account patterns of water accumulation in seasonally freezing road structures operating on an area of 2,451.1 thousand km^2 of Western Siberia.

As a result of our studies, soil moisture content was calculated for the working subgrade layer in terrains of the second and third types. The correlation between experimentally and theoretically determined values of spring moisture was estimated through the linear correlation coefficient, which was equal to $r = 0.82$. For a number of road districts, calculated strength and strain characteristics of clay soils of subgrades were proposed for designing highway pavements. For example, the resilient modulus of loam dusty in climatic conditions of road-climatic zone III, which is recommended in industry-specific standards, is overestimated by about 25–40% compared to the actual one (depending on the relative soil moisture). In contrast, values of angle of internal friction and specific cohesion are underestimated by 5–30% and 10–45%, respectively, compared to experimentally determined values.

Conclusion

Taking the Russian and foreign experience into account, this paper presents the rationale for road-climatic zoning based on the taxonomic scheme “zone – subzone – road district”.

Regardless of the continent and the area of a particular country, road-climatic zoning of territories was found to be based on attributes of the geographic complex.

Theoretical and experimental studies allowed us to separate four road-climatic zones (I, II, III, IV) and three subzones (flat, undulated, mountainous), as well as to localize boundaries of 112 road districts within Western Siberia, reflecting distinctive attributes of the geographic complex for each of them.

A set of calculated moisture content, strength and strain values of clay soils used for designing road pavements was justified and refined. For convenience of practical application, values of the characteristics determined according to the established empirical regression relationships are given in tabular form. The presented results can contribute to reliable functioning of highways in various conditions of the geographic complex.

Further research should be aimed at improving the zoning model for separating areas that are homogeneous in terms of attributes of the geographic complex, taking into account water and heat processes in road structures.

References

1. SP 34.13330.2012. Highways (2013) (in Russian)
2. ODN 218.046–01. Design of flexible pavement (2001) (in Russian)
3. Code of Practice for Highway Routes of the People 's Republic of China: JTG D20-2006, (2006) (in Chinese)
4. Standard of Climatic Zoning for Highway: JTJ 003-86, (1986) (in Chinese)
5. Highway – Specifications for design: TCVN 4054: 2005, (2005) (in Vietnamese)

6. “Filing system of physiographic units helps to resolve local design criteria”, In: *Highway Research News*, 51 (1973)
7. Richlinien fur die Standartisierung des Oberbaues von Verkehrsfiuchen: RStO 01, (2001) (in German)
8. D. Groney, *The design and performance of road pavements* (Transport and road research laboratory, London, 1977)
9. TKP 45-3.03-112-2008, Highways. Flexible pavement. Design rules (2009) (in Russian)
10. VBN B.2.3-218-186-2004, Flexible pavement (2004) (in Ukraine)
11. SNiP RK 3.03-09-2006*, Highways (2014) (in Russian)
12. SNiP RK 32–01:2004, Design. Highways (2004) (in Russian)
13. M. Yinghao, W. Binggang, Journal of Geographical Sciences 14, (2004)
14. V.N. Efimenko, S.V. Efimenko, A.V. Sukhorukov, Sciences in Cold and Arid Regions **7**(4) (2015)
15. V.N. Efimenko, S.V Efimenko, A.V. Sukhorukov, IOP Conf. Series: Materials Science and Engineering **71,** 012049, (2015)
16. V.M. Sidenko, O.T. Batrakov, M.I. Volkov, Highways (Improvement of design techniques and construction), (1973) (in Russian)
17. T.V. Bobrova, E.A. Bedrin, A.A. Dubenkov *Modelirovanie proektnyh reshenij zemljanogo polotna v uslovijah kriolitozony* [Simulating of design solutions of the subgrade in the conditions of the permafrost zone] (2016) (in Russian)
18. C.E. Zapata, W.N. Houston, NCHRP report, 602 (2008)
19. L. Chao, W. Yu-lan, X. Jin-liang, Applied Mechanics and Materials, 353–356 (2013)
20. L. Xu, L. Nai-xing, Journal of Chongqing Jiaotong University, 03 (2006)
21. V.A. Yarmolinsky, Vestnik of TSUAB **5** (2014) (in Russian)
22. V.N. Efimenko, S.V. Efimenko, A.V. Sukhorukov, Key Engineering Materials **683** (2016)
23. I.A. Zolotar, Cold Reg. Res. and Eng. Laboratory Hanover (1972)

Strength of tensed and compressed concrete segments in crack spacing under short-term dynamic load

Zaur Galyautdinov[1,*]

[1]Tomsk State University of Architecture and Building, 634003 Tomsk, Russia

Abstract. Formation of model describing dynamic straining of reinforced concrete requires taking into account the basic aspects influencing the stress-strain state of structures. Strength of concrete segments in crack spacing is one of the crucial aspects that affect general strain behavior of reinforced concrete. Experimental results demonstrate significant change in strength of tensed and compressed concrete segments in crack spacing both under static and under dynamic loading. In this case, strength depends on tensile strain level and the slope angle of rebars towards the cracks direction. Existing theoretical and experimental studies estimate strength of concrete segments in crack spacing under static loading. The present work presents results of experimental and theoretical studies of dynamic strength of plates between cracks subjected to compression-tension. Experimental data was analyzed statistically; the dependences were suggested to describe dynamic strength of concrete segments depending on tensile strain level and slope angle of rebars to cracks direction.

Introduction

Complexity of modern structures, as well as the variety of aspects affecting their operation requires comprehensive approach applied to the developed models of dynamic analysis of reinforced concrete structures. The existing reinforced concrete models enable to take into account basic peculiar features of concrete and reinforcement straining at all stages of structure performance [1-8]. Along with that, a number of issues still require additional study, particularly in the conditions of high-velocity impact.

It is known that in flat-strained reinforced concrete structures cracks occur in the areas of principal tensile stresses; and they can be generally cross-oriented towards the bars of reinforcement mesh. After crack occurrence concrete cannot sustain tensile stresses within the direction which is perpendicular to the crack direction, therefore all tensile stresses within this direction are transmitted through the reinforcement. Stresses in orthogonal direction are withstood by concrete segments in crack spacing.

Strength of such concrete segments under static loading was studied by Russian and foreign scientists [1, 9-12]; a review of the mentioned works was given in [13]. Experimental studies provide the data on influence of the samples strain level in the

* Corresponding author: gazr@yandex.ru

direction which is perpendicular to the crack and the influence of the slope angle of rebars towards the cracks direction. Authors suggested equations for calculation of strength of concrete segments between cracks under static loading that take into account the aspects mentioned above [13]. However, peculiar features of strain of concrete segments in crack spacing under intensive dynamic loading have not been previously studied.

The present paper aims to study strength of concrete segments in crack spacing under short-term dynamic loading.

Materials and methods

Experimental and theoretical analysis of strain of tensed and compressed concrete segments in crack spacing under short-term dynamic loading was conducted. Studies considered both influence of strain level of the sample perpendicular to the crack as well as the slope angle of rebars towards the cracks direction. Reinforced concrete plates with fixed cracks were used for experimental studies. The total amount of tested samples was 72, including: 23 samples under $\alpha = 90°$; 28 samples under $\alpha = 45°$ and 21 sample under $\alpha = 30°$. Celluloid fillings were mounted in all the samples while placement of concrete, which during the experiment lead to formation of cracks separating concrete blocks. Rebars were placed inclined to the crack. The slope angle of rebars to the crack α was accepted as equal to 90°, 45° and 30°. Samples with $\alpha = 30°$ and $\alpha = 45°$ were mounted with metal plates (Pos. 2, Fig. 1) along the exposed face which prevented mutual displacement of rebars. Figure 1 shows the basic sizes and reinforcement schemes of samples.

Experimental studies

Experimental samples were made from heavy weight concrete, of B15 brand. Material used for reinforcement mesh was rebars of A400 brand, 8 mm in diameter.

Simultaneously with these samples 6 reference cubes were made with the edge size of 100 mm, and 6 prisms 100×100×400 mm to define strength and strain properties of concrete. All cubes were subjected to static loading. Prisms were tested under static (3 prisms) and short-term dynamic (3 prisms) loading. Dynamic testing of prisms was conducted under strain rate $\dot{\varepsilon}_b$ which was close to the strain rate of reference samples, which allowed defining the ratio of dynamic strength of concrete segments in crack spacing to dynamic strength of standard prisms.

During the experiment loading was applied to the sample in two mutually orthogonal directions. The sample was subject to tension within one direction and compression within another (Fig. 2). Tensile stresses were completely transmitted through the reinforcement. Compressive stresses were transmitted through the concrete blocks between cracks (Fig. 2). Testing was conducted within two stages, first the tension was applied up to the required level, and then it was followed by compression up to complete failure. Tension was marked by crack occurrence dividing the sample into three concrete blocks connected with each other by reinforcement.

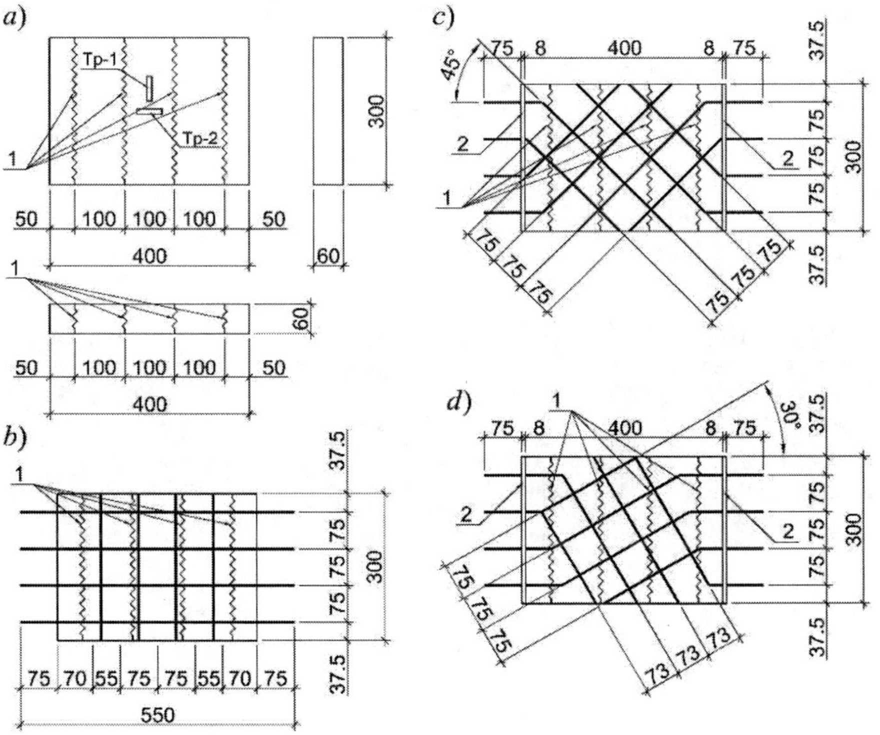

Fig. 1. General view of the samples and placement of strain-gauge sensors on concrete (*a*) and schemes of samples reinforcement under different slope angles of rebars to the cracks (*b*, *c*, *d*): 1 – cracks; 2 – metal plates.

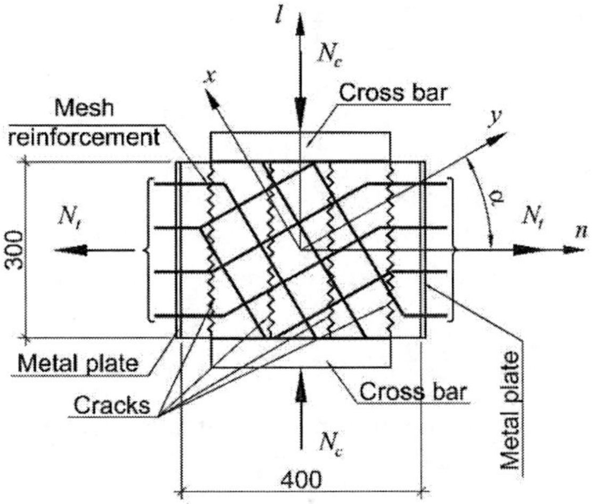

Fig. 2. Testing configuration.

Tensile loading was applied by means of hydraulic jack. Dynamic compressive stresses were provided by falling weight energy. The height of falling weight and its size was selected under condition of elasto-plastic strain of structure up to complete depletion of load-bearing capacity by single loading.

Results

Figure 3 demonstrates fracture patterns of samples with different angles α . As can be seen from Figure 3, with almost the same level of tensile strain $(\varepsilon_l = 111.1\times10^{-5}...184.7\times10^{-5}\text{ u.t.s.})$ for samples with $\alpha = 90°$ fracture of a concrete segment is local, however for samples with $\alpha = 30°$ and $\alpha = 45°$ one can observe almost complete fracture of the concrete segment. Ultimate fracture load for samples was 106.65 kN $(\alpha = 90°)$, 102.9 kN $(\alpha = 30°)$ and 89.89 kN $(\alpha = 45°)$.

a) *b)* *c)*

Fig. 3. Fracture patterns of samples: *a* – under $\alpha = 90°$ and $\varepsilon_l = 184.7\times10^{-5}$ u.t.s. ; *b* – under $\alpha = 30°$ and $\varepsilon_l = 152.41\times10^{-5}$ u.t.s. ; *c* – under $\alpha = 45°$ and $\varepsilon_l = 111\times10^{-5}$ u.t.s.

Figure 4 shows the data on strength of concrete segments in crack spacing under different strain level of samples ε_l, for the accepted slope angles of rebars. It can be seen from the study that the strength of such concrete segments decreases with increase of the strain level perpendicular to the crack ε_l for all values of slope angle of rebars to the crack α . Moreover, for $\alpha = 30°$ and $\alpha = 45°$ larger decrease of strength is observed. This is conditioned by fracture of concrete structure and adhesion of concrete segments in crack spacing with reinforcement under tension. The higher the strain level ε_l is, the more materials structure is fractured.

Experimental studies of strength of concrete segments in crack spacing enabled to estimate qualitatively the influence of the investigated aspects. Quantitative evaluation of the strain influence ε_l under different slope angle of rebars α on the strength of concrete segments in crack spacing was performed using mathematical statistics methods [14].

Theoretical studies

The obtained experimental data testify on the nonlinear change of strength of concrete segments depending on the strain level ε_l (Fig. 4, 5*a*). The largest strength decrease was observed at the initial stage by strain up to $\varepsilon_l = 400\times10^{-5}$ u.t.s. , after that the intensity of strength decrease went down. In this regard, at the initial stage the data was transformed to the linear view, it was followed by the assessment of data samples including screening of gross errors and conformance inspection of strength distribution in concrete segments in crack spacing to the normal distribution law. Results for samples with $\alpha = 30°$ are presented in Figure 5. Screening of gross errors was conducted with the aid of Student's t-distribution. Checking data distribution in accordance with the normal distribution law performed based on Kolmogorov-Smirnov and Shapiro-Wilk criteria showed their correspondence to the normal distribution law (Fig. 5*c*, *d*).

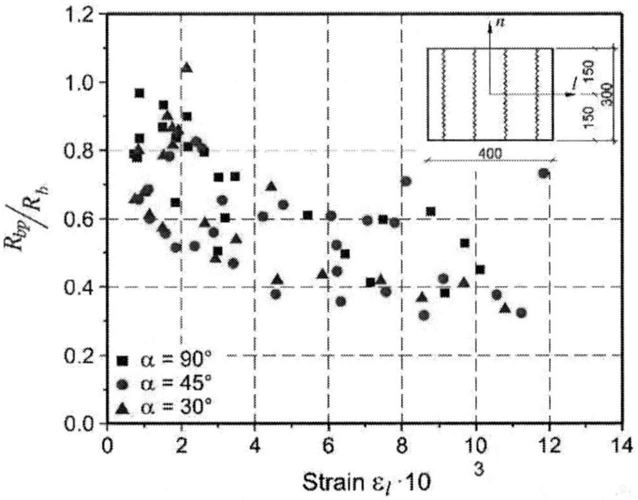

Fig. 4. Change of strength of concrete segments in crack spacing under short-term dynamic loading depending on the slope angle and strain level.

The calculated correlation coefficients characterizing the dependence of strength and strain level ε_l, was equal: to 0.804 for $\alpha = 90°$; to 0.652 for $\alpha = 45°$; to 0.811 for $\alpha = 30°$. This testifies on the average and high degree of connection of the values under study.

Regression analysis was conducted using methodology described in [14]. Based on the least square method the regression equations were derived, which can be given as follows after inverse transformation:

- for $\alpha = 90°$

$$\frac{R_{bpd}}{R_{bd}} = \frac{1}{\left(0.7043 + 0.1639 \cdot \lg(\varepsilon_l)\right)^4}, \tag{1}$$

- for $\alpha = 45°$

$$\frac{R_{bpd}}{R_{bd}} = \frac{1}{\left(0.7529 + 0.1639 \cdot \lg(\varepsilon_l)\right)^4}, \tag{2}$$

- for $\alpha = 30°$

$$\frac{R_{bpd}}{R_{bd}} = \frac{1}{\left(0.65595 + 0.2023 \cdot \lg(\varepsilon_l)\right)^4}. \tag{3}$$

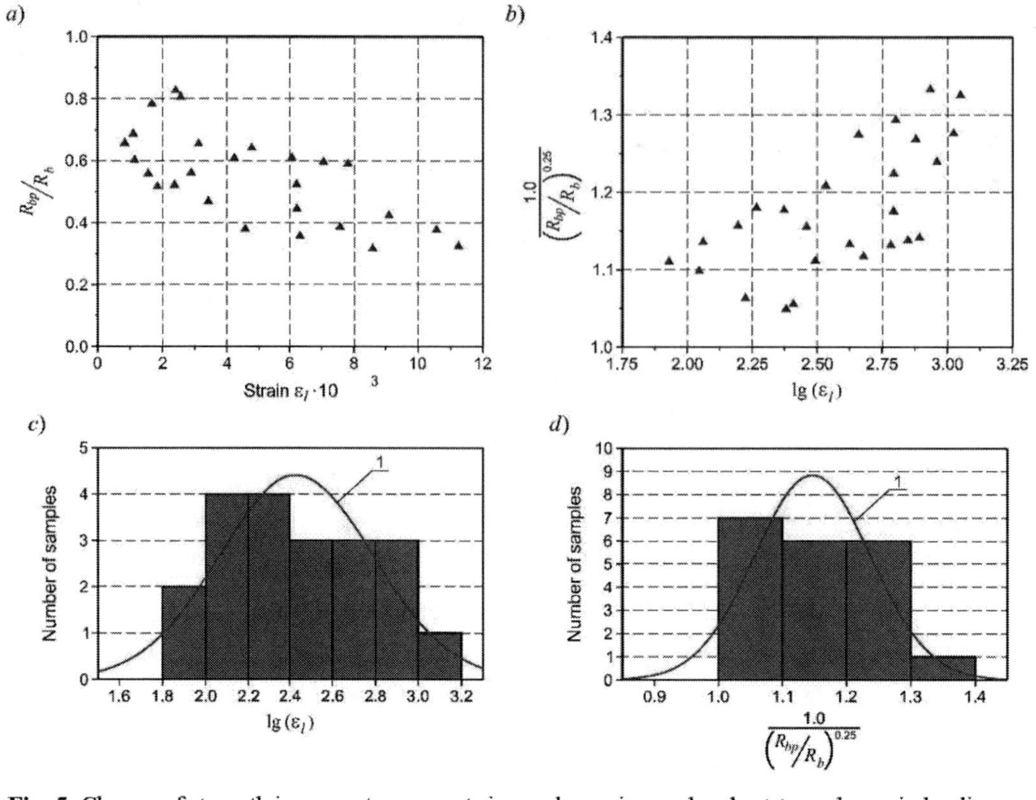

Fig. 5. Change of strength in concrete segments in crack spacing under short-term dynamic loading for $\alpha = 30°$ (*a*, *b*) and experimental and theoretical distribution of the considered values after their transformation to linear view (c, *d*): 1 – normal theoretical distribution.

To estimate significance of the obtained equations Fisher's criteria was used which has the following values: $F_{(1,21)} = 38.523$ for $\alpha = 90°$; $F_{(1,24)} = 17.732$ for $\alpha = 45°$; $F_{(1,18)} = 34.655$ for $\alpha = 30°$. The obtained values were higher than the tabular ones corresponding to 5% probability of the correspondently equal $F^{\text{T}}_{(1,21)} = 4.3248$, $F^{\text{T}}_{(1,24)} = 4.2597$ and $F^{\text{T}}_{(1,18)} = 4.4139$. Thus, the given dependences allow defining dynamic strength of concrete segments in crack spacing with confidence probability of 0.95, as stated in regulatory documents.

Analysis of the obtained theoretical dependences given in Figure 6 shows that the strength of concrete segments of samples whose rebars have slope angle $\alpha = 30°$ and $\alpha = 45°$, is significantly lower than the strength for $\alpha = 90°$. Along with that, the strength values for samples with $\alpha = 30°$ and $\alpha = 45°$ are comparable. This nature of strength variation in concrete segments is conditioned by existence of additional stresses in concrete samples with $\alpha = 30°$ and $\alpha = 45°$ which occur during strain of reinforcement mesh, and do not occur in samples with $\alpha = 90°$.

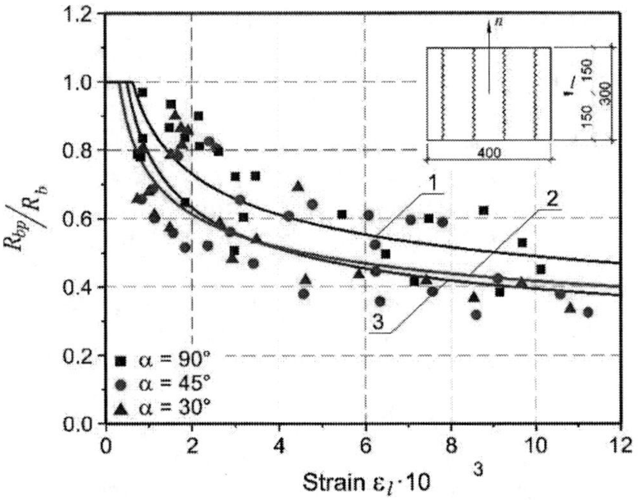

Fig. 6. Experimental and theoretical change of strength of concrete segments in crack spacing under short-term dynamic loading: 1 – according to equation (1); 2 – according to equation (2); 3 – according to equation (3); ■, ●, ▲ – experimental data.

Conclusion

Experimental and theoretical studies revealed significant influence of the strain level of samples in the direction perpendicular to the crack ε_l and the slope angle of rebars to the cracks direction α on the strength of concrete segments in crack spacing. It should be noted that the elements with angles α that differ from 90° demonstrate larger strength decrease in concrete segments compared to the elements with $\alpha = 90°$ within the whole range of tension strain ε_l.

Resulting from data processing obtained during experimental study dependences were suggested to define strength of concrete segments in crack spacing. Assessment of the given equations testifies on the sufficient reliability of the calculated values of concrete segments strength. The obtained results enable to clarify the model of dynamic straining of reinforced concrete; application of this model in calculations will allow obtaining more consistent results characterizing straining of reinforced concrete structures under intensive dynamic loading. Further step of the presented studies will be assessment of deformation property of concrete segments in crack spacing.

References

1. N.I. Karpenko, *Obshhie modeli mehaniki zhelezobetona* [General models of reinforced concrete mechanics] (Strojizdat, Moscow, 1996) (in Russian)
2. N.I. Karpenko, Structural Mechanics and Analysis of Constructions, **1**, 28–32 (1984) (in Russian)
3. J. Jones, C. Wua, D.J. Oehlers, A.S. Whittaker, W. Sunc, S. Marksa, R. Coppola, Engineering Structures, **31**, 2825–2832 (2009)
4. B. Chiaia, O. Kumpyak, L. Placidi, V. Maksimov, Engineering Structures, **96**, 88–99 (2015)
5. Z.R. Galyautdinov, News of higher educational institutions. Construction, **10**, 93–98 (2010) (in Russian)
6. X. Chen, S. Wu, J. Zhou, Construction and Building Materials, **47**, 419 – 430 (2013)

7. F.R. Barrales, Development of a nonlinear quadrilateral layered membrane element with drilling degrees of freedom and a nonlinear quadrilateral thin flat layered shell element for the modeling of reinforced concrete walls, Ph.D. thesis (University of Southern California, 2012)
8. O.G. Kumpyak, D.N. Kokorin, Vestnik of TSUAB, **4**, 101 – 112 (2015) (in Russian)
9. A.F. Yaremenko, Prikladnaja teorija dlitel'nogo deformirovanija i soprotivlenija ploskih zhelezobetonnyh elementov [Applied theory of long-term deformation and resistance of flat reinforced concrete elements], doctoral thesis (Odessa, 1987) (in Russian)
10. A. Belarbi, L.-X. Zhang, T.T.C. Hsu, Sociedad Mexicana de Ingenieria Sismica in Eleventh World Conference on Earthquake Engineering, 1–8 (Acapulco, Mexico, June 23-28, 1996)
11. B.Y. Bahn, T.T.C. Hsu, ACI Structural Journal, **95**, 178–193 (1998)
12. F.J. Vecchio, M.P. Collins, ACI Journal, **83**, 219–231 (1986)
13. O.G. Kumpyak, Z.R. Galyautdinov, D.N. Kokorin, *Prochnost' i deformativnost' zhelezobetonnyh konstrukcij na podatlivyh oporah pri kratkovremennom dinamicheskom nagruzhenii* [Strength and deformability of reinforced concrete structures on yielding supports under short-term dynamic load] (TSUAB Publ., Tomsk, 2016)
14. E.N. L'vovskij, *Statisticheskie metody postroenija empiricheskih formul* [Statistical methods of empirical formulas] (Vysshaya Shkola Publishers, Moscow, 1988)

Computer simulation of yielding supports under static and short-term dynamic load

Oleg Kumpyak[1,*], and *Nikita* Mescheulov[1]

[1]Tomsk State University of Architecture and Building, 634003 Tomsk, Russia

Abstract. Dynamic impacts that became frequent lately cause large human and economic losses, and their prevention methods are not always effective and reasonable. The given research aims at studying the way of enhancing explosion safety of building structures by means of yielding supports. The paper presents results of numerical studies of strength and deformation property of yielding supports in the shape of annular tubes under static and short-term dynamic loading. The degree of influence of yielding supports was assessed taking into account three peculiar stages of deformation: elastic; elasto-plastic; and elasto-plastic with hardening. The methodology for numerical studies performance was described using finite element analysis with program software Ansys Mechanical v17.2. It was established that rigidity of yielding supports influences significantly their stress-strain state. The research determined that with the increase in deformable elements rigidity dependence between load and deformation of the support in elastic and plastic stages have linear character. Significant reduction of the dynamic response and increase in deformation time of yielding supports were observed due to increasing the plastic component. Therefore, it allows assuming on possibility of their application as supporting units in RC beams.

Introduction

Explosive impacts refer to single emergency loads; they are characterized by high intensity and short term of action, which results not only in structural failure but also in production facilities damage and human losses. The existing approaches to designing building structures resistant to explosive impacts are based on higher material consumption, which leads to increasing the cost of objects. Therefore, the alternative economically feasible and reliable approaches are needed in design of structures that are resistant to dynamic loads.

Yielding supports can be applied as one of the active means of prevention or localization of dynamic impact or reducing the intensity of dynamic loading [1-18].

Currently, research results in the field of yielding supports application for protection of buildings and structures subjected to intensive dynamic loading are fragmentary. Experimental, theoretical and experimental-theoretical research [19-22] demonstrate both positive and negative influence of yielding supports on the dynamic response of RC beams.

* Corresponding author: kumpyak@yandex.ru

This paper focuses on the numerical study of the strength and deformation property of yielding supports given as the tubes of annular section under static and short-term dynamic loading. The study aims to obtain the data on load reduction under dynamic loading resulting from the use of yielding supports. The research problem consists in impossibility of conducting experimental study that will provide comprehensive answers to the questions on the stress-strain state of yielding supports. Such situation is caused by the complex geometry of the structure under study and the peculiar features of its straining. Along with that, the solution of the mentioned objective using mathematical package requires deep knowledge in material behavior, taking into account its physical, geometrical and contact nonlinearity under static and dynamic loading.

Research in [1-18] focuses on the elaboration of working models of tubes, and provide comparison of the obtained results with analytical and experimental studies. The works consider the cases of static and dynamic axial and transverse straining of samples made from different materials, including the combined ones (sandwich-tubes). The basic outcomes of the given works are presented as high convergence of the experimental and numerical studies. The present work differs from the above-mentioned ones. Those studies were conducted within the field of machinery and mechanics, while the given research is performed within the studies on increasing the explosion safety of building structures. Therefore, it focuses on variation of the rigidity of yielding supports and reduction of the system dynamic response.

1 Numerical simulation

1.1 Static loading

Program software Ansys Mechanical, v 17.2 was used for solving the set objectives. The research was conducted using finite element method as the most functional and accurate means of the stress-strain state analysis of the solid bodies complex straining processes.

a) b)

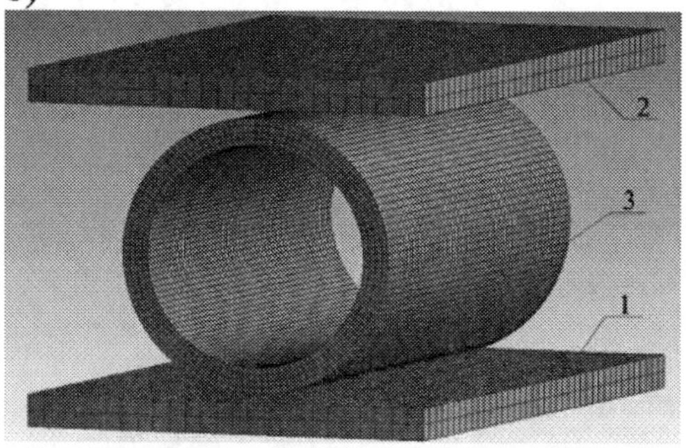

Fig. 1. The general view of yielding supports testing: field experiment (a); numerical simulation (b); 1- supporting slab; 2- force slab; 3- yielding support.

The works [8], [12, 13], [15-18] demonstrate the high degree of convergence between simulation and experimental results. Along with that, numerical studies enable to assess the state of the structure up to its failure and fragmentation of the separate parts, while experimental facilities can record the data mainly in the area of elastic strain. In case of the complex geometry, measurement of the most important data becomes impossible.

The first stage of the work included analysis of yielding supports under quasi-static loading with the constant loading velocity. The design model was idealized and formalized copy of the physical reality and it duplicated the experimental study with a large accuracy [20] (Fig. 1).

Finite element model consisted of the two slabs, one of them was supporting and motionless, the second slab served as a force slab and it had the capacity of vertical displacement. The third element of the model was yielding support. Sample spacing of the finite element mesh of the slabs was 4 mm, of the yielding support – 0.8±0.1 mm. The volume eight-node finite element SOLID185 was used as a finite element.

The analysis was performed for all the considered lengths of yielding supports in the range from 10 to 120 mm (Fig. 2). The dimensions of the solved objective depending on the length of yielding support varied from 5664 to 58464 finite elements.

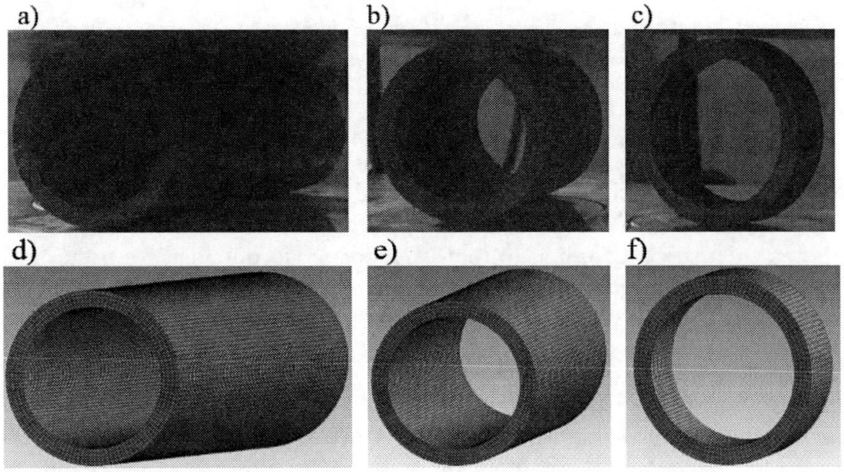

Fig. 2. The general view of yielding supports of various length: experimental studies (a,b,c); numerical finite element model (d,e,f): 120mm (a, d); 60mm (b, e); 10mm (c, f).

The work of material of yielding support was described by multilinear diagram of material straining with isotropic hardening. It was obtained by preliminary field experimental study of material under tension.

The loading of yielding support was performed by means of the given displacement of the upper slab (Fig. 1) to the distance corresponding to the inner diameter of yielding support.

1.2 Dynamic loading

Simulation of dynamic impact was conducted in the module Ansys v17.2 - explicit dynamic. The objectives were solved using Lagrangian approach to the medium motion description.

The simulation of the yielding supports operation under short-term dynamic loading was aimed at assessing the influence degree of the yielding supports on the dynamic response of the system: falling weight - yielding support - supporting slab (Fig. 3). The variable rigidity parameter in the work was changing the length of the deformable element. Cross-section and the material of yielding support were accepted in accordance with the experimental studies [20].

Within the present objective the moment of impact of the falling weight 1 on the yielding support 2 was simulated. The mass of the weight 1 corresponded to the weight used while experimental studies, which is 265 kg. For that purpose, the density of the weight was increased for each of the solved objectives.

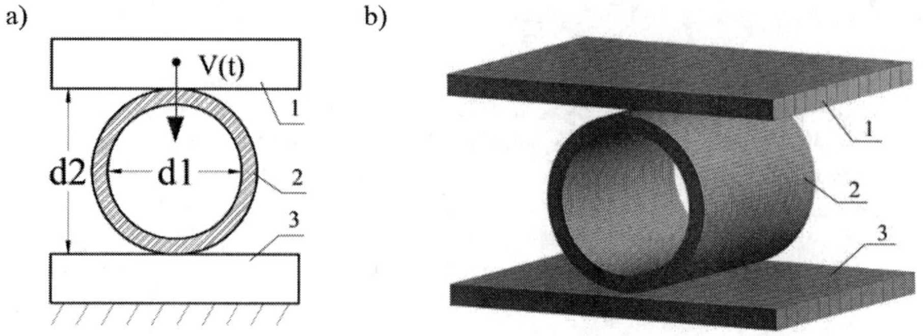

Fig. 3. Design model of the yielding support testing under short-term dynamic loading (a); general view of the finite element model (b); falling weight (1); yielding support (2); supporting slab (3); d1-inner diameter: d2-outer diameter.

The velocity of the weight 1 at the moment of impact on the yielding support 2 corresponded to the velocity of weight dropped from the height of 650 mm during experimental studies. The velocity of the weight 1 at the moment of impact on the yielding support 2 was calculated according to the dependence (1) obtained from the formulas for uniformly accelerated straight line motion and was 3.57 m/s^2.

$$S_x = \frac{V_x^2 - V_0^2}{2_{ax}}, \tag{1}$$

where: S_x- the height of falling weight (0.65 m), $V^2{}_x$- squared velocity at the impact moment of falling weight 1 and yielding support 2, $V^2{}_0$- squared initial velocity of falling weight 1 (0 m/s^2- the weight is suspended), a_x- acceleration of weight 1 (corresponds to gravity acceleration 9.81 m/s^2).

The supporting plate 3 was rigidly fixed along all degrees of freedom. Weight 1 and supporting plate 3, due to their small thickness in the model (Fig. 3 b), were accepted as completely rigid elements.

In the present work, three distinctive types of supports were considered (Fig. 9). a-completely elastic or rigid support; b- yielding support, working in elasto-plastic stage with hardening, c- yielding support, working in elasto-plastic stage.

2 Results

2.1 Static loading

Let us consider the distinctive features of yielding supports straining on the example of inserted element of annular cross-section 40 mm long (Fig. 4). As can be observed, for all the stages of yielding support work the strain pattern during field test corresponds to the strain pattern during numerical simulation.

Geometry of the studied sample during loading significantly changes its shape several times (Fig. 4) and the rigidity accordingly. Considering that fact, the model was supplemented with the option of finite displacements consideration and angular displacements after each equilibrium iteration (geometrical nonlinearity).

During straining process (Fig. 4), the yielding support interacts with the force slabs changing its design model several times. After occurrence of the plastic strain in the yielding support, the contact point with each of the force slabs is divided in two. Then, along with the load increase, the contact points distribute from the center of the support to its edges (Fig. 5). In order to consider the above-mentioned process contacting pairs were established in the model between the inserted element and the force slabs, as well as between the internal surfaces of the yielding support. The blue color in Figure 5 identifies the contact elements, which are significantly remote from the contact surface. Contact elements, which are rather close to the contacting surface, are marked with yellow color. The red color identifies contact elements interacting with the contact surface, the force slab in particular.

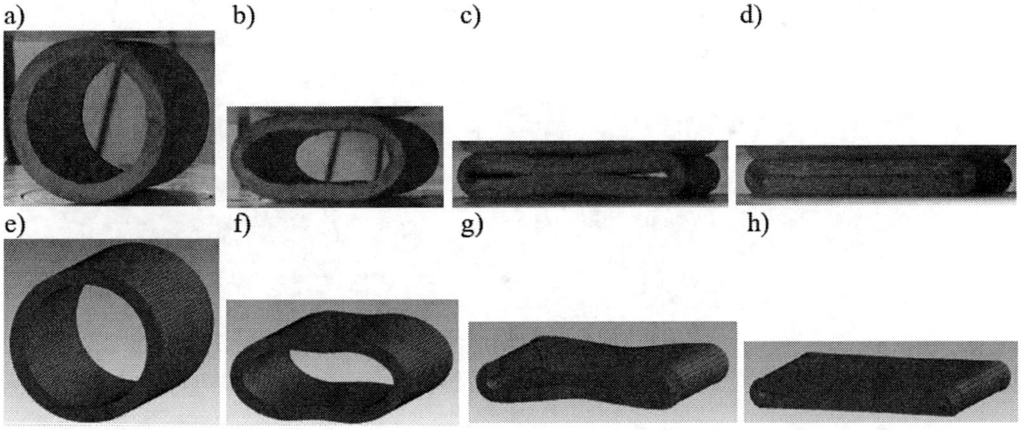

Fig. 4. Peculiar stages of yielding supports deformation: elastic (a, e); elasto-plastic (b, f); elasto-plastic with hardening (c, g, d, h).

Moreover, during yielding supports straining the contact of inner surfaces of the ring occurs (Fig. 4 c, g), which in its turn alters the simulation model one more time. Further, the internal surfaces come into contact (Fig. 4 d, h).

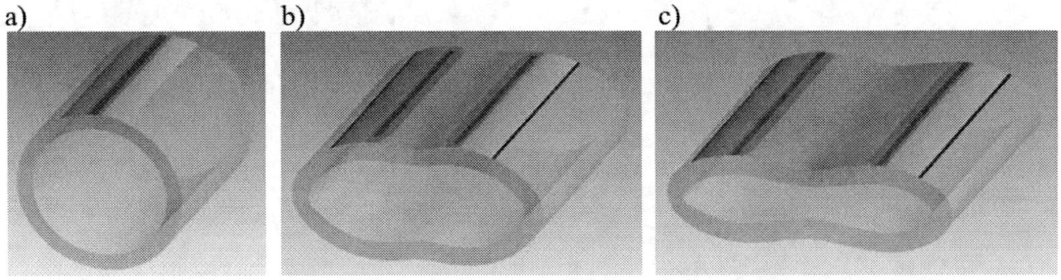

Fig. 5. Alteration of the design model of yielding support in the process of straining: one contact point with the slab (a); two contact points with the slab (b, c).

It should be noted, that during straining process distortion of the ends of yielding support occurs in the areas of plastic hinge. Thus, the contact of internal surfaces of the ring occurs not simultaneously and from the ends to the middle part (Fig. 6).

As a contact interaction interface for all contacting pairs the model of contact with friction $\mu=0.15$ was used. This model includes the possibility to divide contacting pairs after interaction and generally can contain the areas of cohesion and sliding. Advanced Lagrangian method was used as a solution method; stepwise iteration method was used for solution of the system of linear algebraic equations. The checking of non-linear objective solution was conducted using the Newton-Raphson Method. During solving the objective of static loading of the yielding supports, the solution admitted computational error of 5%.

Along with the increase in the number of elements in the solved objective in order to provide high accuracy, the proportional increase in the number of equilibrium iterations was required. Thus, the objectives considered from 500 to 8000 iterations.

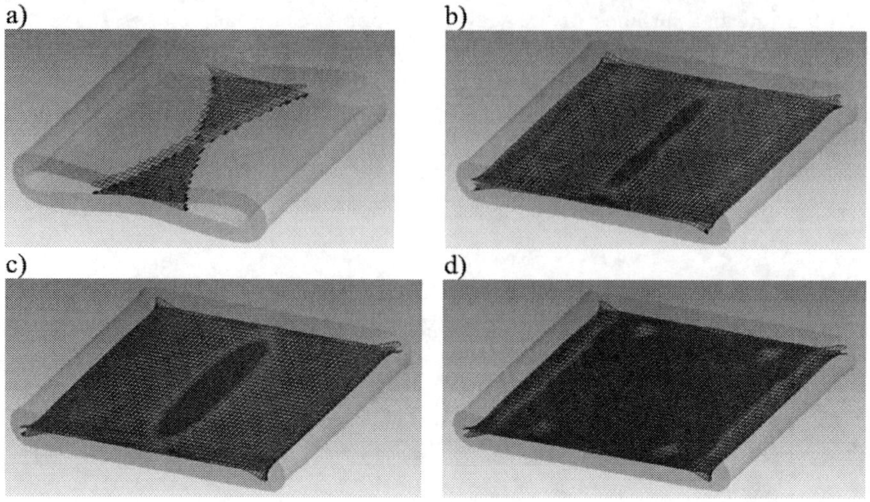

Fig. 6. Alteration in the design model of yielding support during straining: impact of contacting pairs at the edges (a); contact of internal surfaces in the mid-area (b); gradual increase of the contact spot in the middle area (c); full contact of the internal surfaces (d).

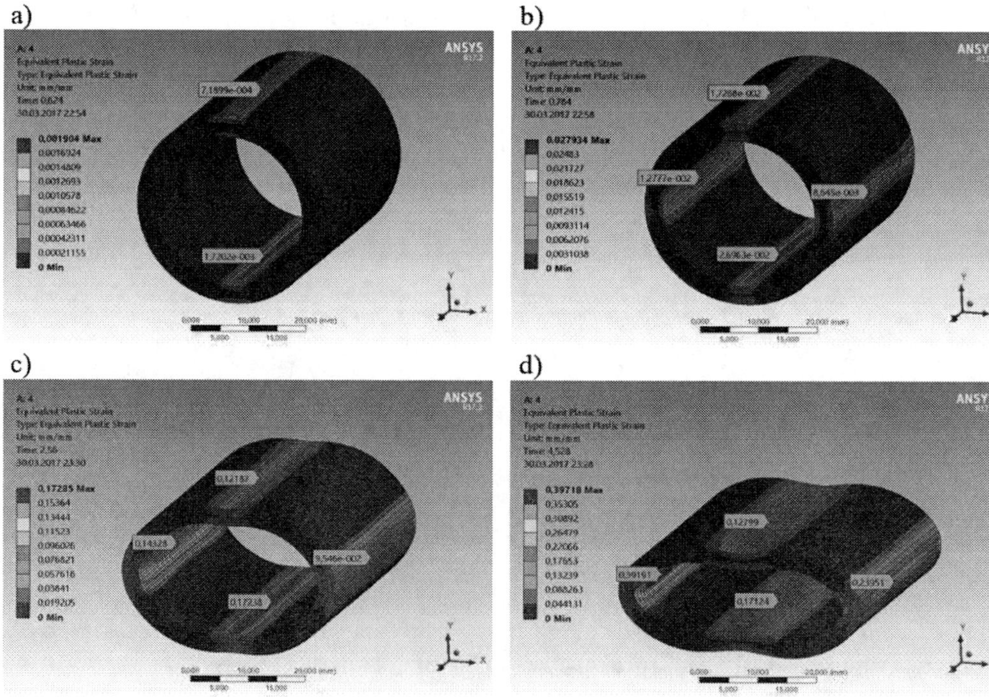

Fig. 7. Distribution pattern of the plastic strain in the yielding support.

Analysis of the results of numerical simulation demonstrated that significant straining of yielding supports occurs due to formation of plastic hinges when the stresses in the material reach the yield stress. These areas are first established in the upper and lower areas of the rings (Fig. 7*a*), and then at the lateral surfaces of the sample (Fig. 7*b*). Further straining is accompanied by the gradual growth of plastic strain and increasing the area of active straining. It should be noted that the value of relative strain at the internal lateral surface of the yielding support ε > 39% (Fig. 7*d*), which exceeds the ultimate strain of the steel 255 $\varepsilon_{s\,255} > 26\%$.

The mentioned area of straining undergoes compressive strain and due to curvilinear geometry of the sample, these locations are marked by the materials self-hardening effect. Analysis of the results of numerical and experimental studies under quasi-static loading can be conducted based on the strain diagram of yielding supports (Fig. 8).

Figure 8 illustrates the diagrams "load-displacement" for the yielding supports 10 and 20- 120 mm long with the span of 20 mm. According to the diagram, it can be judged on the high level of accordance of the experimental and numerical results at all stages of yielding supports performance.

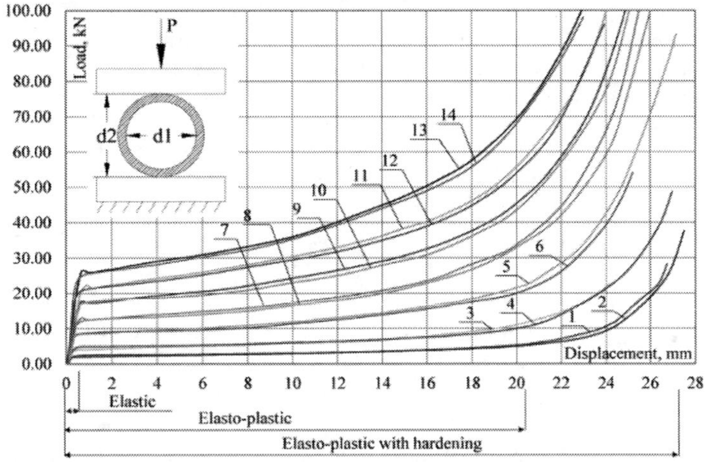

Fig. 8. The strain diagram of yielding supports of various length under numerical and experimental study; d1- inner diameter; d2- outer diameter of the yielding supports; the uneven numbers correspond to the experimental studies; the even numbers present numerical simulation. 1-2- 10 mm; 3-4- 20 mm; 5-6- 40 mm; 7-8 60 mm; 9-10-80 mm; 11-12- 100 mm; 13-14- 120 mm.

Furthermore, with the increase of the yielding supports rigidity during experimental studies the strain diagrams are marked with the "yield drop" during transition of the support work from elastic into plastic stage of hardening, however it was not observed at the diagrams of numerical simulation. This effect applies for the low-carbon steel. The occurrence of the "yield drop" is attributed to the dislocation deformation mechanism. At the initial phase, the density of dislocations is insignificant to provide larger strain degree. After reaching the upper yield stress, the intensive formation of new dislocations begins resulting in decrease of stress. Software package Ansys v 17.2 does not consider the dislocation deformation mechanisms referring to micro-structural steel composition; therefore, it leads to the absence of the "yield drop" in Figure 8.

Along with that, the local inaccuracy of the absolute value of load between the experimental and numerical result does not exceed 3.5 % and does not change significantly the general strain pattern of yielding support within the solution of the given objective.

2.1 Dynamic loading

The degree of influence of the support rigidity can be evaluated from the diagram (Fig. 13). The first case (Fig. 9*a*) presents the support, which possesses the properties of low-carbon steel, which strains according to the Hooke's law. It means that its work does not imply plastic strain occurrence. This support can be conventionally considered as a rigid as the relative strain in it under dynamic loading will be changed according to the linear law with the constant deformation modulus corresponding to the steel elasticity modulus.

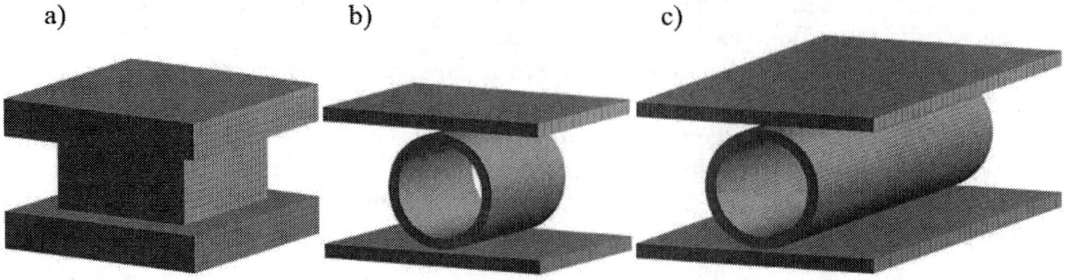

Fig. 9. Computational models of yielding support under short-term dynamic loading; totally elastic (rigid) support (a); yielding support, working in elasto-plastic stage with hardening (b); yielding supports working in elasto-plastic stage (c, d).

The support working in elasto-plastic with hardening stage (Fig. 9*b*) is characterized by the plastic straining with smooth increase in the dynamic response up to the moment when the internal surfaces of the supports contact (Fig. 10) and (Fig. 13*b*, point A). After that sharp increase in the value of dynamic response occurs.

Supports working in elasto-plastic stage (Fig. 9*c*), (Fig. 13 3, 4), depending on their rigidity can have different dynamic response diagrams. Along with the increase in plastic component of such supports increase in time of response occurs. Thus, for instance for the support 120 mm long (Fig. 13 4), compared to the support 1600 mm long (Fig. 13 3), the time of dynamic resistance increased by 2.66 times, which testifies on more plastic work of the first one. Moreover, the support of the larger length does not deplete the whole potential of its plastic properties (Fig. 11), while the support of the less length almost come into contact by the internal surfaces of the ring (Fig. 12). The decrease in the maximum value of dynamic response should be noted (Fig. 13 3, 4) from 407 kN corresponding to the support 1600 mm long to 288 kN for the support 120 mm long which makes 29.2 %.

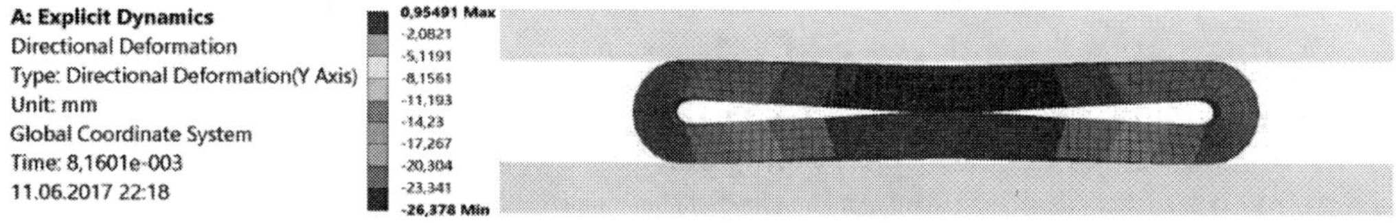

Fig. 10. The overall scheme of the support straining, working in elasto-plastic with hardening stage, the moment of transition of the support into hardening stage point A (Fig. 14*b*).

Having compared the support work in elasto-plastic stage (Fig. 13 3, 4) and the support work in hardening stage (Fig. 13 2), the reduction of the system dynamic response by 6.4 times was observed (Fig. 13).

Conducting the analysis of yielding supports work (Fig. 13 2-4) with totally elastic (rigid) support (Fig. 13 1), a significant reduction in the system dynamic response was revealed, thus, for instance for the elasto-plastic support (Fig. 13 4), compared to the rigid support (Fig. 13 1) the reduction was 23.4 times, and for the elasto-plastic with hardening support it was 6.62 times. Along with that, the increase in the time of dynamic response for yielding supports compared to the rigid ones should be specified (Fig. 13). The time of straining for rigid supports was 0.0005 sec, while for the supports working in the hardening stage it was 0.009, and for elasto-plastic it was 0.012, which is 18 and 24 times more than for the rigid supports accordingly.

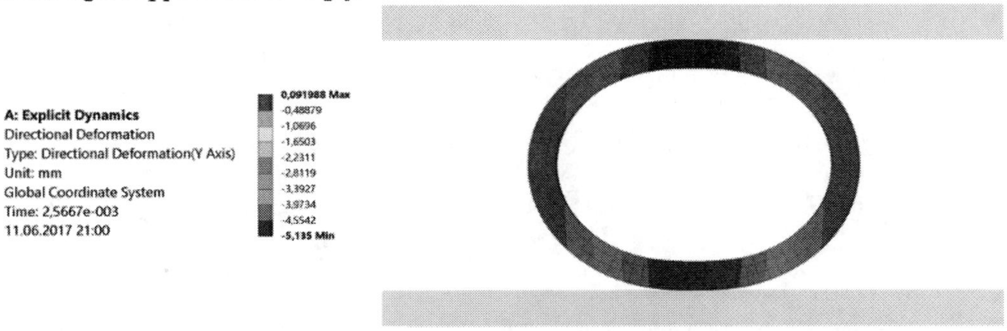

Fig. 11. The general strain pattern of the support working in elasto-plastic stage (the length of the support 1600 mm).

Fig. 12. The general strain pattern of the support working in elasto-plastic stage (the length of the support 120 mm).

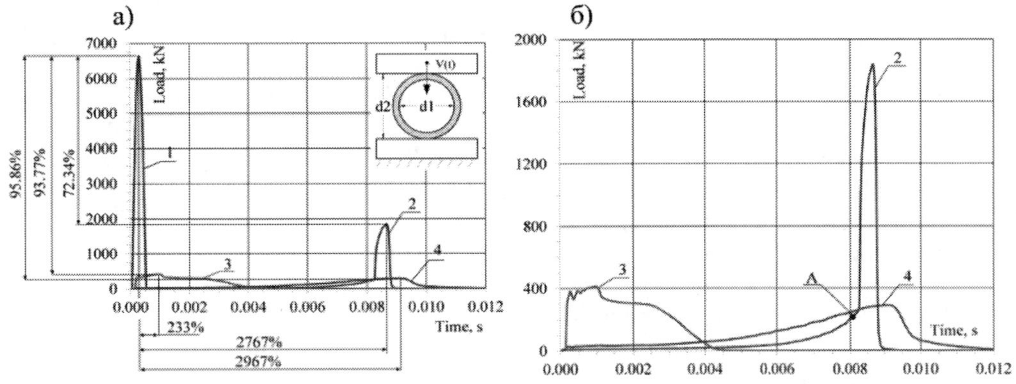

Fig. 13. Diagram of changing in the dynamic response of the supports depending on their rigidity; totally elastic (rigid) support (1); yielding support, working in elasto-plastic with hardening stage (2); the moment of transition of yielding support into the hardening stage (A); yielding supports working in elasto-plastic stage (3, 4).

Conclusions

The findings of the present research enable to conclude on the high efficiency degree of the yielding supports in the shape of annular inserted elements subjected to short-term dynamic loading compared to traditional totally elastic (rigid) supports. The required effect is reached due to the peculiarities of plastic straining of the annular section and increase in the time of dynamic straining. Therefore, the obtained results enable to judge on the possibility of yielding supports application aimed at the increase in the dynamic strength of RC beams.

The reported study was funded by Russian Foundation for Basic Research (RFBR) according to the research project No. 16-38-00133 mol_a, for 2016-2017.

References

1. S. Poonaya, C. Thinvongpituk, U. Teeboonma, Proceedings of WASET, **26**, 329-334 (2007)
2. P. Mata, A. Barbat, S. Oller, R. Boroschek, Arch. comput. methods eng., **15**, 489–539 (2008)
3. G. Nagel, "Impact and energy absorption of straight and tapered rectangular tubes", Ph.D. thesis, Queensland University of Technology, 2005
4. A. Olabi, E. Morris, M. Hashmi, Journal of Thin-Walled Structures, **45**, 706–726 (2008)
5. A. Ahmad Zaidi, G. Ling Lang, Journal of Science and Technology, **2**, 563-580 (2011)
6. J. Marzbanrad, M. Mehdikhanlo, A. Saeedi Pour, Turkish J. Eng. Env., **33**, 159 – 166 (2009)
7. Y. Zhao, J. Fang, International Journal of Pressure Vessels and Piping, **67**, 257-261 (1996)
8. A. Ghaidan, Tikrit Journal of Eng. Sciences, **18** (3), 52-60 (2011)
9. J. DeRuntz, P. Hodge, Journal of Applied Mechanics, **30**, 391-395 (1963)
10. G. Rathnaweera, Y. Durandet, D. Ruan, S. Kinoshita, International Journal of Protective Structures, **2** (4), 465-475 (2011)
11. E. Abdewi, World Academy of Science, Engineering and Technology International Journal of Mechanical, Industrial Science and Engineering, **7** (2), 183-186 (2013)
12. S. Lipa, M. Kotelko, Journal of Theoretical and Applied Mechanics, **51**, 873-882 (2013)
13. T. Chen, J. Ye, Journal of the Chinese Society of Mechanical Engineers, **34** (4), 331 (2013).
14. M. Ghazali, A. Ahmad Zaidi, M. Noh Dalimin, Modern Applied Science, **8** (1), 1-10 (2014)
15. M. Mohamed, A. Kumar, Procedia Engineering, **173**, 1399 – 1406 (2017)
16. A. Olabia, E. Morrisa, M. Hashmia, M. Gilchristb, International Journal of Mechanical Sciences, **50** (1), 104-116 (2008)
17. R. Holloman, V. Deshpande, A. Hanssen, K. Fleming, J. Scully, H. Wadley, Journal of mechanics of materials and structures, **8** (1), 65-94 (2013)
18. Z. Fan, J. Shen, G. Lu, Procedia Engineering, **14**, 442–449 (2011)
19. O. Kumpyak, Z. Galyautdinov, D. Kokorin, AIP Conference Proceedings, **1698**, 070006 (2016)
20. O. Kumpyak, N. Meshcheulov, Vestnik of TSUAB, **6**, 70-80 (2014)
21. A. Kezmane, B. Chiaia, O. Kumpyak, V. Maksimov, L. Placidi, European Journal of Environmental and Civil Engineering, **20**, 1-38 (2016)
22. B. Rastorguev, Building mechanics and analysis of structures, **1**, 50-57 (2009)

Temperature influence on strain state of a flexible plate

Tatiana Treputneva[1,*], *Oleg* Popov[1], *Margarita* Moiseenko[1], *Vladislav* Maksak[1], *Anatoly* Malinovskiy[1], and *Nikita* Mescheulov[1]

[1]Tomsk State University of Architecture and Building, 634003 Tomsk, Russia

Abstract. Developing analytical methods for studying the strain state of flexible plates is a relevant issue, especially for complex problems that recognize nonlinear operation and temperature-dependent changes in mechanical properties of materials. The paper analyzes the stress-strain state of a two-field flexible plate with an initial deflection exposed to a non-uniform temperature field. It presents and analyzes the results of the numerical studies based on an algorithm developed by the authors for designing anisotropic rectangular structural elements with discontinuous parameters, taking into account nonlinearities. The analysis of the design results suggests that initial deflection, physical and geometric nonlinearities, as well as inhomogeneity of the plate material need to be taken into consideration.

Introduction

Plates and shells are the basic structural elements that are exposed to power loads and temperatures during their operation [1–8, 10]. High-temperatures influence leads to a change in mechanical properties of the material and causes a decrease in structural rigidity [4, 5]. Imperfections in the initial shape of the structural surface (initial deflections) lead to qualitative differences in the stress-strain state as compared to a design model [6, 7]. Therefore, it is important to study the stress-strain state of flexible plates with an initial deflection taking into account temperature-dependent changes in mechanical properties of the material. A review of publications in this area also points to the importance of this study.

V. P. Kruglov [3] investigated the load-bearing capacity of flexible steel plates taking into account changes in mechanical properties of the material. A. G. Tamrazyan and A. S. Kozhanova [4] demonstrated high-temperature effects on thermal creep of shallow reinforced concrete shells and flat plates. Pham T. M., et al [11] investigated the influence of eccentric loading on the load-bearing capacity of reinforced concrete columns. Zhang Y. and Karr D. G. [12] developed analytical models to design plates under quasi-static pressure. The pressure in the plate arises from the contact with lake ice, which is caused mainly by thermal expansion and winds. Bhat S.U., Choi S.K., Wierzbicki, Karr D.G. investigated a dynamic response of a ship interacting with an ice floe [13].

* Corresponding author: mmo77@mail.ru

However, there are only a few publications that analyze nonlinear strain of plates with due regard to the combined effect of initial deflection, physical and geometric nonlinearity and temperature-dependent changes in mechanical properties [3, 4, 11]. Therefore, this issue needs to be further studied.

The purpose of this paper is to study temperature-dependent changes in the stress-strain state of the plate. The studies are based on a previously developed mathematical model for designing flexible plates with an initial deflection that are exposed to certain temperatures [9]. The model takes into account anisotropy, heterogeneity and change in physical characteristics of the material under the influence of temperature.

Design Model and Algorithm

The model of the structure under consideration is a two-field plate with thickness $h = 1$ cm that consists of square panels ($a = b = 50$ cm) and is placed in a non-uniform temperature field. The material of the structure is VT6 (titanium wrought alloy) [1]. The plate has a symmetrical initial deflection W_n and rigidly hinged along the contour and the center (Fig. 1).

In order to assess the stress-strain state of the structure more completely, it was designed with due regard to double nonlinearity (physical and geometric), initial deflection and temperature-dependent mechanical properties of the material.

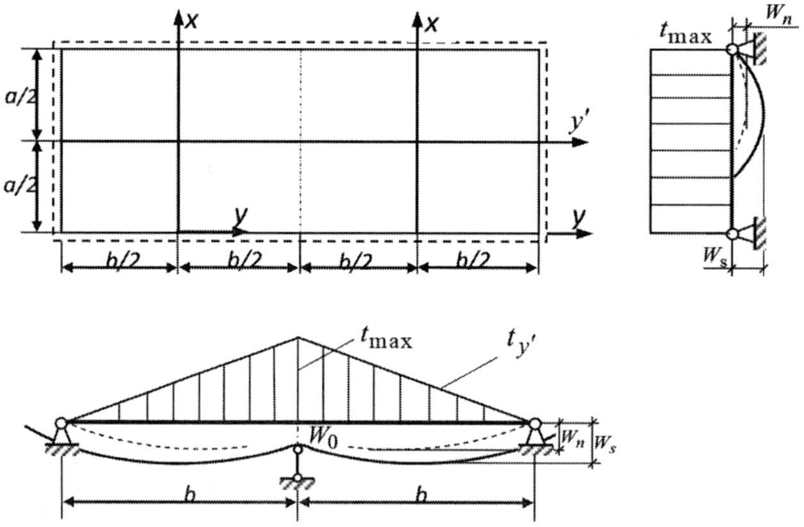

Fig. 1. A two-field flexible plate exposed to a variable temperature field.

This study uses the Goldenblat–Nikolayenko model [1] to take into account temperature-dependent properties of the material by constructing polynomial approximating dependence graphs.

The temperature across the panel field k varies according to the relationship:

$$t_{y'} = (-1)^{k-1} \frac{y + b/2}{b} \cdot t_{max} \quad (1)$$

The initial deflection W_n can be expressed as $W_n(x, y) = W_0 F_1(x) F_2(y)$.

Here $F_1(x) = \sin \pi x$, where $(0 \le \bar{x} \le 1)$; $F_2(y) = \sin 0.25\pi(1+\eta y)$, where $(-1 \le \bar{y} \le 1)$; $\eta = 1$ at $k = 1$; $\eta = -1$ at $k = 2$ (k is the panel number); W_0 is the maximum initial deflection of the plate center: $W_0 = 0$; $h/2$; h. W_s is the total displacement of the plate obtained by summing the initial deflection W_n and additional displacement of the plate W induced by temperature.

In order to take into account changes in physical properties, geometric nonlinearity and additional displacements, the energy method can be applied together with the displacement model [7–9].

Nodal lines are introduced along the lines of intersection between the middle surface of the plate and the vertical inertial plane of the principal edges. In the general case, each nodal line has four functional unknown displacements. The coefficients for functional displacements are chosen so that they ensure compatible displacements of the articulated panels [7–9].

Components of middle panel surface strains ε_x, ε_y, ε_{xy} and curvature χ_x, χ_y are expressed through displacements U, V, W along axes x, y, z , respectively, by the following relationships:

$$\varepsilon_x = \frac{\partial U}{\partial x} + \frac{1}{2}\left(\frac{\partial (W + W_H)}{\partial x}\right)^2 - \frac{1}{2}\left(\frac{\partial W_H}{\partial x}\right)^2 - k_1 W ; \quad \chi_x = -\frac{\partial^2 W}{\partial x^2} . \quad (x \leftrightarrow y) \tag{2}$$

Here: k_1, k_2 are panel curvatures; W_H is the initial deflection of the middle surface.

The relative strains of the panels ε_{xo}, ε_{yo}, ε_{xyo} at the points located at a distance z from the middle surface are written as:

$$e_{xo} = \varepsilon_x + z\chi_x ; \quad e_{xyo} = \varepsilon_{xy} + 2z\chi_{xy} . \quad (x \leftrightarrow y) \tag{3}$$

The centers of gravity of the cross sections and torsion angles of the edges can be expressed using the Kirchhoff-Klebsch hypothesis [6]:

$$W_p = W; \quad U_p = U + e\frac{\partial W}{\partial x}; \quad V_p = V + e\frac{\partial W}{\partial y}; \quad \frac{\partial W_p}{\partial x} = \frac{\partial W}{\partial x} . \tag{4}$$

The strains along the edges ε_{xp}, ε_{yp} are written taking into account eccentricity e of the edges:

$$\varepsilon_{xp} = \varepsilon_{xo} + e\chi_x . \qquad x \leftrightarrow y . \tag{5}$$

The relative strains of the edges $e_{xp}, e_{yp}, e_{zyp}, e_{zxp}$ are written taking into account bending in two planes, longitudinal strain and torsion:

$$e_{xp} = \varepsilon_{xp} - z\frac{\partial^2 W_p}{\partial x^2} - y\frac{\partial^2 V_p}{\partial x^2} ; \quad e_{zyp} = e_{zxp} = \frac{\partial^2 W_p}{\partial x \partial y} . \qquad x \leftrightarrow y . \tag{6}$$

The stresses are expressed through strains, taking into account the temperature load on the panel:

$$\sigma_{xo} = E_{cjx}\left[\frac{e_{xo} + \mu_{cjy} e_{oy}}{1 - \mu_{cjx}\mu_{cjy}} + T^o(\alpha_{jxT} + \mu_{cjy}\alpha_{jyT})\right] . \tag{7}$$

The stresses arising in the edges:

$$\sigma_{xp} = E_{cjx}\left(e_{xp} + \alpha_{cjx}T^o\right); \quad \tau_{zxp} = \gamma_k G_{cjx} e_{zx}. \quad x \leftrightarrow y. \tag{8}$$

Here, γ_k is the coefficient that takes into account the aspect ratio of the prismatic edge; $T^o = T^o(x, y, z,)$ is the temperature; G_{cjx} is the shear modulus; $E_{cjx,}$, E_{cjy} are the secant moduli of elasticity; $\mu_{cjx,}$, μ_{cjy} are the compressibility functions; $\alpha_{cjx,}$, α_{cjy} are the coefficient of linear expansion for the anisotropic material.

The coordinate functions for each individual panel are defined as the sum of two rows [6, 8, 9]:

$$U_k = \sum_n^N U_{ok}^n B_k^{(n)} + \sum_n^N U_{нk}^n Z_j^{(n)}; \; V_k = \sum_n^N V_{ok}^n C_k^{(n)} + \sum_n^N V_{нk}^n Z_j^{(n)}; \; W_k = \sum_n^N W_{ok}^n A_k^{(n)} + \sum_n^N W_{нk}^n Z_j^{(n)}. \tag{9}$$

Here, N is the number of harmonics used in the expansion; $A_к$, $B_к$, $C_к$, Z_i, Z_j are the variable parameters; $U_{ок}$, $V_{ок}$, $W_{ок}$, $U_{нк}$, , $V_{нк}$, $W_{нк}$ are the coordinate functions that satisfy the geometric boundary conditions at the edges of the k-th panel.

The oscillatory functions of the beams that satisfy the boundary conditions in the form of restraints are taken as components of the "homogeneous" series. The functions arising from the forced displacement of the edges along the nodal lines are taken as components of the "heterogeneous" series [2, 8]. The proposed functions are written for each k-th panel.

Based on the main system and functional displacements, an expression for the energy functional of the system can be obtained in Lagrangian form:

$$\delta\Phi = \delta\Pi - \delta A = 0, \tag{10}$$

where Π is the strain energy; A is the external force energy.

By substituting the displacements into the strain expressions and then substituting the latter into the functional, we transform the energy of the system into a multi-parameter function.

Since the functional is cumbersome, the variable parameters can be found by one of the methods designed to search for unconditional extremum – the conjugate gradient method. The calculation algorithm is based on the method of successive approximation. The functional is minimized at each approximation step for a given load level and given boundary conditions. The variable parameters are found to determine displacements, strains, stresses, strain and stress rates.

Testing of the Model

According to the developed algorithm, we developed a design program in the Pascal algorithmic language and tested this program in a number of cases [9]. This paper presents a solution to one of the design problems – the determination of the stress-strain state of the flexible plate with initial deflection which is placed in a non-uniform temperature field and subject to elastic-plastic strains. Figure 2*a* shows diagrams of additional deflections along the axis of symmetry at different temperatures (t = 20°C; 100 °C; 200 °C, 300 °C; 400 °C). Due to the symmetry of the plate and the active temperature load, the diagrams are presented only for the left panel.

Figure 2*b* shows diagrams of total deflections along the axis of symmetry, at different temperatures (t = 20°C; 100 °C – 400 °C).

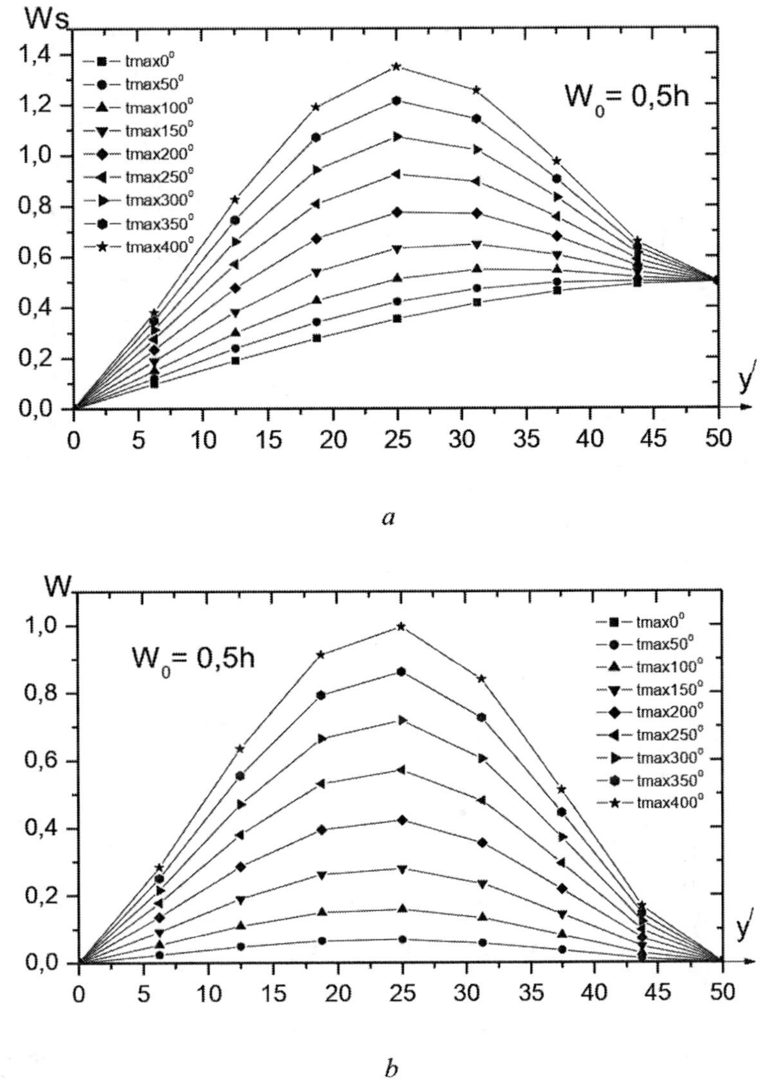

Fig. 2. Diagrams of total (a) and additional deflections (b) of the left panel with initial deflection, under exposure to different temperatures $W_0 = 0{,}5h$.

Figure 3 presents diagrams of strain rates for upper fibers (Fig. 3*a*) and lower fibers (Fig. 3*b*) of the left panel under exposure to different temperatures (t = 20°C; 100 °C – 400 °C).

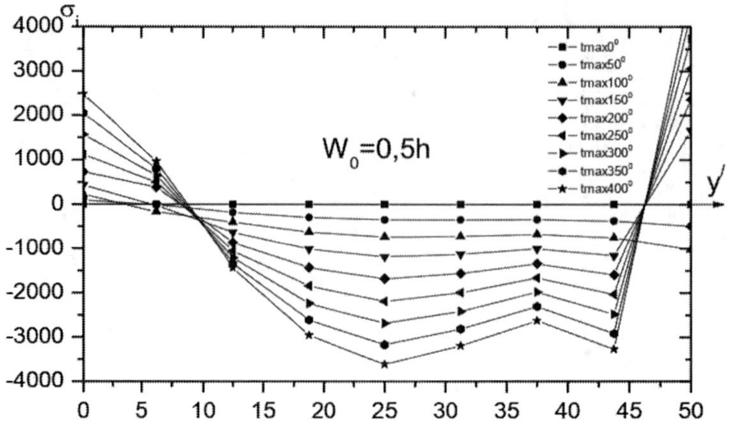

a

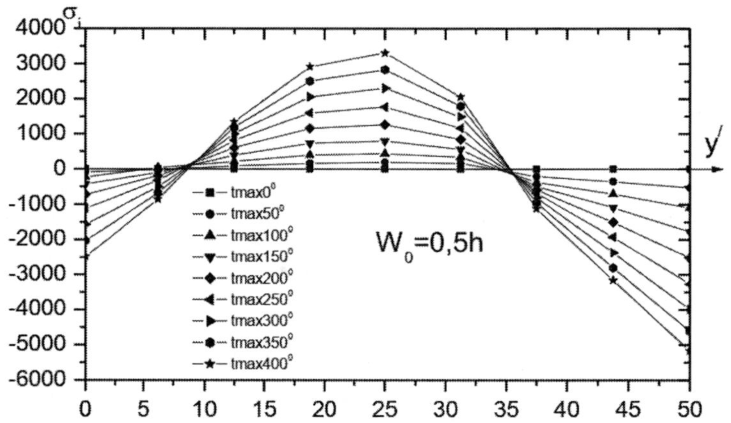

b

Fig. 3. Diagrams of strain rates for upper fibers (a) and lower fibers (b) of the left panel with initial deflection, under exposure to different temperatures $W_0 = 0{,}5h$.

The design results for a two-field flexible plate with an initial deflection ($W_0 = 0{,}5h$) exposed to a non-uniform temperature field shows that additional total deflections and strain rates reach their maximum values in the center of the panel and increase with temperature. At t = 100 °C, the difference between additional displacements of the panels with different initial deflections exceeds 100%. As temperature increases, the difference decreases and, at t = 400 °C, reaches 25.8%.

Conclusion

The developed mathematical model for designing flexible plates with an initial deflection exposed to certain temperatures can be used to perform a comprehensive analysis of the stress-strain state and to estimate safety margin of flexible plates in terms of strength and rigidity.

The further research should be aimed at developing an analytical model and a program for designing thin-walled structural elements with regard to mechanical compliance of the contour.

References

1. N.I. Bezuhov, V.L. Bazhanov, I.I. Gol'denblat, N.A. Nikolaenko, A.M. Sinjukov, Raschety na prochnost' ustojchivost' i kolebanija v uslovijah vysokih temperatur [Calculations for strength, stability and vibrations under high temperature conditions] (Mashinostroyeniye, Moscow, 1965) (in Russian)
2. V.V. Karpov, V.A. Mashkov, V.N. Filatov, *Matematicheskoe modelirovanie, chislennye metody i kompleksy programm* [Mathematical modeling, numerical methods and software], 88-92 (1994)
3. V.P. Kruglov, News of higher educational institutions. Construction and Architecture, **6**, 35-38 (1979)
4. A.G. Tamrazjan, A.S. Kozhanova, Promyshlennoe i grazhdanskoe stroitelstvo [Industrial and civil engineering], **10**, 15-20 (2015)
5. S.I. Trushin, *Stroitel'naja mehanika: metod konechnyh jelementov* [Structural mechanics: the finite element method] (INFRA-M , Moscow, 2016)
6. M.O. Moiseenko, A.P. Malinovsky, O.N. Popov, T.A. Treputneva, Proceedings of the Conference "Construction Science-XXI Century: Theory, Education, Practice, Innovation in the North-Arctic Region", 237-242 (2015)
7. A.V. Aleksandrov, B.Ya. Lashchenikov, N.N. Shaposhnikov, *Stroitel'naja mehanika. Tonkostennye prostranstvennye sistemy* [Structural mechanics. Thin-walled space systems] (Strojizdat, Moscow, 1983)
8. V.N. Zav'jalov, Trudy Tomsk. inzh.-stroit. in-ta: Issledovanija po stroitel'nym konstrukcijam [Reports of Tomsk Engineering and Construction Institute: Studies on building structures], 25-31 (1974) (in Russian)
9. O.N. Popov, A.P. Malinovskij, M.O. Moiseenko, Nelinejnye zadachi rebristyh obolochek [Nonlinear problems of ribbed shells] (Tomsk State University of Architecture and Building, Tomsk, 2004) (in Russian)
10. V.V. Petrov, I.G. Ovchinnikov, V.K. Inozemtsev, Deformirovanie elementov konstrukcij iz nelinejnogo raznomodul'nogo neodnorodnogo materiala [Deformation of structural elements from a non-linear, non-homogenous, inhomogeneous material] (Saratov State University, Saratov, 1989) (in Russian)
11. T.M. Pham, X. Lei, M.N.S. Hadi, Conference on the Mechanics of Structures and Materials (ACMSM22), 139-144 (Sydney, Australia, December 11-14, 2012)
12. Y. Zhang, D.G. Karr, Journal of structural engineering, **5**, 04017003 (2017)
13. S.U. Bhat, S.K. Choi, T. Wierzbicki, D.G. Karr, Journal Offshore Mech. Arct. Eng., **113** (2), 171-178 (1991)

Experimental study of beams on yielding supports with thrust

Oleg Kumpyak[1], *Zaur* Galyautdinov[1,*], and *Daud* Galyautdinov[1]

[1]Tomsk State University of Architecture and Building, 634003 Tomsk, Russia

Abstract. The design of reinforced concrete structures which are subjected to static and intensive dynamic loading requires taking into account the occurrence of thrust reaction resulting in significant increase in strength and crack resistance. Application of yielding supports for the purpose of increasing the energy capacity of structures shall also be considered. The present research aims to define feasibility of yielding supports application in thrust structures. The paper presents the outcomes of experimental studies of reinforced concrete beam structures on yielding supports subjected to thrust under static and short-term dynamic loading. The influence of thrust on the strength, deformation property and crack resistance of reinforced concrete structures under static and short-term dynamic loading was investigated. The combined use of yielding supports and the restriction of horizontal displacement of the support contour were also considered in the study. Research results testify on the positive outcome when yielding supports are applied in the structures subjected to thrust.

Introduction

The basic methods of reinforced concrete (RC) structures analysis under short-term dynamic loading are based on prerequisites which do not always consider the peculiar features of real structure operation to the full extent. All real structures have connections in their bearing member (anchoring of elements on the supports, grouting the gaps between the elements faces), which prevent the horizontal displacement of support sections during loading and result in thrust occurrence. A number of theoretical and experimental studies reflect a significant positive influence of thrust reaction on the stress-strain state of the beams [1–5].

The mentioned influence under static loading is studied rather well [1, 3, 5]. Analysis of the numerical analysis of the structures under static loading subjected to thrust indicate the increase of their load-bearing capacity depending on the reinforcement percentage of the cross-section (μ) and flexibility of the supporting contour (c) [3] (Fig. 1).

The influence of thrust on the structural resistance under dynamic loading is controversial: on the one hand, the presence of the horizontal reaction increases the load-bearing capacity of the element; on the other hand, it reduces its deformation property thus

* Corresponding author: GazR@yandex.ru

leading to reduction of plastic deformation stage. This issue was considered in the works of N.N. Popov and B.S. Rastorguev [6], T.N. Vinogradova [3], and I.N. Tikhonova [5]. The thrust occurrence is necessary to consider as when structures are subjected to the action of emergency loads in a number of cases the work of reinforcement is acceptable beyond the elasticity range. Therefore, it causes increase in the strain of tensile area and the growth of the horizontal reaction accordingly.

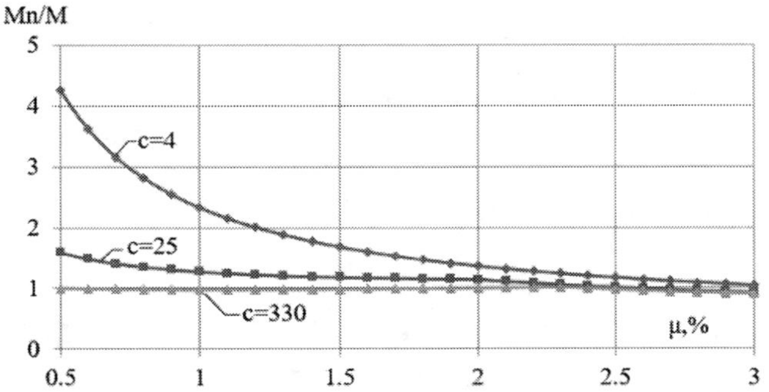

Fig. 1. Alteration of the load-bearing capacity of the beams subjected to thrust under static loading depending on the reinforcement percentage (μ) by various supports ductility (c): Mn – ultimate bending moment taking into account thrust; M – ultimate bending moment without considering thrust.

The dependence of deformation property on the load-bearing capacity under action of the time-constant load on the beam is illustrated in Figure 2.

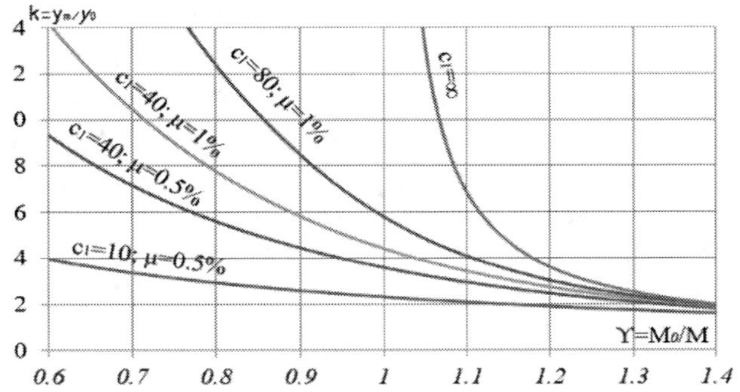

Fig. 2. The dependence of k on Υ for the beams under the action of suddenly-applied time-constant load P_0=const [2]: M_0– the bending moment at the end of elastic stage; M_P– the bending moment of the acting load; y_m – the ultimate deflection of structure; y_0 – the elastic deflection of structure; c – flexibility of the supporting contour.

As can be seen from Figure 2, thrust has significant effect on the deformation property of the structure. Decrease in the flexibility of the horizontal supports (c) and the percentage of longitudinal reinforcement (μ) leads to reduction of the beam strain.

Use of yielding supports contributes to the increase in structural resistance towards dynamic loading. Experimental results of the beams on yielding supports [7, 8] demonstrate great effectiveness of their application under high intensity dynamic loadings. Moreover, the better effect was observed during supports straining in elasto-plastic stage.

Thus, application of yielding supports and consideration of horizontal displacements restriction is a rather relevant issue which requires in-depth study.

Experimental

For the assessment of the stress-stain state of RC beam structures on yielding supports subjected to thrust under static and short-term dynamic loading the program of experimental studies was developed and implemented (Table 1). Six beams were designed and manufactured, their size being 1900x150x220 mm. Heavy concrete of grade B35 was used for the beams production. The beams were reinforced with the spatial frame. Cold-worked reinforcement Ø5 mm grade Bp500 with the span 50 mm in near-support area and 130 mm in the mid-span was used as a lateral reinforcement; while hot-rolled reinforcement Ø 6 mm grade A240 in the compressed area and Ø10 mm grade A500 in tensile area was used as a longitudinal reinforcement. For the reinforcement of front areas of the beams mesh with the cell size of 50×50 mm was made from cold-worked reinforcement Ø5 mm grade Bp500, 7 meshes were installed on each side, and the angle-irons 100×10 mm were placed (Fig. 3).

Table 1. Experimental program.

Loading type	Static load		Dynamic load		
Support type	Rigid supports		Rigid supports		Yielding supports (elasto-plastic stage)
Percentage of reinforcement	μ=0.5%		μ=0.5%		μ=0.5%
The flexibility value of the horizontal contour c_1	∞	25	∞	25	25
Sample code	BS-1	BS-2T, BS-3T	BD-4	BDT-5	BDTY-6
Number of samples	1	2	1	1	1

Notes:

c_1=∞ – corresponds to the lack of the thrust contour;

c_1=25 – corresponds to the presence of the thrust contour presented by two tension bars, 70 mm in diameter;

Notation: "B" – RC beam; "S" – static loading; "D" – short-term dynamic loading (falling weight 450 kg, from the height of 750 mm); "T" – presence of the thrust structure; "Y" – yielding support (made from hollow-section tubes: the outer diameter 32 mm, the inner diameter 26 mm, l=50 mm); "1…6" – the structure No.

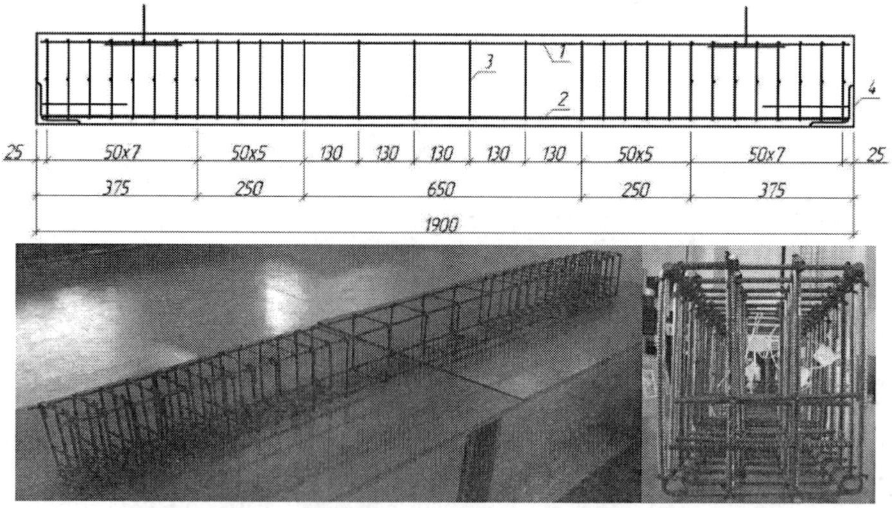

Fig. 3. The scheme of reinforcement of concrete structures under study: 1 – reinforcement Ø6 mm grade A240; 2 – reinforcement Ø10 mm grade A500; 3 – reinforcement Ø5 mm grade Bp500; 4 – equal leg angle 100×10mm.

The whole complex of measuring facilities was placed on the tested samples. Inductive displacement sensor Waycon RL150 was used for displacement measurement. Acceleration gauges DHE 100023 were applied for acceleration measurement. Force measuring strain gauge DST 4126 was used for measuring the system response and the value of the support response was defined by means of dynamometric supports (Russian Federation utility model patent No. 161908). Visualization of the crack development pattern during short-term dynamic loading was conducted using high-speed camera with the speed 2500 shots/sec (Photron Fastcam SA-2). The facilities were connected to the computer program Mic-036R and Mic-400D.

Testing facility was developed and produced (Fig. 4) for implementation of the experimental program, the corresponding patents were obtained for that purpose (Russian Federation utility model patent No.147262, No.148401).

a) b)

Fig. 4. Test facility for RC beams: under static loading subjected to thrust (a); under dynamic loading with thrust on yielding supports (b).

The RC beam under study was placed on dynamometric supports which provide hinge supporting. The supports were fixed on the strong floor, then the initial longitudinal force (thrust) was applied by the hydraulic jack through the traverse system connected by guide ways.

During static loading (Fig. 4*a*) the load was applied by means of hydraulic jack through distributive traverse in one thirds of the span, short-term dynamic loading was applied by means of falling weight of 450 kg, from the height of 750 mm (Fig. 4*b*).

Results

Resulting from static testing for hinged beam not subjected to thrust the ultimate breaking load was 64 kN, and for the beam subjected to thrust it was 168 kN, the increase of load-bearing capacity by the load reached 262.5%, while maximum deflections decreased by 235% (Fig. 5). Occurrence of the first cracks in the beam without thrust was observed at the load of 24 kN, and with thrust – at the load of 36 kN. According to the destruction patterns of the structure and crack formation it can be seen that the presence of thrust leads to increase in crack resistance of beams, and destruction occurs in concrete of the compressed area (Fig. 6).

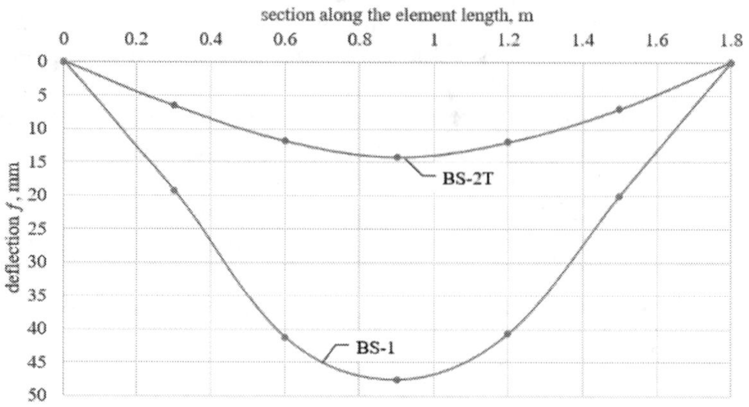

Fig. 5. The ultimate displacements of the beams BS-1, BS-2T under static loading.

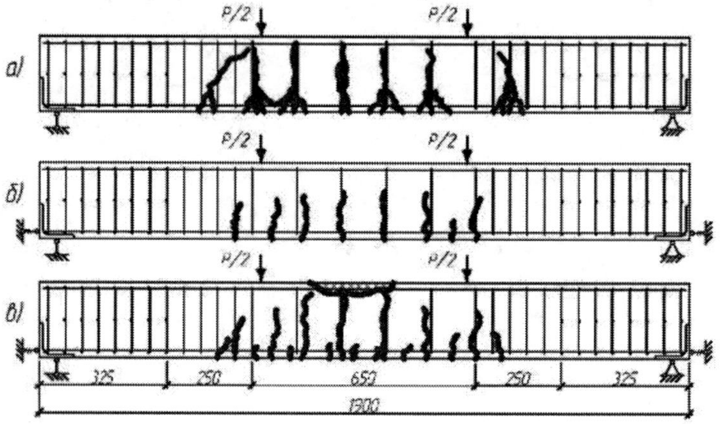

Fig. 6. The schemes of crack formation in the structures under static loading: a) BS-1, by T=64kN; b) BS-2T, by T=64kN; c) BS-2T, by T=168kN.

The results of dynamic testing of the reference samples with equal loading parameters show that application of yielding supports and the restrictive contour (BDTY-6) causes significant increase in crack resistance (Fig. 8).

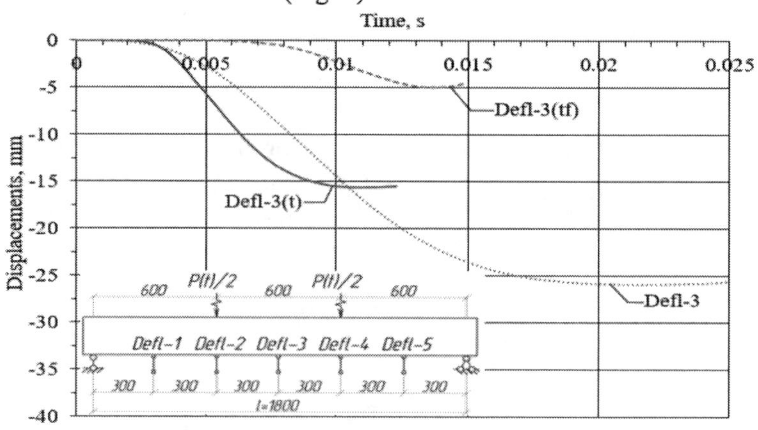

Fig. 7. Alteration in displacements of structures in the mid-span depending on the time: Defl-3 – deflection of RC beam BD-4; Defl -3(t) – deflection of RC beam BDT-5; Defl -3(tf) – deflection of RC beam BDTY-6.

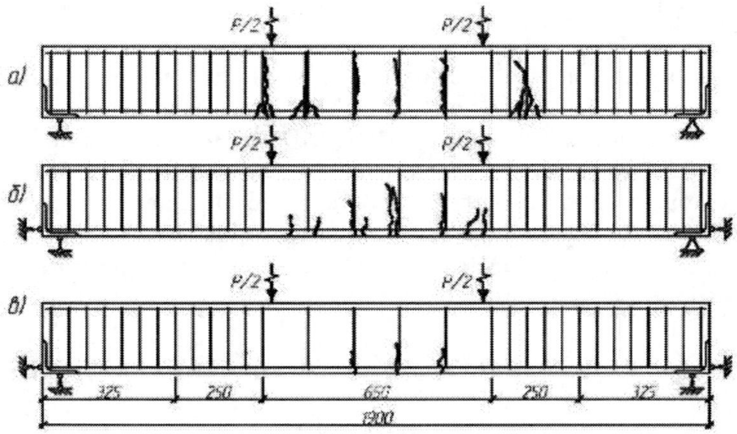

Fig. 8. The schemes of crack formation in structures under short-term dynamic loading: a) BD-4; b) BDT-5; c) BDTY-6.

Displacements analysis demonstrates that occurrence of thrust in the beam BDT-5 results in reduction of deflections in the mid-span by 41% regarding the sample BD-4 (Fig. 7). The reduction of deflection in the structure subjected to thrust indicates higher rigidity of the thrust system. Use of yielding supports in the beams with thrust BDTY-6 reduces displacements by 68% compared to the sample BDT-5, and by 81% compared to the sample BD-4 (Fig. 7). Reduction of displacements of the structure is attributed to the yielding support straining.

Analysis of the system response alteration in time under short-term dynamic loading (Fig. 9) enables to make the following conclusion: presence of thrust in the sample BDT-5 contributes to reduction of the system response by 29% and less time is required for reaching the ultimate load with regard to the beam BD-4. In the structure BDTY-6 the reduction of the system response was 44%, and the time of reaching the maximum value of the system response also slightly decreased due to energy capacity of the yielding supports.

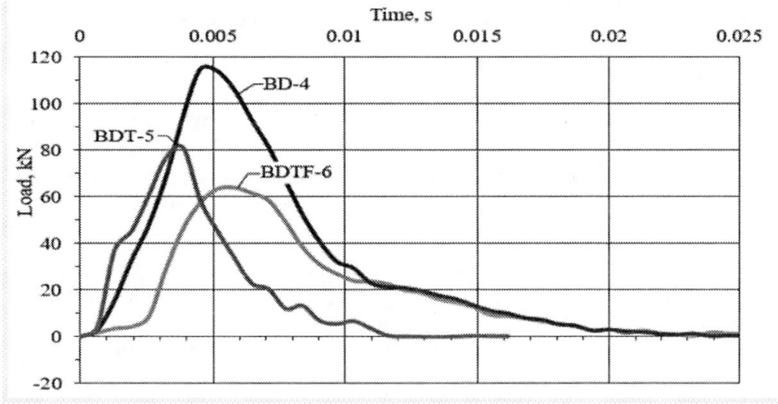

Fig. 9. Alteration in the system response in time during testing of RC beams under short-term dynamic loading.

Conclusion

The conducted experimental studies enable to establish that the presence of thrust in structures results in increase of their load-bearing capacity and reduction of deformation property and crack resistance both under static and intensive dynamic loading. Application of yielding supports in thrust structures provides increase in the system energy capacity (BDTY-6) and increase in the time of the beam plastic straining. A significant lag in reaction response of the structure on yielding supports takes place in this case (BDTY-6) compared to the beam on rigid supports (BD-4, BDT-5), as well as significant reduction of structural response. The results of experimental studies revealed the hided strength reserves of the beams on yielding supports with thrust. Consideration of these parameters will enable to increase the accuracy of structural analysis under short-term dynamic loading. The experimental studies are planned to be extended by changing the work of yielding supports (elastic, elasto-plastic with hardening) and the work of thrust contour by changing its rigidity.

References

1. A.A. Gvozdev, et al., *Novoe o prochnosti zhelezobetona* [Modern studies on reinforced concrete strength] (Strojizdat, Moscow, 1977) (in Russian)
2. L. Huynh, S. Foster, H. Valipour, R. Rendall, Construction and Building Materials **78**, 153-171 (2015)
3. T.N. Vinogradova, Vlijanie raspora na rabotu zhelezobetonnyh balochnyh konstrukcij pri kratkovremennyh dinamicheskih vozdejstvijah [The effect of thrust on performance of reinforced concrete beam structures under short-term dynamic load], PhD thesis (Moscow, 1977) (in Russian)
4. D.R. Galyautdinov, Proceedings of the 62nd Scientific University Conference of Students and Young Researchers, 21–24 (Tomsk, Russia, 26 April, 2016) (in Russian)
5. I.N. Tikhonov, Zhilishhnoe stroitel'stvo [Housing Construction] **2**, 40-45 (2013) (in Russian)
6. N.N. Popov, B.S. Rastorguev, Dinamicheskij raschet zhelezobetonnyh konstrukcij [Dynamic calculation of reinforced concrete structures] (Strojizdat, Moscow, 1974) (in Russian)
7. O.G. Kumpyak, Z.R. Galyautdinov, D.N. Kokorin, *Prochnost' i deformativnost' zhelezobetonnyh konstrukcij na podatlivyh oporah pri kratkovremennom dinamicheskom nagruzhenii* [Strength and deformability of reinforced concrete structures on yielding supports under short-term dynamic load] (TSUAB Publ., Tomsk, 2016) (in Russian)
8. A.R.A. Said, Povyshenie nesushhej sposobnosti zhelezobetonnyh konstrukcij pri vzryvnyh vozdejstvijah [Increase of bearing capacity of reinforced concrete structures under blast load], PhD thesis (Moscow, 1995) (in Russian)

Evolution of the reinforced I-beam strain state

Artem Ustinov[1,*], *Anatolij* Klopotov[1,2], *Aleksandr* Potekaev[2], *Vasilij* Plevkov[1], and *Ekaterina* Marchenko[2]

[1]Tomsk State University of Architecture and Building, 634003 Tomsk, Russia
[2]National Research Tomsk State University, 634050 Tomsk, Russia

Abstract. The paper presents experimental results of the strain state evolution of the steel I-beam reinforced with composite material under four-point bending loading. Digital image correlation technique was used to trace the evolution of plastic strain zones distribution. The images of longitudinal relative strain on the lateral surface of the reinforced I-beam web were obtained and analyzed. It was found that the increase of load-bearing capacity of the reinforced beams can exceed 26 %.

Introduction

The given work presents experimental data on the strain distribution evolution on the surface of the I-beam web without reinforcement and with external reinforcement by carbon fiber under four-point bending.

The real life environment can generate critical conditions for structural elements being exposed to external loads. This state of structural elements is characterized by plastic strain, which in case of emergency results in structural failure. Thus, the mechanisms leading to failure of both structural elements and the systems of external reinforcement are required to be specified.

Digital image correlation technique provides the possibility to receive information in situ about deformation processes on the surface.

When the mentioned technique is applied for investigation of strain impacts on the samples of various geometric shape it makes possible to reveal the common factors required for operation of structure under conditions of ultimate external loads.

The peculiar features attributed to conducting of such experimental studies were previously described in a number of research works [1-5]. Evolutions of plastic strain core areas distribution on the surface of solid bodies under various impacts in various structures are being actively studied nowadays [5-7].

Strain alterations on the surface of solid bodies occur on different structural levels, corresponding to strain mechanisms. Plastic strain is characterized by appearing spatially-temporal hierarchy of structures and phenomena. The basic strain and strengthening mechanisms refer to micro- and meso-structural level of solid bodies strain [8 -9].

Adhesive connections in foreign practice are used in reinforcing and restoring steel structures as an alternative to traditional connecting methods, such as bolted, welded,

* Corresponding author: artemustinov@bk.ru

riveted connections. Thus, the negative aspects of these methods, such as increase of constant load, complexity of work, etc., can be diminished. Experimental and theoretical studies of load-bearing capacity recovery in respect of defective steel structures working in bending reinforced with carbon fiber were conducted by many foreign scientists [11-17].

Materials and methods

Experimental testing was conducted at the Research and Education Center "Testing Building Materials and Structures" in Tomsk State University of Architecture and Building (Tomsk, Russia).

The I-beam was produced from steel with yield stress σ_y=225 MPa, ultimate strength σ_u=360 MPa. The height of the beam cross section is 160 mm, the flange width – 82 mm, the flange thickness – 7.4 mm, the web thickness – 5 mm, the total length – 1800 mm, the span between the centers of supports –1700 mm.

Adhesive used for bonding of the reinforcement elements was FibARM Laminate+; it has shear strength >13 MPa while the thickness of the layer is 1 mm. The external element which is being reinforced is made from carbon fiber lamella FibARM Lamel 12-50. Its tensile strength is >2800 MPa. The thickness of lamella is 1.2 mm; its width is 82 mm, which is defined by the flange width of the I-beam.

The surface of the I-beam web was cleaned by mechanical method using rotating wire brushes. Then, the sample was cleaned from dust and degreased.

The area under study was colored in two layers. The first layer was the base of white color; the second layer was single-point coloring. The wall coloring was performed to measure displacements and the strain on the surface of sample using Vic-3D facility. This system is intended for recording and analysis of stereoscopic speckle images and obtaining the data on spatial straining of the sample surface. The selected methodology enables to obtain the data on spatial straining of the whole area under study. The traditional strain gauge measurements in their turn can only provide local measurements of one strain component restricted in length of measurements. The scheme of recording the evolution of deformation distribution on the lateral surface of the I-beam under study is illustrated in Figure 1.

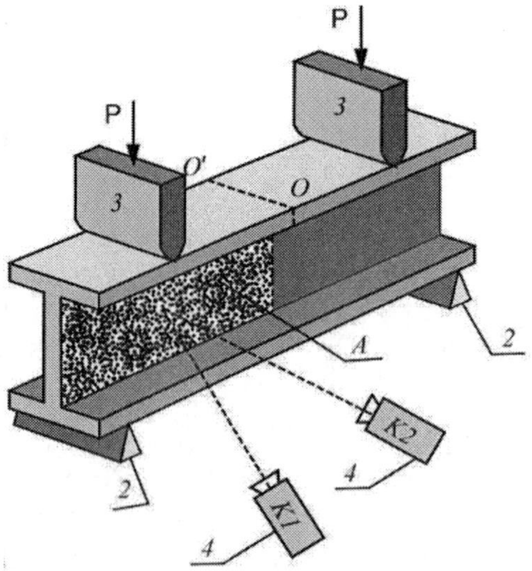

Fig. 1. The scheme of recording the images of the I-beam lateral surface. 1 – I-beam; 2 – supports; 3 – indenter; 4 – digital cameras C1 and C2; A – the area under study; OO' – the axis line passing through the center of the I-beam; P – the applied load.

Bending tests were conducted on a universal testing machine UTM 4500 (GTCS, USA) with maximum strength 4500 kN (458 tf). The loading was carried out at a constant rate of 160 (1/s).

The algorithm of bending tests of the I-beams is given below. The sample was placed on supporting points. One of the supports was hinged-fixed providing only the sample turning; the other was hinged-non-fixed, providing not only the sample turning but also its displacement in the plane of bending. In order to prevent the flange local buckling the steel plates possessing high bending stiffness were mounted above the supports. The load on the sample was transmitted from the testing machine through steel traverse. The traverse transferring load on the tested sample through two hinged supports was mounted on the sample centrally. The scheme of elements location is presented in Figure 2.

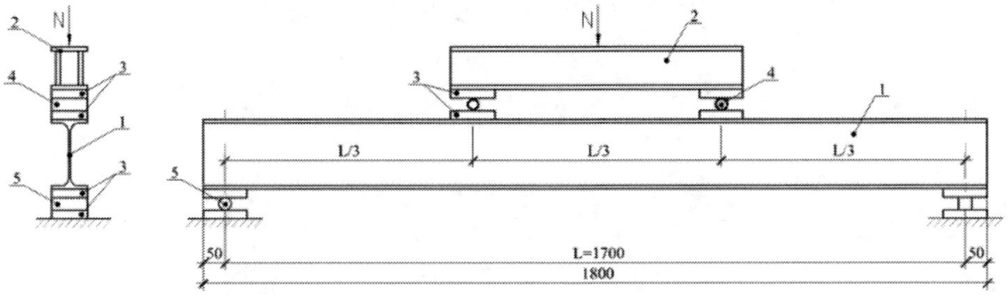

Fig. 2. Loading system for testing of I-beams under four-point bending. 1 – I-beam; 2 – traverse; 3 – distributing plates; 4 – hinged-fixed support; 5 – hinged-non-fixed support.

Results and discussion

Figure 3 presents the strain diagrams in the coordinates "P–f" obtained resulting from bending tests of the reference and reinforced I-beams. Four stages can be observed on the strain diagram of the reference I-beam. The first stage (the interval OA) is characterized by linear nature and corresponds to the elastic stage of material straining. The second stage (the interval AB) – is a transition stage which is characterized by the beginning of elasto-plastic straining. The third stage (the interval BC) at the strain curve "loading-deflection" is conditioned by the peculiar features of the type of external load applied to the I-beam. In this case transition to the third stage is conditioned by the slow flange buckling of the I-beam below the area where the load is applied. The fourth stage (the interval CD) – the total loss of system stability during testing.

In the point C′ of the strain diagram "P-f" of I-beam deflection the experiment was stopped as the system generally loses its stability (Fig. 3, curve 2).

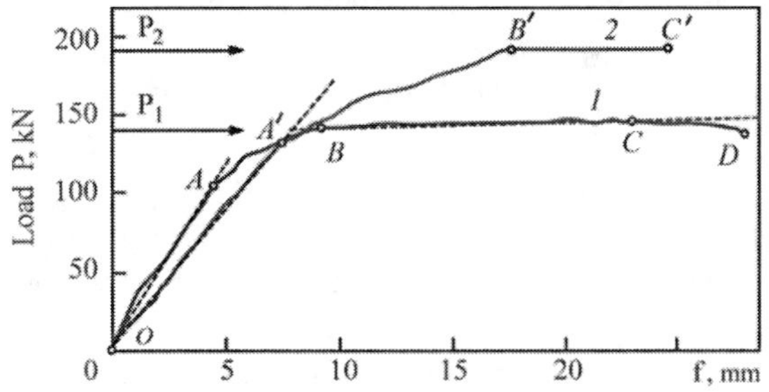

Fig. 3. The diagrams "load-deflection" of the I-beam without reinforcement (curve 1) and the I-beam reinforced with carbon fiber in the top and bottom flange (curve 2)

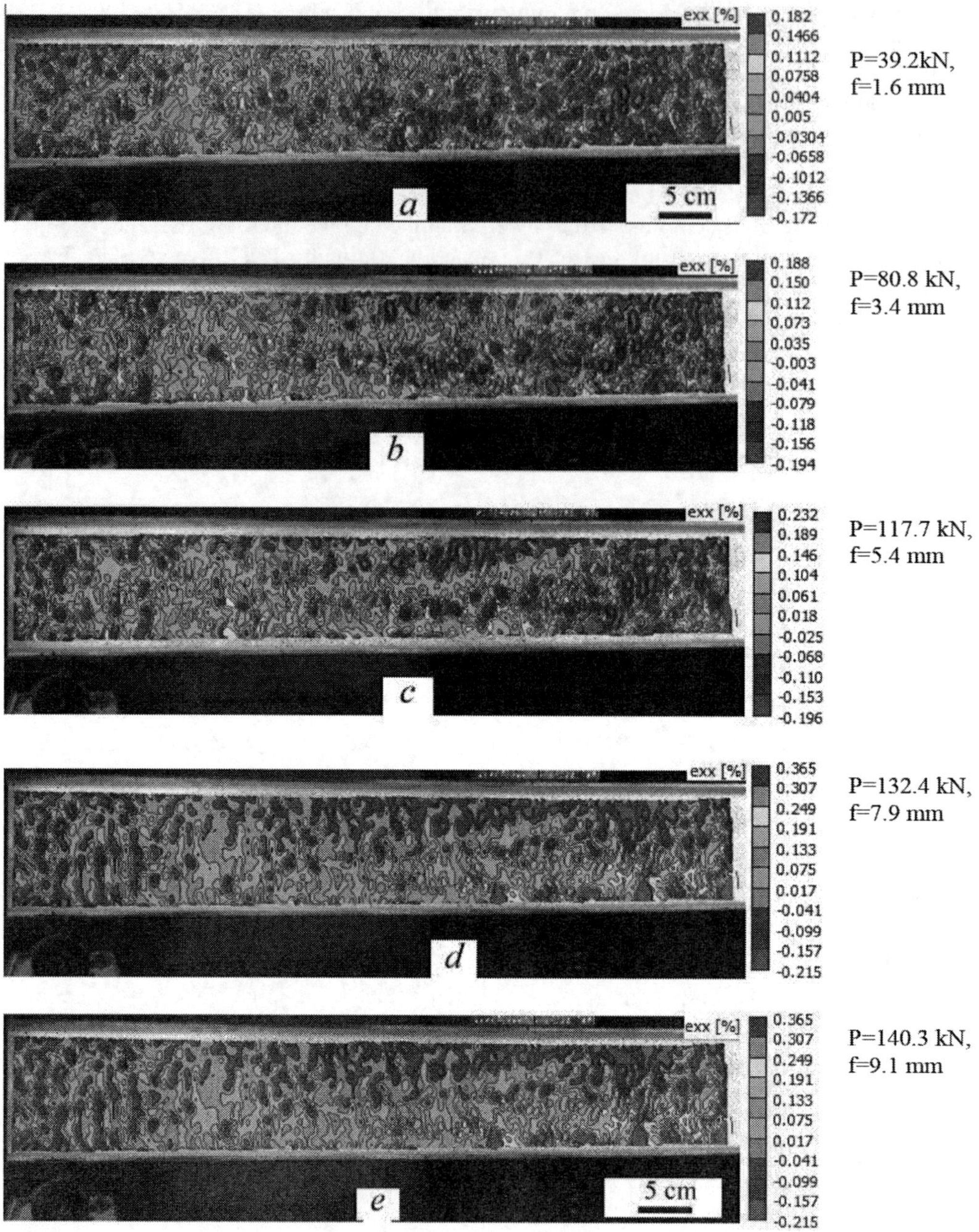

Fig. 4. The images of longitudinal relative strain of the reference steel I-beam under various loads.

The distribution of local strain zones on the lateral web of the reference I-beam at the load ~ 40 kN demonstrates chaotic nature of the cores distribution which corresponds to the first stage (Fig. 4*a*). Plastic strain zones of compression and tension locate chaotically in regards to each other in the central part of the I-beam. Plastic strain zones defined in accordance with the mentioned methods mostly possess the shape of an ellipse elongated along the applied load (perpendicular to the top and bottom planes of the I-beam). At the same time, the local strain areas are of the average sizes from 2 to 10 mm. Along with the increase of the load the image of the local strain zones distribution changes. Depending on the strain the local strain zones in the central part of the beam are located in the following way: compression zones are in the top part, while tension zones are in the bottom part (P ~80 kN, Fig. 4*b*).

During transition from the first stage to the second local strain zones distribution changes significantly. The separate local strain zones start to merge into larger zones which are elongated along the applied load (Fig. 1 and Fig. 4*c–e*). In this case, local strain zones initiate in the top and bottom lateral surface in the central part of the I-beam accordingly, depending on the strain type. In this case, local strain zones are surrounded by the areas with lower values of strain, but of the same type. The other image of the local strain zones distribution is observed in the end of the I-beam where the hinged-fixed support is located (Fig. 2). The local strain zones in this area are located in layers (Fig. 4, *c–e*); the layers of compression zones elongated along the load direction interchange with the tension zones on the beam surface.

The diagram "P–f" (Fig. 3) illustrates the ultimate load (P_1 ~140 kN) with an arrow. After reaching the mentioned load the process of buckling failure occurs with the reference I-beam flange below the area of the applied load (transition to the third stage). Fig. 4*e* demonstrates the image of longitudinal relative strain distribution.

The image of longitudinal relative strain distribution of the reference I-beam flange under ultimate load (P_1) shows wave-like behavior of plastic strain zones distribution (Fig. 4 *c–e*). Along with that, in the top part of the I-beam lateral surface the maximum value of compressive plastic strain in local plastic strain zones reaches the values of 0.215%. At the same time, in the bottom part of the I-beam lateral surface the maximum value of tensile plastic strain in local strain zones is larger than in the top part of the lateral surface (it reaches the values of ~ 0.36 %).

The other image of longitudinal relative strain distribution at the same external load applied is observed on the lateral web of the reinforced I-beam (Fig. 5). The local strain zones with high compressive strain values are basically located on the top plane in the central part of the I-beam at the load P ~40 kN (Fig. 5*a*). Along with the increase of applied load but within the range of the first stage (P ~80 kN) the density of local strain zones with high values of compressive strain increases on the lateral web of the reinforced I-beam in its top part. In the bottom part of the lateral beam surface the decrease in density of local compressive strain zones is observed (Fig. 5*b*).

During transition from the first stage to the second in the central part of the I-beam surface plastic strain zones in the middle part are also observed. However, the distribution of strain zones in the area of the I-beam end behaves the other way as compared to the case of the reference I-beam.

Strain zones distribution in the end of the I-beam has its own nature. There are no clearly distinguished local strain zones. There are strain zones which are significant in size with low values of strain and they are elongated along the action of the applied load (Fig. 5*b*). The expansion deformation occurs in these zones, they interchange as the layers: strain zone with the values 0.005%, and its neighboring strain zone with higher values (~0.092 %).

The diagram "P–f" (Fig. 3) shows the ultimate load of the reinforced I-beam with an arrow (P_2 ~190 kN). After reaching the mentioned load the process of buckling failure of the flange occurs below the location of applied load (transition to the third stage). However, the zones on the top and bottom parts of the I-beam lateral surface which are occupied by local strain zones with the maximum values are significantly lower than in the reference I-beam with the same values of the applied load. Also, the density of local strain zones location is significantly lower.

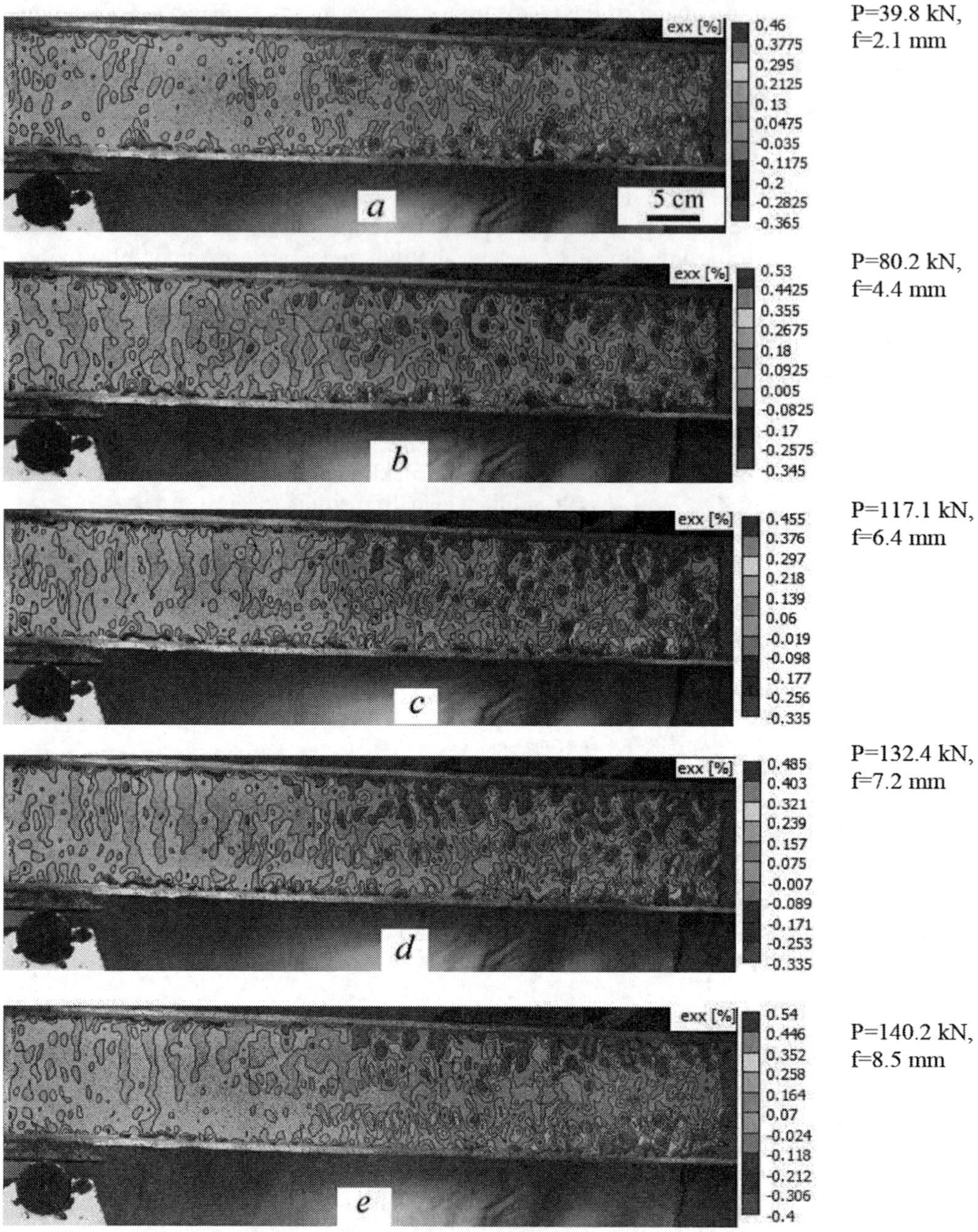

Fig. 5. The images of longitudinal relative strain of the steel I-beam web reinforced with carbon fiber lamellas glued to the top and bottom flanges of the beam under various loads.

Conclusion

Based on the conducted research the following conclusions can be made.

1. On the strain diagrams in the coordinates "P–f" during testing of the reference I-beams and the I-beams reinforced with carbon fiber lamellas the four stages can be distinguished: the first stage is characterized by the linear dependency and elastic strain of the material; the second stage is the transition stage, it is characterized by the initiation of

the plastic strain in the areas which are the most strained; the third stage is conditioned by the slow buckling failure of the I-beam flange below the area of the applied load; the fourth stage reflects structural failure (the loading system and the beam) during the process of testing and it is defined not by mechanical properties of the material but by the structure itself and the type of strain impact.

2. It was defined that the ultimate load of the I-beams reinforced with carbon fiber lamellas equals to 190 kN; that is the load that initiates the process of buckling failure of the reinforced I-beam top flange; and that value is by 26 % more than for the reference I-beam.

3. It was established that on the reference and the reinforced I-beams the distribution of strain zones and local plastic strain core areas on the lateral planes evolve in a different manner under bending at the same external actions.

The obtained data can be used during reinforcement of the operating steel structures which are subjected to static loads. Application of carbon fiber will enable to extend the operation lifetime of the steel structures, to increase their load-bearing capacity in case of increasing the design loads, and to recover the load-bearing capacity of defect elements.

The scheme of the steel I-beams reinforcement with carbon fiber in the area of maximum normal stresses contributed to the increase in load-bearing capacity by a quarter. The application area of the elaborated reinforcement type is restricted by the fact that temporary load on beams should be excluded.

Introduction of composite materials provides good technical and economic perspectives in case of reinforcing the linear structural elements of buildings and structures under operation.

References

1. V. Vildeman, T. Tretyakova, M. Tretyakov, Journal of Machinery Manufacture and Reliability **5**, 106 (2010) (in Russian)
2. D. Kopanitsa, A. Ustinov, A. Potekaev, A. Klopotov, G. Kopanitsa, AIP Conf. Proc. **1698**, 030014 (2016)
3. A. Potekaev, A. Klopotov, AIP Conf. Proc. **1683**, 020233 (2015)
4. A.M. Ustinov, D.G. Kopaniza, Y.A. Abzaev, A.A. Klopotov, B.O. Koshko, G.D. Kopaniza, AIP Conf. Proc. **1800**, 040007 (2017)
5. A.F. Ioffe, *Fizika kristallov* [Physics of crystals] (OGIZ, Moscow, Leningrad, 1929) (in Russian)
6. V.P. Alekhin, *Fizika prochnosti i plastichnosti poverhnostnyh sloev materialov* [Physics of Strength and Plasticity of Surface Layers in Materials] (Nauka, Moscow, 1983) (in Russian)
7. V.S. Ivanova, *Sinergetika. Prochnost' i razrushenie metallicheskih materialov* [Synergetics. Strength and fracture of metallic materials] (Nauka, Moscow, 1992) (in Russian)
8. V. Panin, V. Fomin, V. Titov, Physical Mesomechanics **6**(2), 5 (2003) (in Russian)
9. V. Panin, Physical Mesomechanics **2**(6), 5 (1999) (in Russian)
10. M. Tavakkolizadeh, H. Saadatmanesh, J. Struct. Engl. **129**(2), 186–196 (2003)
11. A. Taoum, H. Jiao, D. Holloway, International Journal of Steel Structures **15**(1), 125-134 (2015)
12. S.S.J. Moy, A.G. Bloodworth, Structures & Buildings **160**, 81-93 (2007)
13. M. Elchalakani, D. Fernando, Thin-Walled Structures **53**, 58-71 (2012)
14. Y.J. Kim, J. LaBere, I. Yoshitake, Composites: Part B 51, 233–245 (2013)
15. K.M. El-Sayed, N.N. Khalil, I.M. El-Shennawy, International Journal of Advanced Engineering, Management and Science **2**(1), 9-17 (2016)

16. E.Y. Sayed-Ahmed, Advanced Composite Materials in Bridges and Structures **IV**, 1-8 (2015)
17. M.H. Seleem, I.A. Sharaky, H.E.M. Sallam, Materials and Design **31**, 1317–1324 (2010)

Basics of recreation of ancient ceramics production technology

Elza Sutakova[1,*], and *Alexei* Mestnikov[1]

[1]M.K. Ammosov North-Eastern Federal University, 677000 Yakutsk, Russia

Abstract. The paper investigates possibility of ancient ceramics technology recreation based on materials of archaeological excavations and experimental simulation and describes results of the research. The structure and properties of ceramic samples found during archaeological excavations of the neolithic age and clay raw materials from different deposits were studied. The fragments under study were shown to consist mainly of light-melting clays, possessing polimineral composition with additives of fire clay, sand, loose sedimentary rock, and in some cases wool and vegetation. Clays under study are characterized by variability of chemical and mineral composition, by their combining it is possible to obtain optimal moulding compositions to produce ceramic products of artistic and architectural-structural purpose. Overview of scientific and technical information in archaeology and the obtained experimental results enable to reveal compositions and peculiarities of ancient technology of ceramics production and technical decoration application. Peculiar features of experimental simulation of the ancient ceramics technology considering chemical and mineral composition of clay raw materials are described.

Introduction

Ancient ceramics was initially used as houseware made from clay and it was the most popular and information-bearing material. For revealing the secrets of ancient ceramics technology latest results of archeological studies and analysis of existing materials may be useful [1].

It is known that talented craftsmen turn simple things into true pieces of applied arts. Ceramic art reached its highest development in Ancient Egypt, Assyria, Babylon, Greece, China and Hungary. Various museums of the world exhibit houseware made by ancient potters. The old masters were able to mould houseware of enormous sizes sometimes. Greek pithoi – vessels for water and wine which can be two meters high, strike by their highly technical implementation [2].

Ceramic pots and vessels in the Neolithic culture were made on a rather high technology level of that time. Items of different shape and size can be found, as different production techniques were used [3]. The round-bottomed ceramics of open-type with marked neck are generally met [4].

* Corresponding author: em.sut@mail.ru

According to the existing archaeological remains and data of petrographic analysis moulding composition of ancient ceramics was defined [5]. The basic components of the composition were clay, sand of various fracture, rotted rock, wool, vegetation and fire clay. Ancient craftsmen used specially selected varieties of clay raw materials or mixtures on their basis for production of quality products [6]. Regulation of the ready products' properties was performed using additives of mineral and organic origin, as well as by applying different processing techniques [7].

From the viewpoint of construction engineering ceramic products attract by their durability, ease of manufacturing various products, availability of raw materials and low products cost, and possibility to organize small production. Therefore, studying ceramic fragments of ancient times is one of the most crucial and relevant objectives of ancient ceramics technology revival and its enhancement on a modern level. Revival of ancient ceramics production technology can be put into action in production of ceramic tiles and hardscape with decorative coating and exclusive painting which are of special demand for green construction [8-10].

Thus, experimental simulation of ancient technology requires conducting a complex of research of the existing samples of ancient ceramics and clay raw material for their production, as well as systematization and analysis of the obtained data. Defining the possibility for revival of ancient ceramics technology on the basis of archaeological excavations and experimental simulation using modern methods and facilities is the basic goal of the present study.

Materials and methods

The object of experimental study is samples of ancient ceramics and clay raw materials used for their production. The most difficult objective in experimental simulation of ancient ceramics technology is searching and obtaining permission for research of ceramic fragments being museum specimen. Moreover, the largest interest for research is ceramic samples of special purpose, e.g. large-size vessels with thin wall.

Ceramic fragment under study has waffle drawing performed on light-burning clay, peculiar for "Ymyyakhtatskaya culture" (archaelological culture of Siberia). "Ymyyakhtatskaya culture" is characterized by round-bottomed ceramics with the distinctive surface, waffle and ribbed imprint, dated II-I thd. B.C. [2].

Experimental simulation apart from ancient ceramic fragment was also applied to clay raw materials of different deposits of North Eastern Region of the Russian Federation.

Modern physical-chemical research methods enable to study structure and properties of materials, as well as define moulding composition of ancient ceramics and clay raw materials with high accuracy [11-12]. The studies were conducted with the help of X-ray spectral analysis using facility SRS-3400, X-ray powder diffraction apparatus ARL X TRA (Thermo Fisher Scientific, Switzerland), diffraction data bases ICDD PDF-2, search system Cristallografica Search – Match (Oxford Cryosystems), scanning electron microscope JEOL JSM-7800F.

Results

The sample of ceramic fragment under study contains aluminum oxides – 23.61 wt.%, it is quite possible that it contains kaolinite minerals. Chemical analysis makes possible to confirm that ceramics under study was made from semi-acid clay with average grade of water-soluble salts. Considering ferrous oxide content it refers to clays with low content of colouring oxides. Apart from oxides given in Table 1, the following ones were also observed: V_2O_5, Cr_2O_3 and MnO in the amount of 0.01-0.02 wt.%.

Table 1. Results of chemical analysis of ceramic fragment, wt.%.

Na_2O	MgO	Al_2O_3	SiO_2	P_2O_5	SO_3	Cl	K_2O	CaO	TiO_2	Fe_2O_3
1.41	1.45	23.61	67.43	0.99	0.12	0.04	0.66	1.04	0.67	2.39

X-ray phase analysis identified variety of quartz, ferric hydroxysulfate, clinochrysotile, kaolinite, and muscovite in the ceramic sample under study. Diffraction angles of quartz, clinochrysotile, and kaolinite are quite well observed (Fig. 1). Clinochrysotile refers to the group of zeolite minerals.

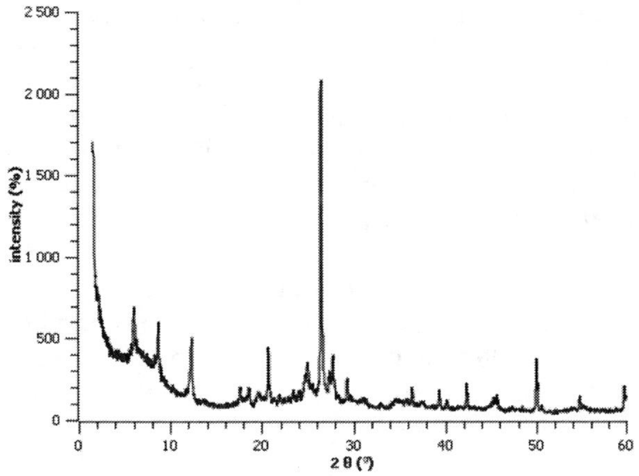

Fig. 1. X-ray pattern of the ceramic fragment.

Cross-sectional view of the ceramic fragment shows inhomogeneous composition, large size particles are visible, i.e. sand, possibly with fire clay additives. The arrangement of structure is close-packed; firing was probably conducted at the temperature close to 900°C (Fig. 2).

(a) (b)

Fig. 2. Microstructure of ceramic fragment: a – surface; b – chip.

Chemical composition was investigated (Table 2) along with mineral composition of the clay raw material from the four deposits of North Eastern region of the Russian Federation (Kangalasskoe *1*, Kempendyaiskoe *2*, Namtsyrskoe *3*, Sannikovskoe *4*).

Table 2. Chemical composition of the clay raw material, wt.%.

No	SiO_2	Al_2O_3	Fe_2O_3	CaO	MgO	P_2O_5	K_2O	TiO_2	Na_2O
1	60.28	34.38	0.95	0.69	0.49	0.07	0.78	1.49	0.33
2	42.62	11.22	5.68	16.69	14.41	0.19	4.31	0.83	2.77
3	63.57	22.38	1.71	0.54	0.55	0.04	2.49	0.81	0.95
4	61.44	14.23	14.39	4.28	2.14	0.13	2.63	0.73	2.11

The clays of Namtsyrskoe deposit possess complex polimineral composition, minerals of quartz subgroup were revealed (quartz 56.5%), potassium sodium feldspar (white feldspar – 11.5%, orthoclase feldspar – 12.6%), kaolinite (4%), as well as hydromicaceous minerals were observed – 4.9%. Considering refractory properties the samples refer to low-melting clays of polymineral composition, considering Al_2O_3 content they refer to semi-acid clays, depending on the content of coloring oxides they refer to the group of average content of Fe_2O_3 and TiO_2, and is characterized by the average content of water-soluble salts.

Sannikovskoe deposit contains the following minerals, wt.% : quartz – 17.1; potash feldspar – 24.4; muscovite – 57.5; vermiculite – 10.3 and cronstedtite – 1.3 (Fig. 3). The revealed minerals refer to lamellar hydrosilicates. The deposit was defined as belonging to the group of moderate plastic, coarsely dispersed semi-acid clay raw with high content of silty particles, coloring oxides and soluble salts.

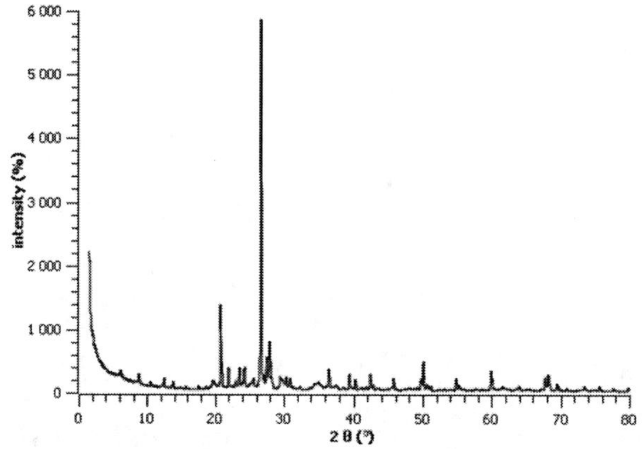

Fig. 3. X-ray pattern of clay raw from Sannikovskoe deposit.

Kempendyaiskoe deposit is characterized by high content of water-soluble salts. According to chemical composition it is characterized by low content of aluminum oxides – 11.22%, silicon oxides content is 42.62%, considering content of coloring oxides it refers to clays with high content of: ferrous oxides (5.68%), titanium oxide (0.83%). With regard to mineral composition it is characterized by quartz content (16.4%), muscovite (41.4%), calc spar (25.9%), dolomite (13.7%), traces of hematite are observed (2.6%).

Authors consider that the largest interest for manufacture of ceramic products of architectural construction and artistic purpose is presented by clay raw materials from Kangalasskoe deposit (Fig. 4). Clays are characterized by not high content of Al_2O_3 (34.38 wt.%), low content of coloring oxides Fe_2O_3 – 0.954, TiO_2 – 1.49%. The basic minerals are presented by kaolinite – Al_4 [Si_4O_{10}] $(OH)_8 \cdot nH_2O$ (69.8%), quartz SiO_2 (19.7%), according to X-ray phase analysis presence of halloysite $Al_2Si_2O_5(OH)_4$ (10.5%) is observed. Clay mineral of lamellar silicates subclass is close in its composition to kaolinite and refers to light-burning clays.

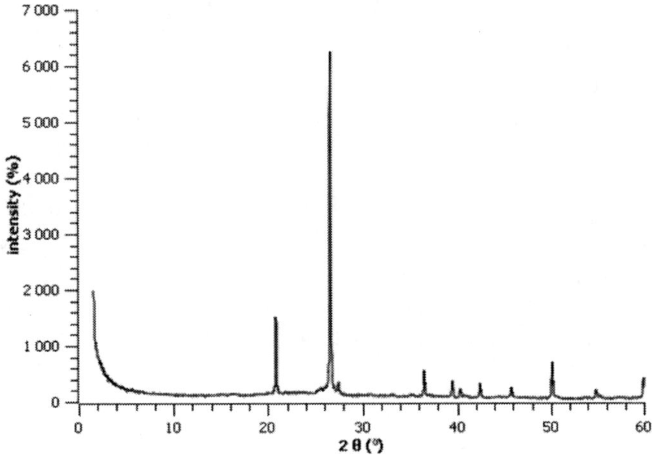

Fig. 4. X-ray pattern of clay raw material from Kangalasskoe deposit.

Conclusions

Overview of the literature data and results of experimental studies enable to make the following conclusions:

1) ancient masters used the simplest moulding composition for making household items, i.e. "clay + mineral additives (sand, rock pieces, fire clay) + organics (vegetation, wool)";

2) the basic secret of ancient technology lies in chemical-mineral composition of clay raw materials, thus proving comparison of experimental results of ancient ceramic fragment with clay samples from different deposits: for production of coarse ceramics (vessels of everyday use) ancient masters used the available kinds of clay, while for fine ceramics (regular and large-size vessels with thin walls) they looked for special kinds of clays that would satisfy the technology requirements.

Further research will be devoted to studying the ancient ceramics technology relating to large-size items production, recreation of their production technology and firing in non-industrial conditions.

Authors acknowledge Ruslan A. Zakharov for providing materials for research, and Vasilii V. Mukhin, senior teacher of the Department of General, Analytical and Physical Chemistry, for valuable comments.

References

1. V.M. D'iakonov, Izvestija Laboratorii drevnih tehnologij [Reports of the Laboratory of ancient technologies], **1 (4)**, 104-119 (2006)
2. A. Cercone, M. Golitko, Poster at the SAA 81st Annual Meeting (Orlando, Florida, April 6-10, 2016)
3. S.A. Fedoseeva, *Ymyjahtahskaja kul'tura Severo-Vostochnoj Azii* [Ymyiakhtakh culture of Northeast Asia] (Nauka, Novosibirsk, 1980)

4. V.M. D'iakonov, *Drevnie kul'tury Mongolii i Bajkal'skoj Sibiri* [Ancient cultures of Mongolia and Baikalian Siberia], Proceedings of the International Scientific Conference, **2**, 163-170 (Irkutsk, Russia, May 3-7, 2011)
5. M. Romagnoli, American Ceramic Society Bulletin, **4**, 9301-9304 (2005)
6. N.D. Yatsenko, É.O. Ratkova, Glass and Ceramics, **3**, 93-94, (2009)
7. V.Z. Abdrakhimov, E.S. Abdrakhimova, News of higher educational institutions. Construction, **7**, 12-16 (2006)
8. M. Golitko, European Journal of Archaeology, **14**(3), 419-440 (2011)
9. H. Falih, Anbar Journal of Engineering Science, **3**(2), 72–77 (2010)
10. T. Takeuchi, K. Nakamura, L.N. Mylnikova, S.P. Nesterov, N.A. Kulik, Ethnology and Anthropology of Eurasia, **37** (1), 39-51 (2009)
11. A. Mestnikov, S. Semenov, V. Strokova, V. Nelubova, *AIP Conference Proceedings*, **1698**, 070010 (2016)
12. T.A. Khoroshun, I.M. Summamen, The role of natural science methods in archaeological research of ancient ceramics in Karelia. Institute of Language, Literature and History, Karelian Research Centre, Russian Academy of Sciences, **8**, 17-27 (2015)

Material and field impacts on cement compositions

Aleksandr Rubanov[1,*], *Nikolay* Gorlenko[1], and *Vladimir* Vergasov[1]

[1]Tomsk State University of Architecture and Building, 634003 Tomsk, Russia

Abstract. The paper studies the effect of combined field impact on cement compositions. The priority research direction involves creation of cement compositions and technologies that meet the requirements of technological availability and efficiency, environmental safety and sustainability, biocompatibility, economic efficacy and energy and resource saving. The paper proves that these criteria can be affected by material and field impacts on cement systems, which leads to considerable increase in strength of hardened cement paste, mortars and concretes. With this purpose, the paper presents detailed study on combined impact of ultrasound and magnetic field on grouting fluid of cement systems and strength of hardening systems in normal conditions. Compressive strength of cement and sand mortar was found to increase on average by 30-45%.

Introduction

Development of new cement types, modifying chemical admixtures, external impacts on cement systems enhance the ability to control physico-mechanical and operational characteristics of such systems, both on the stages of cement and water interaction and ultimate strength formation in a system [1-3].

According to the universal framework of environmental studies and modern building materials science, all newly obtained materials, including those based on cement systems, must meet the criteria previously formulated in [4-5].

Among all cement system components, grouting fluid still remains the most available one for direct modification of hardened cement paste properties [6-10]. However, if we take water, in spite of remarkable success in this field, its modification by physical and chemical impacts meets the above mentioned criteria to a greater extent, especially when combined methods of material and field impacts are applied [3].

This study is aimed at increasing strength of cement and sand mortar by affecting the grouting fluid by ultrasound with given frequency and by static magnetic field with given intensity.

To date, no previous studies have been performed on the topic of combined impact on cement systems grouting fluid.

* Corresponding author: rubal06@yandex.ru

Materials and methods

When performing the study, USU-0707 ultrasound machine with the acoustic frequency of 125±6 kHz and power demand of 9W was used. Static magnetic field was produced by a system of NdFeB magnets with the magnetic field strength of 0.10±0.02 Tesla on the magnet surface. System of magnets created areas with magnetic field gradients, through which distilled water was put at the velocity of 0.7 m/s. Grouting fluid was affected in cycles in a closed-loop system. Different variants of combined impact were applied, including separate and successive exposure to ultrasound and magnetic field, as well as simultaneous and combined impact of these material-field modifiers.

The impact types were chosen based on multidirectionality of their effect on water and electrolyte water solutions.

Changes in physico-chemical properties of water and cement compositions were studied with the help of LAMBDA 265 UV-spectrophotometer and SDTQ 600 thermal analyzer.

To define compressive strength cubic specimens sized $(70{,}7\times70{,}7\times70{,}7)\times10^{-3}$ m were made from cement and sand mortar composed of 1 part of M400 Portland cement and 3 parts of sand, with water to cement ration of 0.45. The specimens mixed with activator liquid and distilled water were cured in a moist chamber and tested for compressive strength within given periods according to the standard methods (GOST 5802-86) [13].

The number of specimens in each series of compressive strength tests was no less than 6. All obtained test results were statistically analyzed.

Results and discussion

Exposure of both distilled and natural water to static magnetic field results caused shift of pH medium towards the alkaline one up to the values of 8.5–9.0 compared to the corresponding values in test water (pH=6.86). At the same time, UV spectroscopy analysis showed that density of hydrogen bonds under cyclic magnetic treatment increases by 25–50 % depending on the number of cycles (Fig.1).

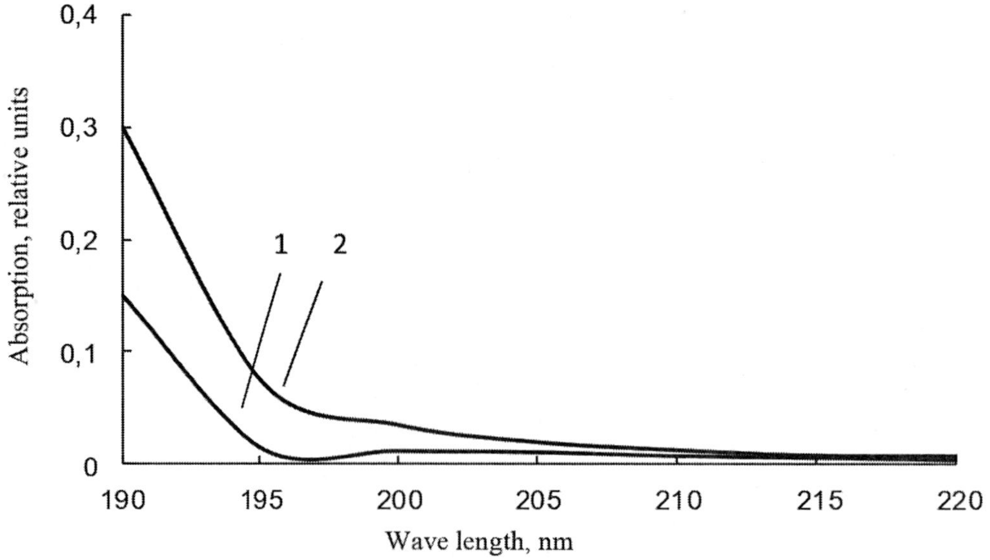

Fig. 1. UV-absorption spectra in distilled water: 1 – test specimen; 2 – after magnetic field treatment.

Mixing of hardened cement paste with such water ensures higher bond density between water molecules and Portland cement clinker, which is confirmed by comparison of thermograms of hardened cement paste in the studied test specimens of the same age [9]. Shift of endo effects of water dehydration towards higher temperature values (by 20–30 %) confirms the previously made assumptions [11].

Treatment of water with ultrasound of various energy and with various exposure time results in dispersion of large water clusters into smaller fragments and, as a consequence, leads to increase of its electrical conductivity and hydrogen ions concentration in liquid phase. Based on this, a conclusion can be drawn that the cement-water system must be very sensitive to succession of combined impact methods application for water as grouting fluid for cement systems.

The results of compressive strength tests on cement and sand mortar specimens under various impacts are given in Table 1.

Table 1. Strength of cement and sand mortar.

No.	Type of impact	Time and mode of impact on grouting fluid, min	Hardening time, days	
			3	28
			Specimens strength, MPa	
1	Without treatment	–	3.4±0.2	14.9±0.2
2	Ultrasound (US)	15	7.9±0.4	18.4±0.2
		30	6.1±0.3	20.3±0.4
		45	7.8±0.4	19.1±0.3
		60	6.7±0.3	20.1±0.4
3	Magnetic field (MF)	7 cycles	6.6±0.3	18.6±0.3
4	Successive impact by US+MF	30 minutes + 7 cycles	7.2±0.3	21,6±0,4
5	Successive impact by MF+US	7 cycles + 30 minutes	6.2±0.2	16.7±0.2
6	Simultaneous impact y US and MF	30	5.8±0.2	15.4±0.2

As can be seen from the Table 1, strength value of hardening structures in cement-water system after impact by only ultrasound at the age of 28 days varies between 18.4 and 20.3 MPa depending on the exposure time, which is 23–35 % higher than that in test specimens.

After impact by static magnetic field only the strength of specimens increased by 25 % at the same age. When combined impact of ultrasound + magnetic field was applied to specimens, their compressive strength increased by 45 %, and after impact by magnetic field + ultrasound – by 12%.

Test results analysis shows that succession of impact leads to different strength parameters of cement and sand mortar. According to the authors' opinion, it can be caused by the reasons as follows. Under initial ultrasound impact, dispersion of water medium comes with formation of various short-lived particles and states in water, cavitation effects and other related phenomena which become sensitive to magnetic field impact. As a result, water reactive capacity rapidly increases in relation to cement and leads to increase in strength parameters. Under magnetic field impact followed by ultrasound impact, the latter appears less effective for the same values of energy and exposure time, due to formation of stronger structures of liquid phase clusters. As a consequence, lower strength values are observed (compared to those obtained for impact type No.4, Table 1), which are still higher than those in test specimens.

At the same time, under simultaneous impact by ultrasound and magnetic field, the strength values of hardening structures fluctuate and come with interchange of increase and decrease of strength in a cement system.

Additional tests showed that the water subjected to a certain type of impact retains its maximum reactive capacity in relation to cement within first 4 hours. However, even 7 days after water treatment, the impact effectiveness is still high, and the strength of hardening structures is higher than that of test specimens [9].

Conclusion

The performed study showed that the suggested method of grouting fluid successive treatment first with ultrasound and then with magnetic field can be recommended for production of cement-based building materials.

Further investigations in this field will be aimed at increasing efficiency of the studied technologies for cement systems hardening not only in normal conditions, but also in extreme ones. The examples of fields for further studies can be the following: hardening at negative temperatures due to complex modification of grouting fluid with antifreeze admixtures [12] which are widely used in cold weather concreting of precast structures.

References

1. S.V. Samchenko, *Formirovanie i genezis struktury cementnogo kamnja* [Formation and genesis of the structure of hardened cement paste] (MISI - MGSU Publishing House, Moscow, 2016) (in Russian)
2. Y.M. Bazhenov, *Modificirovannye vysokokachestvennye betony* [Modified High Performance Concrete] (Association of Civil Engineering Universities, Moscow, 2006) (in Russian)
3. Y.S. Sarkisov, News of Higher Educational Institutions. Construction **2**, 106–109 (1993) (in Russian)
4. Y.S. Sarkisov, N.P. Gorlenko, I.A. Rakhmanova, Polzunovsky vestnik **3**, 168–171 (2016) (in Russian)
5. N.O. Kopanitsa, V.N. Safronov, O.V. Demyanenko, A.A. Shpilekova, Efficient techniques and technologies in building materials science, Proceedings of Scientific Conference (Novosibirsk, February 14–17, 2017) (in Russian)
6. J.C. Smith, F. Merzel, C.S. Verma, S. Fischer, J. Mol. Liq. **101,** 27–33 (2002)
7. T. H. Plumridge and R. D. Waigh, J. Pharm. Pharmacol. **54**, 1155–1179 (2002)
8. C. H. Cho, J. Urquidi, S. Singh, S. C. Park, G. W. Robinson, J. Phys. Chem. **A 106**, 7557–7561 (2002)
9. N.P. Gorlenko, Y.S. Sarkisov, *Nizkojenergeticheskaja aktivacija dispersnyh sistem* [Low-energy activation of dispersion systems] (TSUAB Publishing, Tomsk, 2011) (in Russian)
10. A.N. Gul'kov, Y.A. Zaslavsky, P.P. Stupachenko, *Primenenie magnitnoj obrabotki vody na predprijatijah Dal'nego Vostoka* [Application of magnetic water treatment at industrial enterprises of the Far East] (Far East University Publishing, Vladivostok, 1990) (in Russian)
11. Y.S. Sarkisov, N.P. Gorlenko, V.N. Safronov, S.A. Kugaevskaya, M.A. Kovaleva, T.A.Ermilova, D.A. Afanasyeva, Vestnik of Tomsk State University. Chemistry **2**, 20–

29 (2015) (in Russian)
12. A.V. Lagojda, N.A. Romanova, O.S. Ivanova, O.E. Koroleva, *Rukovodstvo po primeneniju betona s kompleksnymi protivomoroznymi dobavkami* [Handbook on the use of concrete with complex antifreeze admixtures] (Publishing house of Reinforced Concrete Research Institute, Moscow, 1987) (in Russian)
13. GOST 5802-86. Mortars. Test methods, 16 (Standartinform, Moscow, 2010) (in Russian)

Macromodel of interfacial transition layer in ceramic matrix composites

Andrey Stolboushkin[1,*], *Vadim* Syromyasov[1], *Vladimir* Vereschagin[2], and *Oksana* Fomina[1]

[1]Siberian State Industrial University, 654007 Novokuznetsk, Russia
[2]National Research Tomsk Polytechnic University, 634050 Tomsk, Russia

Abstract. The purpose of the study is to create a macromodel of interfacial transition layer in ceramic matrix composites. Chemical and mineralogical compositions were investigated by means of X-ray fluorescence analysis and X-ray diffractometry, ceramic and technological properties of raw materials were defined using standard test methods for argillaceous raw materials. Phase composition and structure of ceramic specimens were studied using a complex of modern physico-chemical analysis methods. The layer-by-layer model of shell-core transition in ceramic matrix material was suggested. Boundary conditions for obtaining specimens were defined in terms of number of layers, thickness of such layers and pitch of core-to-shell material ratio. Forced air supply was organized while burning for directed heat and mass transfer inside the specimens. Mineral composition of layers was defined for ceramic specimens with the core of iron ore waste and the shell of clay. The study enabled to determine dependences between qualitative and semi-quantitative variation of new mineral formations content in transitional layers of a composite, which is the evidence of interaction between the core and the shell products while burning a ceramic matrix material.

1 Introduction

Scarcity of high quality clay resources generates the need for search of alternative raw materials for ceramic building products. Off-spec silica raw materials and industrial mineral waste can be used as additional raw material source [1-5]. At the same time, slow development of technogenic deposits for ceramic products manufacturing is connected with insufficient previous studies on physico-chemical processes that occur while burning such raw materials [6-10].

New approaches to directed structure formation at the stages of raw materials and molding composition preparation and unburnt brick compaction are required to obtain ceramic bricks with specified physico-mechanical properties from off-spec natural raw materials and industrial mineral waste [11-13]. Technological principles for obtaining ceramic matrix composites were developed by the authors of the study [14]. Modeling of layer-by-layer transition between core and shell in ceramic matrix composites enables to control structure formation processes and, in the end, to obtain tailored ceramic products.

* Corresponding author: stanyr@list.ru

Based on the previous research, the authors developed a framework for formation of matrix transition layers on a macrolevel and prepared model specimens. Critical conditions for obtaining macromodels were defined and factors to increase their performance were considered [15, 16].

Specialists emphasize the necessity to develop and use new research methods, along with standard ones, for studying ceramics made from technogenic raw materials with new structural and mechanical characteristics [17-21]. The novelty of this research compared to other studies is selectivity while studying interphase boundaries in ceramic matrix composites using a transition layer macromodel.

The purpose of this research is to study phase composition and structure of interfacial transition layers between core and shell of ceramic matrix composite made of iron ore waste and argillaceous raw materials.

2 Framework for macromodel development

The framework to obtain ceramic specimen models consisted in formation of successive layers with various percentage of raw material components (Fig. 1).

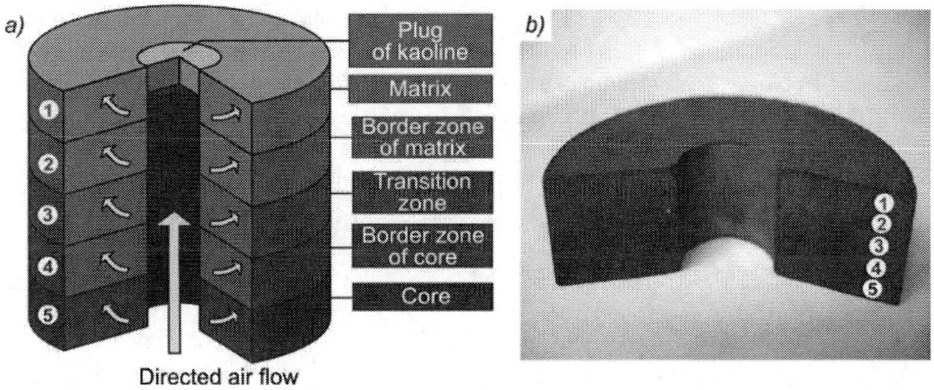

Fig. 1. Scheme of transition layer formation between shell and core in a matrix composite (a) and its specimen model after burning at 1000 °C (b): 1 – matrix (shell); 2 – border zone of matrix (shell); 3 – transition zone; 4 – border zone of core; 5 – core.

According to the developed method for obtaining ceramic matrix composite products, argillaceous batch components roll up onto the surface of waste granules in the process of their formation. At the same time, both raw materials partially mix in the border zone. Distribution of both components at the waste↔clay border in the total mass changes integrally from 100 to 0 wt.%, and conversely. By differentiating this smooth transition into several zones with a certain components ratio, one can approximately model a phase boundary (dispersed phase↔dispersion medium) in a ceramic matrix composite. A specimen model is obtained by forming successively several different layers (Fig. 1, b).

Optimum conditions for obtaining a macromodel were defined using design and experimental methods. Previously, the number of transition layers varied from 1 to 5. The experiment showed that one or two transition layers provide a rough approximation of the model in relation to the real matrix structure condition. While forming specimens with five transition layers a problem of obtaining uniform layer thickness appears. That is why the optimum number of transition layers in a model is 3, while the total number of layers is 5. Investigation was performed to decrease the layer thickness to the minimum possible one enabling to perform material sampling for further X-ray diffraction analysis [15]. Such investigation allowed approximating the model to the real conditions of boundary transition between core and matrix. To intensify the processes of heat and mass transfer and interaction between layers, it was suggested to organize forced air change while burning. Special molds were designed for this purpose, using which cylinder specimens with end-to-end holes can be produced [16].

3 Materials and Methods

3.1 Raw materials for macromodel preparation

The suggested model (Fig. 1) was tested using the following raw materials. Iron ore waste from the Abagursky ore-dressing plant (Novokuznetsk city, Russia) was used as core material, and slightly plastic Novokuznetsk loam was used as matrix material.

Mineral part of iron ore waste consists of feldspars, quartz, mica, pyroxenes, amphiboles, ferruginous chlorites, and contains small amount of mixed-layer argillaceous formations. The material is fine-grained (with the average grain size of 15-18 microns), slightly plastic (with plasticity index less than 4), non-sintering and has low drying sensitivity.

Novokuznetsk loam contains less than 1% of coarse-grained particles (>0.5 mm). The amount of clay fraction particles (<0,001 mm) is less than 20-30%. Loam has high drying sensitivity and belongs to hydromicaceous-montmorillonitic type based on its argillaceous minerals characteristics. Non-plastic part of loam is represented by quartz, carbonate, feldspars, chlorite and amphiboles.

Chemical composition of raw materials is given in Table 1.

Table 1. Chemical composition of raw materials.

Raw material	Mass fraction of components, % (on a dry basis)											
	SiO_2	Al_2O_3	FeO	Fe_2O_3	CaO	MgO	TiO_2	MnO	R_2O	P_2O_5	SO_3	Loss on ignition
Iron ore waste	32.74	8.99	5.70	13.59	14.37	10.88	-	0.59	1.15	0.24	1.45	10.3
Novokuznetsk loam	62.03	14.71	5.07		4.56	2.49	0.93	0.21	3.94	-	0.47	5.59

3.2 Macromodel design method

According to the section 2, ceramic hollow cylinder specimens with the diameter of 60 mm and the height of 20-25 mm were prepared using semidry pressing method (Fig. 1, b). Amount and composition of raw materials components of the layers are given in Table 2. Moisture of ceramic batch components was 9-9.5 wt.%, pressure value was 15 MPa. The specimens were burnt at 1000 °C and cured for 1 hour at maximum temperature.

Table 2. Composition of raw materials for model specimens.

Raw material	Content of batch components depending on the layer number, wt. %				
	1st layer	2nd layer	3rd layer	4th layer	5th layer
Iron ore waste	-	25	50	75	100
Novokuznetsk loam	100	75	50	25	-

While preparing multilayer specimens, accurate batching of mold powder was ensured for each layer on the volume basis. The layer thickness was 4-5 mm. Molding material moisture control was performed for each layer. Finely fragmented components were dried to constant mass. Accurate amount of water was calculated for prepared batch compositions. Mold powders were simultaneously kept in a desiccator for 6 hours for moisture leveling.

In order to organize directed heat and mass transfer inside a specimen and trigger layers interaction, one side of the end-to-end cylindrical hole was tightly plugged with a combination of kaolin wool and clay. On the opposite side of the model specimen, directed air flow was supplied into the hole (Fig.1, a).

3.3 Macromodel research methods

Radiographical studies of the layers which modeled the layer-by-layer transition from matrix to core were performed using XRD-6000 X-Ray diffractometer (Shimadzu, Japan). Powders obtained from each level were studied in the range of 2θ from 4° to 80° at room temperature using FeKα radiation (Wavelength 1.936040 Å). Diffractometer settings were as follows: 40 kV, 30 mA, continuous scan mode, 2θ step width 0.02° and speed 2 deg/min. Phase composition analysis was performed with the help of Match!3 (Crystal impact, Germany) software and the PDF-2 data of The International Centre for Diffraction Data.

Microstructure and chemical composition of the matrix and core layers of the specimen were investigated using JSM-6460LV (JOEL, Japan) scanning electron microscope and a device for energy dispersive X-ray analysis (INCA Energy-350, Oxford Instruments).

4 Experiment

X-ray diffraction patterns of the model specimen layers showed presence of the following solid mineral phases: quartz (4.255; 3.343; 2.128 Å, etc.), hematite (2.69; 2.25; 1.843; 1.697 Å), feldspars (4.03; 3.78; 3.484 Å, etc.), spinel (4.67; 2.85; 2.43 Å), augite (3.204; 2.989; 2.917; 2.550 Å), diopside (2.512; 2.985; 2.943 Å) and olivine (1.748; 2.510; 2.457 Å). Their intensity and semiquantitative content are presented in Figure 2.

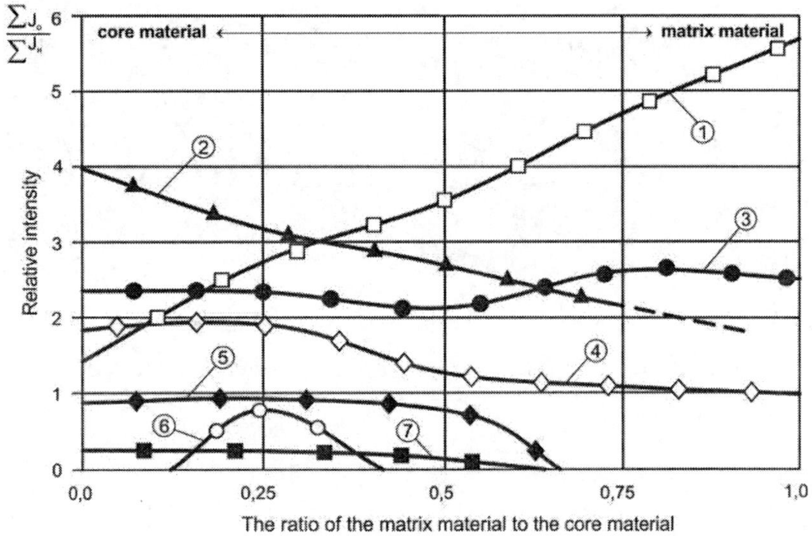

Fig. 2. Variation of total intensity of the minerals while core to matrix transition in a ceramic matrix composite: 1 – quartz; 2 – hematite; 3 – feldspar; 4 – augite; 5 – spinel; 6 – diopside; 7 – olivine.

Analysis of the curvatures of intensity variations showed that quartz crystal amplitude increases with the increase of content of matrix material in relation to core material from 0 to 1, which is explained by higher content of quartz crystal in loam. Layer-by-layer decrease of hematite intensity towards the matrix is caused by lower content of ferrous oxide in argillaceous component compared to that of iron ore waste. With the increase of loam amount in layers content after burn, spinel and olivine become less intense and are almost absent in the layer formed of loam only. The intensity of feldspar has almost no change in transition from core to matrix of ceramic matrix specimen, irrespective of layers initial content. The line of augite intensity indicates decrease in reaction of pyroxene mineral formations while transition to matrix of the composite material.

Therefore, the main crystal phases of transition layers and the core are quartz, hematite, feldspar and augite. The same minerals except hematite are present in the matrix. The amount of quartz rises from core to matrix due to quartz component in loam. Spinel and olivine are contained in the core in small amounts, but they disappear in layers that are close to the matrix.

Studies of microstructure and chemical composition of core and matrix layers of a model specimen are presented in Figure 3 and in Table 3.

By using average values, high content of Fe, Ca and Mg was found in the model core layer compared to the matrix layer. Such phenomenon is caused by their predominance in iron ore waste and confirms formation of augite *(Ca, Fe, Mg)*$_2Si_2O_3$ and diopside $CaMgSi_2O_6$ while burning. Higher content of *Na* and *K* was found in the matrix layer compared to the core layer.

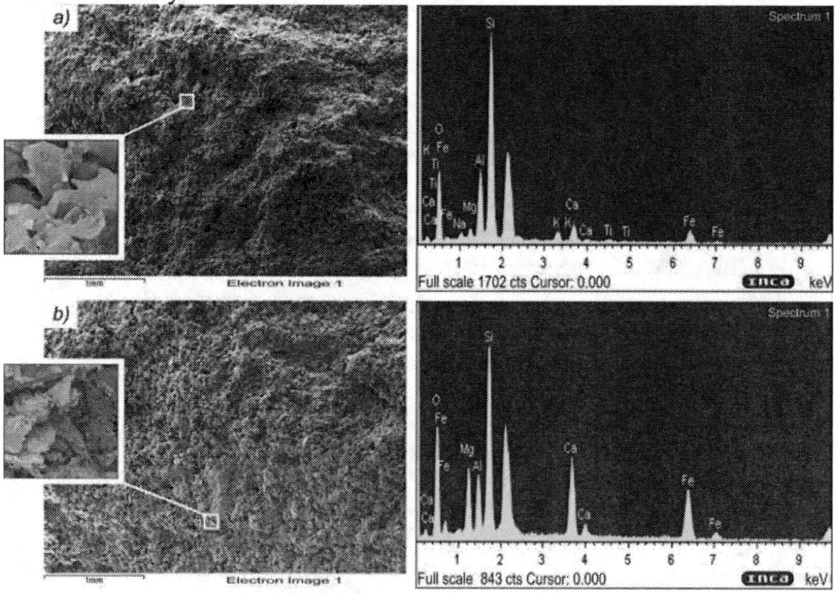

Fig. 3. X-ray microanalysis of matrix (a) and core (b) layers in model specimen burnt at 1000 °C.

Table 3. Element composition of core layer and matrix of model specimen.

Zone	Element composition of crystal phase	Chemical element								
		O	Na	Mg	Al	Si	K	Ca	Ti	Fe
Matrix	weight, %	39.56	1.13	1.57	9.65	33.23	2.30	4.12	0.62	7.83
	atomic, %	55.67	1.11	1.45	8.05	26.64	1.32	2.31	0.29	3.16
Core	weight, %	34.62	0.74	6.81	6.34	19.97	0.46	10.58	0.00	20.49
	atomic, %	53.24	0.79	6.89	5.78	17.49	0.29	6.49	0.00	9.03

Using the investigation data on microstructure and chemical composition of matrix and core layers of a model specimen, we can conclude that alkali elements interact with Si, Al, Mg, Fe and ensure increased content of liquid phase in the boundary layer (Fig. 3).

5 Results

The determined dependences of new mineral formations variation in transition zone between core and matrix are the evidence of interaction between the core and the matrix products while burning a ceramic matrix material.

Presence of identical crystal phases in the core, transition layers and the matrix of sintered loam excludes occurrence of thermal stresses on the core↔matrix boundary.

The obtained experimental data of X-ray diffraction analysis coincide with the forecast of suggested solid phase reactions behavior.

6 Conclusion

The conducted research proved the relevance of the developed scientific hypothesis of modeling core to matrix transition in ceramic matrix composite in the form of discrete layers with different content of raw components. According to the authors, evaluation of layer-by-layer change of phase composition of mineral formations in model specimens' layers enables to forecast the minimum required amount of argillaceous batch components to provide the required operational properties of the product. Such evaluations shall be done along with studying physico-mechanical properties of materials, as well as chemical, structure and phase transition while burning. Further investigation in this field will include development of method of batch component design for obtaining ceramic matrix composites depending on the type, chemical and mineral compositions and ceramic and technological properties of technogenic and natural raw materials. This way, relevance and prospects for this research direction is provided.

The research was performed in the framework of State Assignment of the Ministry of Education and Science of the Russian Federation, project code No. 7.7285.2017/8.9 "Basic research on ceramic matrix composite building materials based on technogenic and natural raw materials".

References

1. B.K. Kara-Sal, D.H. Sat, Sh.V. Seren, D.S. Mongush, Stroitel'nye Materialy [Construction Materials], **4**, 33-36 (2016) (in Russian)
2. E. Wolff, W.K. Schwabe, S.V. Conceição, J.A. Santanna-Greco, M. Greco, R.R.Machado, Clean Technologies and Environmental Policy, **19**(2), 379-389 (2017)
3. T.V. Vakalova, V.M. Pogrebenkov, I.B. Revva, Siberian Journal of Science, **1**(2), 339-347 (2012) (in Russian)
4. A.M. Salakhov, K.A. Ariskina, R.A. Ariskina, Georesources, **18**(3), 236-239 (2016) (in Russian)
5. L. Zhang, Construction and Building Materials, **47**, 643-655 (2013)
6. V.A. Guryeva, V.V. Prokofyeva, Stroitel'nye Materialy [Construction Materials], **4**, 55-57 (2014) (in Russian)
7. L.F. Zhang, S.J. Xiao, S. Chen, X.G. Chen, C.Z. Dong, Y.B. Zhang, Advanced Materials Research, **652-654**, 294-297 (2013)
8. A.F. Khuzin, A.A. Lamberov, S.R. Egorova, O.V. Sroyanov, M.G. Gabidullin, Herald of Kazan Technological University, **18**(16), 89-91 (2015) (in Russian)

9. R.M. Nafikov, G.R. Faseeva, A.M. Salakhov, A.I. Hatsrinov, Herald of Kazan Technological University, **8**, 224-229 (2010) (in Russian)
10. T. L. Lazareva, E.S. Kulikova, Technical Sciences – theory and practice, **55**, 135-140 (2016) (in Russian)
11. J. Martın-Marquez, J.M. Rincon, M. Romero, Ceram. Int., **34**, 1867-1873 (2008)
12. V.F. Pavlov, Stroitel'nye Materialy [Construction Materials], **8**, 28-30 (2003) (in Russian)
13. V.P. Tarasevich, Stroitel'nye Materialy [Construction Materials], **5**, 5-8 (1992) (in Russian)
14. A.Yu. Stolboushkin, G.I. Berdov, V.I. Vereshchagin, O.A. Fomina, Stroitel'nye Materialy [Construction Materials], **8**, 19-23 (2016) (in Russian)
15. A.Yu. Stolboushkin, V.A. Syromyasov, A.I. Ivanov, N.V. Petrova, Dolgovechnost' stroitel'nyh materialov, izdelij i konstrukcij [Durability of building materials, products and structures], Proceedings of the All-Russian Scientific and Practical Conference, 161-166 (Saransk, 2016) (in Russian)
16. A.Yu. Stolboushkin, V.A. Syromyasov, A.I. Shcherbina, N.V. Petrova, Proceedings of the II Bryansk International Innovative Forum – *Construction-2016*, **1**, 153-158 (Bryansk State University of Engineering and Technology, Bryansk, Russia, December 1, 2016) (in Russian)
17. M.D. Herd, Materials and Manufacturing Processes, **8**(2), 259-264 (1993)
18. A.Walter, R. Herber, C. Hess, T. Ressler, Chemistry Central Journal, **4**(December), 4-3 (2010)
19. J. Malaiskiene, R. Maciulaitis, Procedia Engineering, **57**, 739-745 (2013)
20. V.D. Kotlyar, Yu.V. Terekhina, A.V. Kotlyar, Stroitel'nye Materialy [Construction Materials], **4**, 24-27 (2014) (in Russian)
21. N.D. Yatsenko, A.P. Zubekhin, Stroitel'nye Materialy [Construction Materials], **4**, 28-31 (2014) (in Russian)
22. A.Y. Stolboushkin, O.A. Fomina, News of Higher Educational Institutions. Construction, **7**, 23-35 (2016) (in Russian)

Effect of fine-grained components on concrete properties and structure formation

Galina Zimakova[1,*], *Valentina* Solonina[1], *Marina* Zelig[1], and *Viktor* Orlov[1]

[1]Tyumen Industrial University, 625000 Tyumen, Russia

Abstract. The paper investigates the influence of fine-grained components on the synthesis of new formations and structure of cellular concrete. Grain size composition was defined with the aid of a laser diffraction analyzer of particles ANALYSETTE 22 NanoTecplus. The composition of the obtained hydrated phases was studied using electron microscopy and X-ray phase analysis. The introduction of silica and aluminosilicate components with the specific surface area >350 m^2/kg enabled to directly influence the mechanism of the hydrated phases formation. Complex hydrated calcium aluminosilicates of the frame structure of zeolite type and non-permanent composition were identified in the hydration products. The formation of the total set of properties is attributed to the morphological features of the hydrated phases, the nature of connections between the crystalline hydrates, and their location in the material structure. X-ray phase analysis showed that ultra-micro-dispersed components have the stimulating impact on the processes of silicate formation. With the increase in the amount of hydrated new formations and decrease in the liquid phase volume the conditions for building of strong crystal intergrowth contacts were created.

Introduction

The latest developments in the field of construction materials science open wide opportunities to control the structure and properties of cement concretes due to introduction of mineral additives, chemically reactive components and modifiers into the binding matrix [1-3]. Such composites as cellular concretes are characterized by composition complexity. Requirements in energy efficiency of building envelopes stimulated the research in the field of autoclaved aerated concrete. However, obtaining of cellular concrete with low density structure predetermines extension of the research on effective control of the structure formation processes. As reported by some of the authors [4-7], the required conditions for minimizing defects in the formed structure of concretes along with provision of the required technical properties are connected with the increase of dispersion of the components and mineral additives. By means of optimal binding content the thickness of intergranular layers is decreased, porosity is also decreased up to a minimum volume and macro-pores transit into micro- and ultrapores.

* Corresponding author: zimakovaga@tyuiu.ru

The peculiar features of cement systems with mineral components consist in the variety of chemical and interphase interaction processes in which the additive substances participate along with the cement minerals. Reactive capacity of the mineral additives and components is defined by their chemical and mineral composition, dispersive capacity and the conditions of concrete composite hardening.

Scientific potential of understanding the mechanisms of reaction interaction in multi-component and multi-level systems (considering their dispersive capacity) which vary in their structure and functional organization grows rapidly [8-11]. However, a number of issues are still being discussed, even though they explain comprehensively the interconnection between the hydration processes and the formation of spatial structures.

Application of fine-grained silica as a component of rheological matrix of type I [12] is considered for powder and powder-activated concretes in modern concrete study. Inclusion of fine-grained silica into the composition of conventional compact-grained concretes is not considered as a relevant issue. Basically, the properties of compact-grained concretes are controlled by introduction of microsilica. Thus, the growth of interest to this material from the side of researchers and manufacturers has lead to the groundless growth of its cost. Application of industrial low-lime ashes in this regards can serve as a technically and economically justified solution.

Previous studies [13, 14] have shown that the condition for obtaining heavy cement concrete with high water impermeability (up to grades W16-18) and the strength 45-55 MPa was the method of structure control by means of mechanically activated ash intcorporation in the presence of effective water-reducing additive. High degree of hydration interaction and the formation of monolithic structure are reached resulting from the establishment of strong epitaxial intergrowth contacts, also from formation of the new hydrated phases due to increasing of the pozzolanic activity of ashes. High hydration degree is also attributed to the synthesis of crystals and cluster crystals of hydrated new formations which are naturally connected with each other by the phase intergrowth contacts.

The given paper is aimed to thoroughly study the processes of hydration interaction in cellular cement systems in the presence of mineral additives and fine-grained chemically reactive components in the conditions of hydrothermal treatment.

Hydrothermal hardening conditions allow full accomplishment of reactive capacity of silica components and provide the conditions of interphase interaction. High degree of porosity is peculiar for cellular concrete and therefore justification of the rational composition and technical properties of the mixture is integral in obtaining concretes with the given technical properties. However, description of the hydration hardening processes of the cellular cement systems basically follows the scheme which is indicative of the clinker minerals. Thus, the end products of hydrate formation are generally given as hydrated silicates of various basic capacities. As supposed by the authors of this paper, in a complex system which is subjected to hydrothermal treatment depending on crystal-chemical particular features and the dispersity of components the process of the new phases synthesis will have its own peculiar features. A.E. Sheikin shrinkage theory [15] is based on a two-phase model; according to that model the hardened stone is presented by crystal joint and sub-microcrystalline gel.

When clarifying the influence of composition-dispersive properties of the mixture on the processes of hydration interaction and morphological peculiarities of the new hydrated phases we could get the opportunity to control the properties of cellular composites to a greater extend.

Materials and methods

The materials used in the given study are as follows: Portland cement with the low content of C_3A (<4.7%); high-calcium lime; silica sand with SiO_2 content > 98%; ash from Reftinsky State District Power Plant [13]; triethanolamine (TEA).

The whole number of aspects influences the structure formation processes and the properties of cellular concretes: concrete composition, water-solid ratio, temperature, the mode and duration of mass maturation, heat performance of autoclaved curing. In order to reach the set objective the following properties were kept permanent: rheological properties of gas-concrete mixture, temperature conditions and the duration of concrete cellular structure formation, the modes of hydrothermal treatment, as well as the consumption of binding material and gas-forming agent.

Silica sand and ash were grinded in a laboratory mill. Silica powder with specific surface area 200, 290 and 350 m^2/kg was preliminary obtained. The specific surface area 350 m^2/kg was reached by grinding of silica sand in the presence of TEA. The dispersity of grinded ash was 700 m^2/kg. These conditions allowed obtaining micro-heterogeneous and ultra-micro-heterogeneous systems.

The research was conducted using a laser diffraction analyzer of particles size ANALYSETTE 22 NanoTecplus, the facility PSKH-10(SP) (the measuring instrument of the specific surface area and the mean particles size of powders), X-ray diffractometer DRON 7M c α_{cu}, inverted electron optical microscope GX-51.

The new formations of cement composites were identified using the crystallographic and crystal-chemical data bases for minerals and their structural analogues [16, 17].

Results

Results of analytical estimation of the obtained fine-grained silica powders and ash grain composition can be seen in Figure 1.

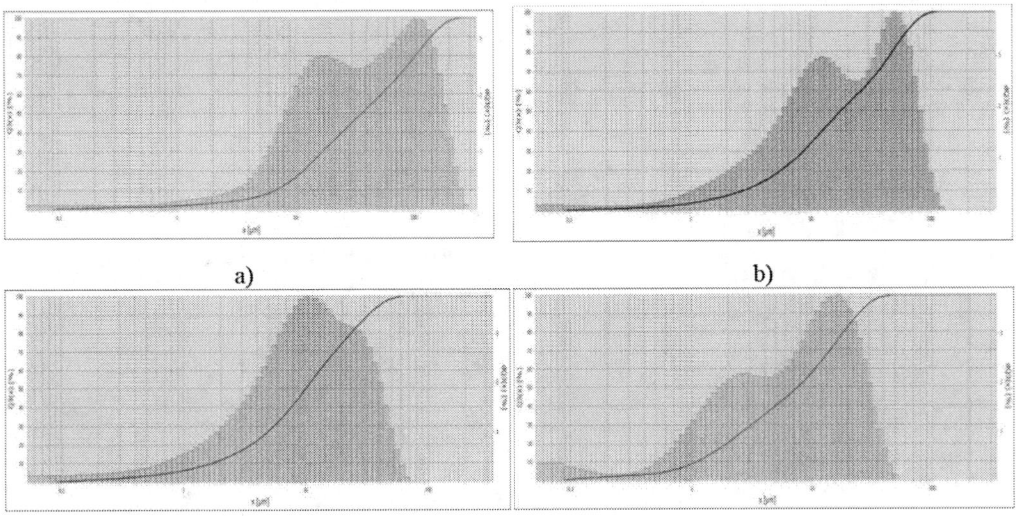

a) b) c) d)

Fig. 1. Grain size composition of fine-grained mineral powders: a) silica powder with the specific surface area of 200 m^2/kg; b) silica powder with the specific surface area of 290 m^2/kg; c) silica powder with the specific surface area of 350 m^2/kg; d) grinded ash with the specific surface area of 700 m^2/kg.

Silica powder with the specific surface area of 200 m^2/kg (Fig. 1*a*) has 20% of its grains with the sizes of more than 100 μm. Integral characteristic of the grain composition indicates predominance of fractions sized 70-150 μm, the content of ultra-dispersed grains is insignificant. Every further trial shows that fineness of silica grinding is increasing and the integral curve is shifting towards ultra-dispersed grains. The maximum grain size in silica powder with the specific surface area of 350 m^2/kg (Fig. 1*c*) does not exceed 80 μm; the grains with size 8-40 μm are prevailing. The integral curve of grain size composition of grinded ash given in Figure 1*d* illustrates that the maximum grain size does not exceed 55 μm; the content of particles of nanometer range reaches 10%. Table 1 provides summary analysis of the particles size of the obtained dispersed materials.

Table 1. Particles size of dispersed materials.

Material	Specific surface,m^2/kg	The mean particle size, μm in trial bulk		
		10%	50%	90%
Silica powder	200	6.74	36.43	139.79
	290	2.68	17.58	64.03
	350	1.68	9.97	32.83
Grinded ash	700	1.09	7.59	24.39

5 set of samples were produced and further tested: compositions No.1-3 differed in the dispersion degree of silica powder, compositions No.4 and No.5 had 18% of silica powder with the specific surface area of 350 m^2/kg replaced with grinded ash, while TEA was also added into composition No.5 while batching.

Resulting from physical-mechanical testing of samples it was established that concrete strength of the composition No.3 demonstrates almost two-times increase compared to the concrete strength of the composition No.1.

Mineralogy of the new formations was studied in detail with the aim of results justification. Figure 2 presents results of X-ray phase analysis of the obtained cellular concrete samples.

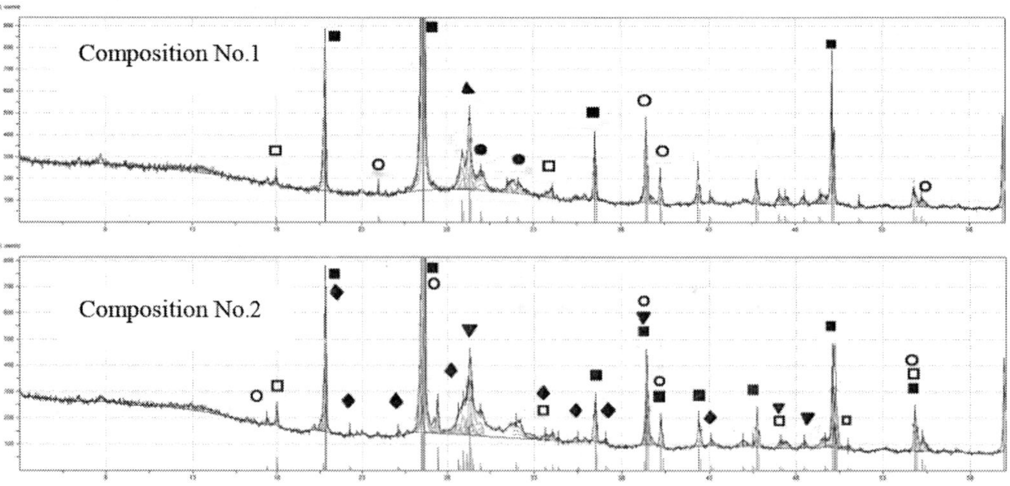

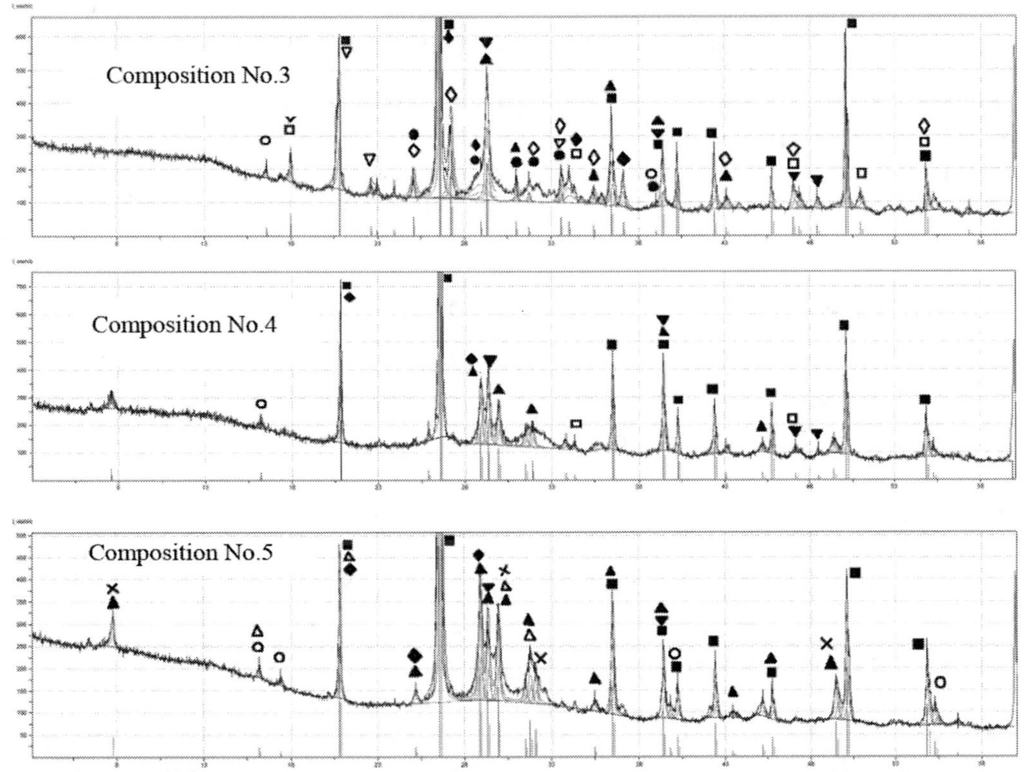

Fig. 2. X-ray phase analysis of the cellular concrete composition: ▪ - silica; ▫ - Portlandite; ∇ - gismondite; ◊ - xonotlite; ▲ – monoclinic crystal system tobermorite; ▼ - calcite; ● – triclinic tobermorite; ♦ - orthorhombic hydrosilicate; o - hydrated calcium aluminate; × - clinotobermorite; Δ – scolecite.

Resulting from X-ray phase analysis crystal chemical peculiarities of the hydrated phases were defined in micro-heterogeneous system (composition No.1) and ultra-micro-heterogeneous system (compositions No.3 and No.5). The basic peculiar feature of ultra-micro-heterogeneous system was two-times increase of the amount of silica participating in hydrated silicate and hydroaluminosilicate phases synthesis. In the products of hydrothermal hardening of the composition No.1 orthorhombic hydrated silicates $2CaO \cdot SiO_2 \cdot 2H_2O$ were detected, along with monoclinic crystal system tobermorite and Portlandite. Polysilicates of various structures, like triclinic tobermorite, xonotlite were crystallized in the composition No.3 due to high reactive capacity of silica-oxygen anions.

In the presence of TEA solubility of the basic mineral constituents of the cement was changed. As noted in [18, 19] TEA mainly increases the solubility of aluminate phases. In this system in the hydration products the formation of frame hydroaluminosilicates of gismondite type $CaO \cdot Al_2O_3 \cdot 2SiO_2 \cdot 4\text{-}5H_2O$ of monoclinic crystal system was revealed.

In the presence of grinded ash in the composition No.4 full binding of Portlandite into hydrated silicates could be observed. In this case hydrated silicates formed a continuous series of xonotlite-tobermorite type, including both high-basic and low-basic xonotlite phases. Also, the presence of hydrated calcium aluminates was established.

Composition No.5 differed by replacement of 18% silica sand with ash and additional introduction of TEA. The synthesis of new phases revealed frame hydrated silicates of scolecite type $CaO \cdot Al_2O_3 \cdot 3SiO_2 \cdot 3H_2O$.

It can be seen from the X-ray patterns that decreasing the particles diameter not only intensified the participation of silica in new phases synthesis, but also it had lead to changing of their morphology with the formation of monoclinic, triclinic, orthogonal hydrated silicates and hydrated calcium aluminates. Application of fine-grained silica in combination with TEA stimulates the increase of the balance of hydrated phases with formation of the minerals which are similar in their structure to natural zeolites. Crystal-chemical properties of the revealed new phases are given in Table 2 and illustrated in Figures 3, 4.

Table 2. Crystal-chemical properties of the new phases.

Name of the identified phase	Chemical formula	Cell parameters, Å:			Cell volume, V_c, $Å^3$
		a	b	c	
Orthorhombic tobermorite	$2CaO \cdot SiO_2 \cdot 2H_2O$	5.5860	3.6960	22.7790	470.29
Monoclinic crystal system tobermorite	$2CaO \cdot SiO_2 \cdot 2H_2O$	6.6900	7.3900	22.7790	939.21
Triclinic tobermorite	$5CaO \cdot 6SiO_2 \cdot H_2O$	11.1560	7.3030	9.5660	763.87
Xonotlite	$6CaO \cdot 6SiO_2(OH)_2$	16.5300	7.3300	7.0400	853.00
Gismondite	$CaO \cdot Al_2O_3 \cdot 2SiO_2 \cdot 4\text{-}5H_2O$	10.0200	10.6200	9.8400	1046.16
Scolecite	$CaO \cdot Al_2O_3 \cdot 3SiO_2 \cdot 3H_2O$	18.5080	18.9810	6.5270	2292.79

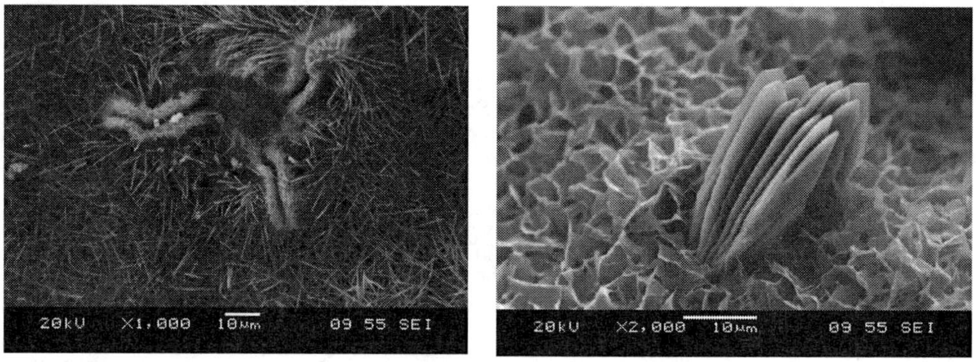

Fig. 3. Hydroaluminosilicates of frame structure.

Electron microscopy data showed that in pores with sizes 50 μm and larger, ettringite-like crystals were formed (Fig. 4). During microscopic studies of the obtained samples multiple repetition of their formation was observed, however these crystalline hydrates cannot be revealed by X-ray diffraction analysis.

Fig.4. Ettringite-like crystalline hydrates in the pores of cellular concrete.

The gel structural component is subjected to shrinkage, the crystalline joint prevents its free shrinkage, and therefore, while increasing the crystallinity degree of matrix, shrinkage will be defined by elastic properties of the crystalline joint. The presence of tobermorite gel was not established according to the X-ray phase analysis.

Increase in the concrete strength was provided resulting from the formation of cementing components of complex composition with the higher condensation degree of silica-oxygen anions of scolecite, gismondite type, referring to framework silicates of zeolite type.

Conclusion

Based on the outcome of the present research one can conclude that by means of mechanical activation during grinding up to the specific surface area of >350 m^2/kg the potential activity of silica and aluminosilica additives and concrete components increases. Use of TEA allows changing directly the mechanism of hydrated phases formation along with building complex hydrated calcium aluminates of frame structure of zeolite type with non-permanent composition and similar structure. When increasing dispersion and partly replacing silica with ash, up to 70% of silica and up to 100% $Ca(OH)_2$ participate in new phases synthesis. Hydrated silicates of xonotlite-tobermorite type build a continuous series of crystalline structures.

Further investigation will involve studying the influence of fillers dispersion capacity and mineral additives on the morphology of crystalline hydrates which are formed during hydration of cement and silicate systems. The study will also address the development of new compositions of dry building mixtures and special purpose concrete.

References

1. J.W. Bullard, H.M. Jennings, R.A. Livingston, A. Nonat, G.W. Scherer, J.S. Schweitzer, K.L. Scrivener, J.J. Thomas, Cem. Concr. Res. **41**, 1208 (2011)
2. N. van Tuan, G. Ye, K. van Breugel, O. Copuroglu, Cem. Concr. Res. **41**(11), 1104 (2011)

3. S. Pashkevich, A. Pustovgar, A. Adamtsevich, A. Eremin, Appl. Mech. Mater. **584-585**, 1659 (2014)
4. Yu.M. Bazhenov, V.S. Demyanova, V.I. Kalashnikov, *Modificirovannye vysokokachestvennye betony* [Modified high-quality concretes] (Association of universities for civil engineering, Moscow, 2006) (in Russian)
5. V.G. Batrakov, *Modificirovannye betony. Teorija i praktika* [Modified concretes. Theory and practice] (Strojizdat, Moscow, 1998) (in Russian)
6. I.N. Akhverdov, *Osnovy fiziki betona* [Fundamentals of concrete physics] (Strojizdat, Moscow, 1981) (in Russian)
7. S.S. Kaprielov, Beton i zhelezobeton [Concrete and Reinforced Concrete] **4**, 16 (1995) (in Russian)
8. H. Taylor, J. Amer. Ceramic Soc. **69** (6), 464 (1986)
9. B. Lothenbach, G. Le Saout, M. Ben Haha, R. Figi, E. Wieland, Cem. Concr. Res. **42**(2), 410 (2012)
10. S. M. Leisinger, B. Lothenbach, G. Le Saout, C. Annette Johnson, Cem. Concr. Res. **42**, 158 (2012)
11. M.K. Takhirov, Proceedings of MIIT: *New in building materials science* **902,** 48 (1997) (in Russian)
12. V.I. Kalashnikov, Beton i zhelezobeton [Concrete and Reinforced Concrete] **1**, 82 (2012) (in Russian)
13. G.A. Zimakova, V.A. Solonina, M.P. Zelig, *IOP conf. series: Earth and Environmental Science* **50**, 012007 (2017)
14. G.A. Zimakova, V.A. Solonina, M.P. Zelig, International Research Journal **12-3** (54), 90 (2016) (in Russian)
15. A.E. Shejkin, Ju.V. Chehovskij, M.I. Brusser, *Struktura i svojstva cementnyh betonov* [Structure and properties of cement concretes] (Strojizdat, Moscow, 1979) (in Russian)
16. WWW-MINCRYST, Crystallographic and crystallochemical database for minerals and their structural analogs, available at http://database.iem.ac.ru/mincryst/rus/ (1997) (in Russian)
17. V.S. Gorshkov, V.V. Timashev, V.G. Savelyev, *Metody fiziko-himicheskogo analiza vjazhushhih veshhestv* [Methods of physico-chemical analysis of binders] (Vysshaja shkola, Moscow, 1981) (in Russian)
18. V. Ramachandran, *Shestoj mezhdunarodnyj kongress po himii cementa* [The 6th International Congress on Chemistry of Cement] **2**(2), 37 (Strojizdat, Moscow, 1976)
19. E. Gartner, D. Myers, J. Am. Ceram. Soc. **76** (6), 1521 (1993)

Naturally cured foamed concrete with improved thermal insulation properties

Nikolay Mashkin[1], *Ekaterina* Bartenjeva[1,*] and *Rustam* Mansurov[1]

[1]Novosibirsk State University of Architecture and Civil Engineering (Sibstrin), 630008 Novosibirsk, Russia

Abstract. The paper is dedicated to investigation on improvement of thermal insulation properties of non-autoclaved concrete by increasing aggregate stability of foamed concrete mixture. The study demonstrates influence of mineral admixtures on the foam stability index in the mortar mixture and on decrease of foamed concrete density and thermal conductivity. The effect of mineral admixtures on thermal conductivity properties of non-autoclaved concrete was assessed through different ways of their addition: to the foam and to the mortar mixture. The admixtures were milled up to the specific surface area of 300 and 600 m^2/kg using an АГО-9 centrifugal attrition mill with continuous operation mode (Institute of Solid State Chemistry and Mechanochemistry, Siberian Branch of the Russian Academy of Sciences, Novosibirsk). Laboratory turbulent foam concrete mixer was used to prepare foamed concrete. Thermal conductivity coefficient was defined by a quick method using "ИТП-МГ 4 "Zond" thermal conductivity meter in accordance with the regulatory documents. The impact of modifiers on the foam structure stability was defined using the foam stability index for the mortar mixture. The research demonstrated the increase in stability of porous structure of non-autoclaved concrete when adding wollastonite and diopside. Improvement of thermal and physical properties was demonstrated, the decrease of thermal conductivity coefficient reaches 0.069 W/(m×°C).

1 Introduction

The growth of interest in cellular concrete investigation is connected with the change of requirements to thermal protection of buildings and structures, particularly those located in harsh climate conditions. Thermal conductivity of foamed concrete can be decreased by creating pores of smaller size [1-3], however, other research show that the size and shape of pores for low-density foamed concretes have no particular impact [4]. There are some research dedicated to mineral content control in hardened cement paste [5], the result of which define the decrease of thermal conductivity coefficient of interpore partitions in foamed concrete by 6.5-24 % with the maximum allowable content of calcium aluminate (C_3A) and calcium alumoferrite (C_4AF) in Portland cement and with the minimum allowable content of alite (C_3S). Natural and artificial microsilica [6] is widely used to

* Corresponding author: ek.bartenjeva@yandex.ru

improve thermal insulation properties of materials, investigations are conducted on the use of carbon nanotubes [7] and geopolymers [8]; however, such technological methods can considerably raise the price of the end product. Thermal conductivity was found to decrease with addition of silica sol (up to 35%) and wood fiber (up to 30%) [9]. Use of components with lower thermal conductivity and high density like ash, slag, peat, expanded perlite, vermiculite, also allows improving thermal insulation properties [10-15]. Use of fine montmorillonite clay enables to obtain ultralight foamed concrete with the density of 200 kg/m^3 due to high adsorption capacity [10].

The key factor for thermal conductivity decrease in cellular concretes is the increase of total porosity. Thus, the decrease of mean density by 100 kg/m^3 leads to thermal conductivity reduction by 20%. In this manner, decrease of cellular concrete mean density up to 200 kg/m^3 will cause reduction of thermal conductivity up to 0.06 W/(m×°C) and lower, which corresponds to that of highly effective thermal insulation materials, such as mineral wool and cellular plastics.

It is difficult to obtain non-autoclaved foamed concrete with low density and thermal conductivity in a production environment. Increase in porosity of foam concrete, especially non-autoclaved one, reduces its structure stability, which goes along with considerable shrinkage, deformation and disturbance of optimum porosity [4, 16-20]. Stability of foamed concrete mixture can be increased by controlling water temperature [17], another effective method for improving aggregate stability is use of fiber admixtures [21-23]. Use of viscosifiers, such as gelatin, carboxymethylcellulose, soluble sodium silicates, also increases the foam structure stability [24-25].

This paper is aimed at investigating technological principles for obtaining non-autoclaved foamed concrete with lower thermal conductivity.

2 Experiment

2.1 Materials

Portland cement of type ЦЕМ I 42.5 (Iskitim town) with the following chemical composition (wt.%) was used to obtain foamed concrete: Na_2O_3 – 0.41, MgO – 1.94; Al_2O_3 – 4.29; SiO_2 – 20.46; P_2O_5 – 0.08; K_2O – 0.97; CaO – 63.63; TiO_2 – 0.23; MnO – 0.09; Fe_2O_3 – 3.69; SO_3 – 1.77; loss on ignition (LOI) – 2.44. Specific gravity of Portland cement is 3.060 g/cm^3, bulk density – 1.083 g/cm^3. Acidic fly-ash from thermal power station No.5 located in Novosibirsk obtained after burning Kuznetsk hard coal was used as silica filler. Chemical composition of the filler is as follows (wt.%): SiO_2 – 60.77; Al$_2$O3 – 19.45; Fe_2O_3 – 5.16; CaO – 5.12; MgO – 2.10; Na_2O – 0.89; K_2O – 2.01; SO_3 – 0.54; P_2O_5 – 0.39; TiO_2 – 0.82; BaO – 0.20, MnO – 0.07. Bulk density of fly-ash is 0.885 g/cm^3, specific gravity – 1.870 g/cm^3, sieve residue 008(by weight) – 4.49 %. «FoamCem» protein-based foaming agent (Italy) was used in the production process.

Wollastonite and diopside were used as mineral admixtures. Wollastonite is a mineral pertaining to silicates group of the Altai deposit, its specific gravity is 2.455 g/cm^3, specific surface area was 90 m^2/kg and 574 m^2/kg. Chemical composition of wollastonite is: SiO_2 – 46.1; Al_2O_3 – 2.93; Fe_2O_3 – 4.44; CaO – 45.12; MgO – 0.9; LOI – 0.51. In the production process diopside from the Slyudyanskoe deposit with the following chemical composition was used (wt.%): CaO – 25.03, MgO — 20.01; SiO_2 — 51.33, Al$_2$O3 – 1.88; Fe_2O_3 – 0.84; MgO – 20.01; K_2O –0.17, TiO_2 – 0.14; LOI – 0.61; specific gravity – 2.778 g/cm^3, specific surface area – 116 m^2/kg.

2.2 Methods of experiment

Experimental research was performed using samples of naturally cured foamed concrete produced according to the traditional technology with the use of turbulent foam concrete mixer. The foamed concrete composition was selected in compliance with the known techniques [26], the ratio of silica component to binder was 0.6; water/solid ratio = 0.49.

The influence of mineral admixtures on the foam stability was investigated using foam stability index in the mortar mixture. This indicator characterizes foam shrinkage while mixing it with mortar fraction of foamed concrete. It can be related to the processes of coalescence and synaeresis with insufficient structural strength of interpore partitions caused by variation of pH medium or redistribution of surfactants in disperse system [27-28].

The value of foam stability index in the mortar mixture is one of the crucial factors for foamed concrete production, as foam destruction during mineralization causes porous mixture degradation. It was calculated by the formula [29]:

$$C^{\text{p}}_{\text{cp}} = \frac{V^{por}_{\text{c.p.}}}{V_{c.p.}+V_{foam}}, \tag{1}$$

where $V^{por}_{\text{c.p.}}$ – volume of porous cement paste, ml; $V_{c.p.}$ – volume of cement paste, ml; V_{foam} – foam volume, ml.

Introduction of mineral admixtures was performed in two different ways: to the foam and to the mortar mixture. The admixtures were milled up to the specific surface area of 300 and 600 m^2/kg using АГО-9 centrifugal attrition mill with continuous operation mode (Institute of Solid State Chemistry and Mechanochemistry, Siberian Branch of the Russian Academy of Sciences, Novosibirsk). Mean density (ρ) was defined after 28 days according to the GOST 12730.1-78. Thermal conductivity coefficient (λ) was defined by a quick method using "ИТП-МГ 4 "Zond" thermal conductivity meter in accordance with the GOST 30256-94. Quick method to define thermal conductivity is based on the dynamic method, i.e. on the change rate of averaged temperature of the heat probe (as the first probe temperature derivative with respect to time) in the tested material. The lower thermal conductivity of material is, the slower the probe cools down.

3 Results

The values of foam stability in the mortar mixture are given in Table 1.

Table 1. Influence of type, amount and method of mineral admixture introduction on the foam stability index in mortar mixture.

Type of admixture	Specific surface area, m^2/kg	Method of admixture introduction	Amount of admixture, wt. %			
			0	1	2.5	4
Wollastonite	100	Foam	0.74	0.79	0.91	0.71
		Mortar mixture		0.81	0.89	0.73
	300	Foam		0.84	0.84	0.85
		Mortar mixture		0.96	0.93	0.96
	600	Foam		0.67	0.84	0.56
		Mortar mixture		0.81	0.87	0.74
Diopside	100	Foam		1.14	0.77	0.66
		Mortar mixture		0.91	0.93	0.66
	300	Foam		0.90	0.86	0.91
		Mortar mixture		0.89	0.90	0.87
	700	Foam		0.96	0.87	0.84
		Mortar mixture		0.87	0.90	0.93

The highest values of foam stability index in the mortar mixture correspond to specific surface area of the admixture equal to 300 m^2/kg and the amount of added wollastonite equal to 1-2.5%. When adding diopside in the amount of 1% to the foaming agent solution the value of foam stability index increases up to 1.14.

The foam stability index in porous mortar mixture is higher for foamed concrete with mineral admixtures introduced in the mortar mixture.

For diopside, the best values of mean density correspond to the specific surface area of 100-300 m^2/kg when adding it to the mortar mixture in the amount of 1 and 2.5%. The optimum amount of admixture for wollastonite is 1-2.5%. Increase of the admixture amount causes increase of foam concrete density (Table 2). Introduction of diopside has bigger effect on decrease of mean density of foam concrete, than introduction of wollastonite. When increasing specific surface area of admixtures the variation of mean density of samples are inconsiderable.

Introduction of diopside and wollastonite allows increasing aggregate stability of foam concrete mixture. Positive effect of wollastonite on foamed concrete structure is possible by means of reinforcement of interpore partitions, which occurs due to its acicular particles. This effect allows reducing plastic shrinkage, limiting sedimentation of the mixture components, which agrees with other research [30-31].

Introduction of admixtures reduces thermal conductivity of foamed concrete (Table 2). When using diopside, thermal conductivity coefficient is slightly lower than when using wollastonite. The lowest values correspond to the specific surface area of admixtures of 100 m^2/kg.

Table 2. Influence of type, amount and method of mineral admixture introduction on the mean density and thermal conductivity of foamed concrete.

Type of admixture	Specific surface area, m^2/kg	Method of admixture introduction	Amount, wt. %							
			0		1		2.5		4	
			ρ	λ	ρ	λ	ρ	λ	ρ	λ
Wollastonite	100	Foam	547	0.122	415	0.119	435	0.105	638	0.122
		Mortar mixture			375	0.070	395	0.107	449	0.114
	300	Foam			420	0.082	502	0.097	512	0.117
		Mortar mixture			420	0.075	435	0.103	444	0.085
	600	Foam			545	0.093	404	0.104	783	0.134
		Mortar mixture			361	0.083	406	0.095	436	0.094
Diopside	100	Foam			345	0.088	560	0.108	664	0.130
		Mortar mixture			274	0.069	377	0.097	520	0.105
	300	Foam			401	0.073	450	0.096	563	0.079
		Mortar mixture			433	0.074	467	0.091	477	0.091
	700	Foam			488	0.101	463	0.086	530	0.110
		Mortar mixture			434	0.084	461	0.090	424	0.076

The lowest values of density and thermal conductivity coefficient of foamed concrete are obtained for the minimum amount of introduced admixtures. Apparently, it is connected with the positive effect of mineral admixtures, particularly diopside, on aggregate stability of foamed concrete mixture. Considering that mineral admixtures have reinforcing effect on interpore partitions of foamed concrete, the reinforcement process speeds up when introducing the maximum amount of admixture. This can be observed by the high values of foamed concrete density (500-700 kg/m^3). The minimum amount of admixture provides stability of foamed concrete mixture and reinforces interpore partitions to such extent that while mixing mortar and foam the air entrainment occurs, due to which optimum porous structure of foamed concrete is obtained.

As a result of the experiment, dispersity of the admixture was noted to influence the properties of foamed concrete. It is preferable to introduce admixtures with the specific surface area from 300 m^2/kg and lower.

Adding wollastonite with the specific surface area of 100 m^2/kg to mortar in the amount of 1% allows obtaining foamed concrete with the density of 375 kg/m^3 and thermal conductivity coefficient of 0.070 W/(m×°C). For diopside, the best parameters are as follows: specific surface area is 100 m^2/kg, admixture amount is 1%, type of introduction is to the mortar mixture, foamed concrete density is 274 kg/m^3, thermal conductivity coefficient is 0.069 W/(m×°C).

4 Conclusion

It was found that due to directed control of porous structure of cellular concretes by introducing mineral modifying admixtures it is possible to considerably improve the stability of foam concrete mixture and structure of foamed concrete. This, in turn, is a step to decrease density and thermal conductivity of a material and to reach technical and economical effect due to saving cement. The outcomes of this research allow using production waste for manufacturing of low-density foamed concrete, thus contributing to solving environmental issues.

References

1. Y.P. Gorlov, *Tehnologija teploizoljacionnyh i akusticheskih materialov i izdelij* [Technology of thermal insulation and acoustic materials and products] (Vysshaya Shkola Publishers, Moscow, 1989) (in Russian)
2. A.S. Kolomatsky, S.A. Kolomatsky, Construction Materials **1**, 38-39 (2003) (in Russian)
3. T.E. Kobidze, V.F. Korovyakov, A.Y. Kiselev, Construction Materials **1**, 26-29 (2005) (in Russian)
4. T.V. Anikanova, S.M. Rakhimbaev, *Penobetony dlja intensivnyh tehnologij stroitel'stva* [Foam concretes for intensive construction technologies] (BGTU Publ., Belgorod, 2015) (in Russian)
5. A.F. Kosach, I.N. Kuznetsova, M.A. Rashchupkina, News of higher educational institutions. Construction **8**, 26 – 31 (2009) (in Russian)
6. V.P. Selyaev, V.A. Neverov, L.I. Kupriyashkina, P.V. Selyaev, M.A. Okin, News of higher educational institutions. Construction **4**, 30-37 (2014) (in Russian)
7. G. Yakovlev, J. Kerienė, A. Gailius, I. Girnienė, Materials Science **12**(2), 147-151 (2006)
8. Z. Zhang, J.L. Provis, A. Reid, H. Wang, Cement and Concrete Composites **62**, 97-105 (2015)
9. A.M. Sycheva, E.A. Popova, D.I. Drobyshev, I.P. Filatov, *Aktivirovannoe tverdenie penobetonov* [Activated hardening of foam concretes] (PGUPS, Saint Petersburg, 2007) (in Russian)
10. L.B. Svatovskaya, V.Y. Solovyeva, V.A. Chernakov, *Poluchenie monolitnogo penobetona s uchetom osobennostej prirody zapolnitelja* [Obtaining cast foam

concrete with regard to the nature of filler] (PGUPS, Saint Petersburg, 2001) (in Russian)

11. S.V. Aleksandrovskiy, *Prikladnye metody teorii teploprovodnosti i vlagoprovodnosti betona* [Applied methods in theory of concrete thermal and moisture conductivity] (Kompanija Sputnik+, Moscow, 2001) (in Russian)
12. N.V. Pavlenko, P.P. Pastushkov, A.N. Kharkhardin, E.V. Voytovich, Vestnik SIBADI **6** (52), 80-86 (2016) (in Russian)
13. E.G. Velichko, Construction Materials **1**, 9-13 (2009) (in Russian)
14. A. Kudyakov, I. Prishepa, M. Tolchennikov, *IOP Conf. Series: Materials Science Engineering* **71**, 012016 (2015)
15. R. Demirboğa, R. Gül, Cement and Concrete Research **5**(33), 723-727 (2003)
16. V.V. Kondrat'ev, N.N. Morozova, V.G. Khozin, Construction Materials **11**, 35-37 (2002) (in Russian)
17. V.N. Morgun, *Penobeton* [Foam concrete] (RGSU, Rostov-on-Don, 2012) (in Russian)
18. A.I. Kudyakov, A.B. Steshenko, V.V. Konusheva, O.O. Syrkin, Vestnik of TSUAB **5**(58), 129-139 (2016) (in Russian)
19. A.K. Abdulin, V.N. Feklistov, Construction Materials **6**, 14-17 (2005) (in Russian)
20. K. Wan, G. Li, Sh. Wang, Ch. Pang, Cement and Concrete Composites **82**, 217-226 (2017)
21. A.I. Kudyakov, A.B. Syrkin, Vestnik of TSUAB **2**(43), 127-133 (2014) (in Russian)
22. H.R. Shah, J. Weiss, Materials and Structures **9**(39), 887–899 (2006)
23. D.-Y. Yooa, N. Banthia, Cement and Concrete Composites **73**, 267-280 (2016)
24. V.G. Bezborodov, V.F. Zavadskiy, T.Y. Nikulina, News of higher educational institutions. Construction **12,** 29-33 (2002) (in Russian)
25. V.V. Ivanitskiy, A.V. Bortnikov, Y.V. Gudkov, N.A. Sapelin, A.F. Buryanov, patent no. 2213717 Russian Federation (2003) (in Russian)
26. V.F. Zavadskiy, A.F. Kosach, P.P. Deryabin, *Stenovye materialy i izdelija* [Wall materials and products] (SIBADI Publ., Omsk, 2005) (in Russian)
27. P.M. Kruglyakov, D.R. Ekserova, *Pena i pennye plenki* [Foam and Foam Films] (Himija, Moscow, 1990) (in Russian)
28. S.M. Rakhimbaev, V.N. Tarasenko, D.V. Tverdokhlebov, Bulletin of BSTU named after V.G. Shukhov **5**(1), 119-122 (2003) (in Russian)
29. E.G. Velichko, A.G. Komar, Construction materials **3**, 26-29 (2004) (in Russian)
30. N.M. Khasanov, Bulletin of Civil Engineers **3**, 181-186 (2016) (in Russian)
31. G.I. Berdov, L.V. Il'ina, V.N. Zyrianova, *Vlijanie mineral'nyh mikronapolnitelej na svojstva kompozicionnyh stroitel'nyh materialov* [Influence of mineral microfillers on the properties of composite building materials] (NGASU (Sibstrin) Publ., Novosibirsk, 2013) (in Russian)

Modifying admixtures to cement compositions produced from local raw materials

Daria Vasileva[1,*], *Egor* Protodiakonov[1], *Anastasia* Egorova[1], and *Svetlana* Antsupova[1]

[1]Ammosov North-Eastern Federal University, 677000 Yakutsk, Russia

Abstract. Durability of hardened cement paste depends on chemical and mineralogical composition of Portland cement. The main factor for hardened cement paste is higher content of calcium aluminate and free calcium hydroxide, binding of which into water-insoluble compounds causes increase in resistance to water, frost and corrosion. The purpose of this research is to develop modifying admixtures to cement compositions based on local raw material - rock sand. Chemical and mineralogical properties of the source materials were studied using X-ray spectroscopy and X-ray diffraction analysis. Standard methods were used for defining physico-mechanical properties of sand and binder. Influence of the degree of mechanochemical activation of modifying admixture on the properties of binder and hardened cement paste made on its basis was studied. Research methods of scanning electron microscopy and spectral measurements were applied. The possibility of using admixture based on rock sand as a modifier was determined, its usage providing increase of strength, sulphate and frost resistance, which causes higher durability of cement concrete.

Introduction

Portland cement is one of the most demanded binders used for various operational conditions. In the Far East region (Russia), with its high tariffs and seasonal supply of materials, the transportation costs for 1 ton of cement is three times higher than its actual cost [1]. This fact indicates the need to search for opportunities to extend the sphere of Portland cement usage by adding in its composition mineral or complex admixtures. At the same time, the stock of local natural raw materials are abundant, they are potentially suitable for being used as admixtures to Portland cement. It is known that in production of cement with mineral admixtures specific fuel consumption is reduced by 10-35 % , and electricity consumption - by 12 % [2, 3]. These data with the account of transportation costs for long-haul organomineral cement indicate relevance of the research on development of modifying admixtures to Portland cement on the basis of local raw materials.

The second direction for research is investigation of the influence of admixture on sulphate resistantce of hardened cement paste. Observations demonstrate that buildings and structures in Sakha Republic (Yakutiya, Russia) during their operation often become unfit for further exploitation prior to the schedule. This entails additional costs on reconstruction

* Corresponding author: omsvdv2910@mail.ru

and construction of new buildings. It can be explained by higher content of salts Na_2SO_4 and $CaSO_4$ in the soil. Similar problems can occur in other regions with higher soil salinity (South Kazakhstan, Middle Asia, West of the USA, South America, North Africa, in Russia - middle and southern Volga region, northeast Caucasus, south of Western and Eastern Siberia) [4].

One of the solutions to this problem can be the use of sulphate-resistant Portland cement. There were some previous studies on influence of admixtures [5] and industrial raw materials [6, 7] on increase of sulphate resistance of hardened cement paste. The purpose of this study is development of modifying admixtures from mineral and organic components. Local silica raw material - rock sand - was studied as a mineral component.

Materials and Methods

Raw materials used in the research were: Portland cement of the brand ЦЕМ I 32,5 Б (М400 Д0) based on alite clinker manufactured at "Yakutcement" production plant and meeting the requirements of GOST 31108-2003 [8]; red and orange rock sand (Kildyamsky open pit, Yakutsk city); superplasticizer C-3 meeting the requirements of TU 5870-005-58042865-05 [9]. Properties of Portland cement were defined according to the requirements of GOST 10178 [10] and GOST22266-94 [11]; for rock sand - according to GOST 8735-88 [12]; samples were produced from mortar and were tested according to GOST 5802-86 [13]; frost resistance of the samples was defined in compliance with GOST 10060.0-95 [14].

Experiment

The following scheme of the admixture preparation and introduction in binder was accepted for the research:

- mechanochemical activation of rock sand by milling with superplasticizer C-3 up to the optimum value of specific surface area using the laboratory ball mill Retch PM-400;
- mixing of Portland cement and activated rock sand in the laboratory mixer.

The research includes mathematical planning of factorial experiment with two variables: X_1 – amount of admixture (activated rock sand), %; X_2 – rock sand activation time, min. Variation step for X_1 – 15 % of Portland cement mass, and for X_2 – 2 minutes. The following optimization parameters were applied: standard consistency of cement paste, %; initial and final setting time, min.; compressive strength as the age of 14 + 14/28 days of curing in moist cabinet, MPa; compressive strength at the age of 14 (in moist cabinet) + 14/28 days of curing in 5 % solution of Na_2SO_4, MPa; sulphate resistance coefficient at the age of 28 and 56 days. Equation representing dependence of optimization parameters on variable factors looks as follows:

$$Y_n = b_0 + b_1 X_1 + b_0 X_2 + b_{11} X_1^2 + b_{22} X_2^2 + b_{12} X_1 X_2. \quad (1)$$

According to the accepted plan, binder compositions were prepared and beam specimens with the size of 4x4x16 cm were produced from sand-cement mortar. Having reached standard age the reference samples of each composition were cured in a moist cabinet for strength development and the experimental samples were put in sulphate environment (5 % solution of Na_2SO_4). Then, after 28/56 days both types of samples were tested to define the ultimate compressive strength for finding sulphate resistance coefficient.

Results

X-ray spectral analysis of rock sand using SRS-3400 showed the following chemical composition, %: SiO_2 – 76.32; Al_2O_3 – 13.30; Fe_2O_3 – 6.88; K_2O – 4.94; Na_2O – 3.41; CaO – 1.08; P_2O_5 – 0.52; MgO – 0.37; TiO_2 – 0.24, etc. Rock sand samples were investigated using X-ray diffraction analysis using D8 Discover with GADDS. The following spectrum analysis results were obtained: the studied sample consists of quartz SiO_2 – 87.9 % and anorthoclase $(Na,K)(Si_3Al)O_8$ – 9.67 %. Physical properties of rock sand are given in Table 1. The sand composition is close to the research outcomes [15-16].

Table 1. Physical properties of rock sand.

Title	Bulk density, kg/m^3	Specific gravity, kg/m^3	Fineness modulus	Void ratio, %	Water absorption capacity, %
Rock sand from Kildyamskoe deposir	1.03	2.23	2.73	52	8.7

After that, according to the accepted plan, Portland cement samples were prepared with a rock sand admixture of various activation degree. Properties of binders and samples produced on their basis and cured in different conditions are presented in Table 2.

Results of the obtained experimental data analysis were used to make equations to represent dependences of optimization parameters on variable factors:

- for milling fineness, %:

$$Y_{mn} = 14{,}31 + 1{,}18 * X_1^2 + 0{,}54 * X_2^2 + 3{,}58 * X_1 - 2{,}66 * X_2 - 1{,}57 * X_1 * X_2 \quad (2)$$

- for standard consistency of cement paste, %:

$$Y_{sc} = 24.61 + 0.14 * X_1^2 + 1.39 * X_2^2 - 0.685 * X_1 + 0.061 * X_2 + 0.062 * X_1 * X_2 \quad (3)$$

- for initial setting, minutes:

$$Y_{is} = 210 - 20 * X_1^2 - 5 * X_2^2 + 16.66 * X_1 + 15 * X_2 + 15 * X_1 * X_2 \quad (4)$$

- for final setting, minutes:

$$Y_{fs} = 297.77 + 8.33 * X_1^2 - 1.66 * X_2^2 + 15 * X_1 + 5 * X_2 \quad (5)$$

- for ultimate compressive strength of samples at the age of 28 days:

$$Y_{Rc} = 29{,}4 - 12.07 * X_1^2 - 2.63 * X_2^2 - 3.01 * X_1 - 1.49 * X_2 - 1.16 * X_1 * X_2 \quad (6)$$

- for ultimate compressive strength of samples at the age of 56 days:

$$Y_{Rc} = 35.37 - 14.88 * X_1^2 - 2.24 * X_2^2 - 3.45 * X_1 - 2.05 * X_2 - 2.87 * X_1 * X_2 \quad (7)$$

- for sulphate resistance coefficient of samples at the age of 28 days:

$$Y_{CSR} = 0.816 + 0.03 * X_1^2 + 0.01 * X_2^2 + 0.203 * X_1 + 0.016 * X_2 - 0.012 * X_1 * X_2 \quad (8)$$

- for sulphate resistance coefficient of samples at the age of 56 days:

$$Y_{CSR} = 0.808 + 0.02 * X_1^2 - 0.01 * X_2^2 + 0.05 * X_1 + 0.011 * X_2 - 0.007 * X_1 * X_2 \quad (9)$$

Table 2. Results of investigation of modifying admixture to Portland cement.

No. of composition	X_1, amount of admixture of binder mass, %	X_2, milling time, min.	Amount of superplasticizer in admixture, %	Milling fineness residue on sieve No. 008, %	Standard consistency, %	Setting time, min.		Ultimate compressive strength at the age of 28 days, MPa			Ultimate compressive strength at the age of 56 days, MPa		
						Initial setting	Final setting	Samples cured in moist cabinet	Samples cured in sulphate environment	Sulphate resistance coefficient C_{SR}	Samples cured in moist cabinet	Samples cured in sulphate environment	Sulphate resistance coefficient C_{SR}
1	0	4	0.6	12.10	26.4	170	290	23.23	19.28	0.83	28.06	22.17	0.79
2	15	4	0.6	19.72	24.5	190	280	26.21	26.00	0.99	31.80	29.09	0.91
3	30	4	0.6	23.20	22.3	170	320	18.51	18.99	1.02	21.02	19.79	0.94
4	0	6	0.6	12.10	26.4	170	290	23.23	19.28	0.83	28.06	22.17	0.79
5	15	6	0.6	14.28	25.0	210	300	18.52	17.47	0.94	25.68	27.54	1.07
6	30	6	0.6	19.02	22.8	210	320	18.68	19.50	1.04	20.10	20.48	1.02
7	0	8	0.6	12.10	26.4	170	290	23.23	19.28	0.83	28.06	22.17	0.79
8	15	8	0.6	10.02	25.5	220	310	21.43	21.10	0.98	26.39	22.14	0.85
9	30	8	0.6	16.90	23.0	230	320	20.36	22.50	1.10	24.20	21.86	0.90

The obtained equations were used to build Mathcad models of dependences in the form of response surfaces. Based on the analysis of experimental data from Table 2 and and equations (2-9) the following conclusions can be made:

- Residue on sieve No. 008 increases as the amount of admixture in binder (X_1) grows, increase in time of admixture activation (X_2) causes reduction of difference between cement without admixtures and cement with maximum amount of admixture. It can be explained by high content of quartz in sand and its higher milling resistance.
- It is found that higher density of sand particles is the reason for decrease of standard consistency of cement paste as long as the amount of admixture in binder grows.
- Admixture with longer activation period have more obvious influence on the setting time, particularly on initial one, which can be explained by increase of specific surface area of the admixture and chemical activity of such elements as CaO and Al_2O_3. This can cause formation of initial ettringite in bigger amount.
- Ultimate compressive strength decreases almost in all cases when admixture is introduced. The exception is composition No.2 in which rock sand activated for 4 minutes was added in the amount of 15 % of Portland cement mass. This composition demonstrated increase of ultimate compressive strength by 13 % compared to pure Portland cement.
- Samples are considered to be tested for sulphate resistance in case coefficient of sulphate resistance is higher than 0.8. Conducted research demonstrates that sulphate resistance of hardened Portland cement paste fails the test on 56th day of experiment, whereas introduction of admixture considerably improves this parameter.

Figure 1 presents the model of dependence for the most important parameter of the developed admixture - its effect on sulphate resistance of hardened cement paste of standard age. Analysis of the model shows that sulphate resistance of hardened cement paste is mostly affected by variation of the amount of admixture and is almost independent of milling fineness of rock sand. Based on the parameters of strength and sulphate resistance the optimum amount of admixture is 15 % with its activation time of 4 minutes, which corresponds to 19 % of binder residue on sieve.

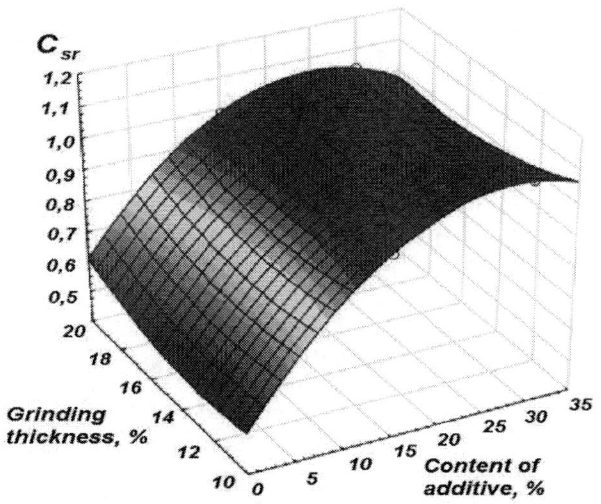

Fig. 1. Dependence of coefficient of sulphate resistance (C_{sr}) of samples at the age of 28 days on the amount of admixture and its milling fineness (by residue on sieve No. 008).

In order to study the structure of obtained hardened cement paste, samples of Portland cement with (15 %) and without admixtures were investigated using JEOL JSM-7800F scanning electron microscope, the results are presented in Figure 2. Analysis of obtained micropictures allows making a conclusion that introduction of activated rock sand admixture causes densification of its structure. At the picture with 60-fold zoom (Fig. 2 a, b) the modifying admixture is seen to densify the structure of hardened cement paste.

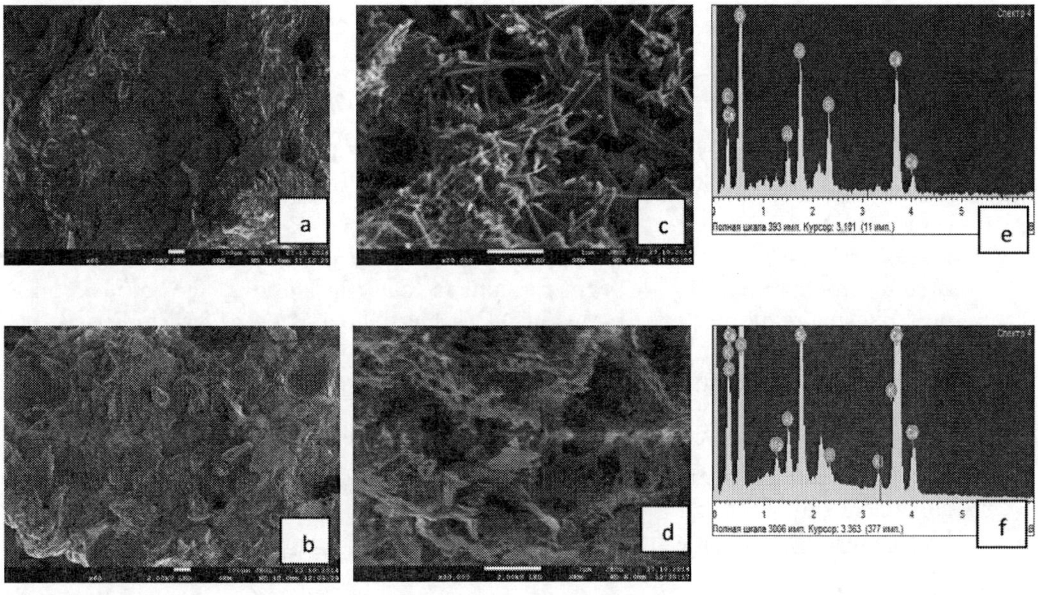

Fig. 2. Micropictures and spectrograms of hardened cement paste without admixture (a, c, e) and with 15 % of admixture (b, d, f): a,b - 60-fold zoom; c, d - 20000-fold zoom, e,f - spectrograms.

20,000-fold zoom (Fig. 2 c, d) demonstrates the structure of new formations, it is visible that space between structural crystals of hydrosilicate and calcium aluminate sulfate hydrate is filled with other microcrystalline formations, which provides dense and solid hardened cement paste and has positive effect on its strength. Such structure formation can be referred to technogenic metasomatism [17]. Data from spectrograms (Fig. 2 e, f) confirm presence of greater amount of silicates and calcium aluminates in hardened cement paste with activated rock sand.

The influence of the developed admixture on frost resistance properties of hardened cement paste was determined using accelerated method. During the experiment cube specimens with the size of 70x70x70 mm produced from fine-grained concrete based on Portland cement with optimized amount of rock sand (15 %) activated for 4 minutes were subjected to 20 freeze-thaw cycles with thawing in 5 % water solution of sodium chloride, which corresponds to at least F200 brand of concrete frost resistance.

Conclusion

Due to adding to Portland cement the developed modifier based on rock sand of Kildyanskoe deposit (Sakha Republic, Yakutia) having quartz-feldspar composition the sulphate resistance coefficient of hardened cement paste was increased up to the value of 1. This fact proves that local silica raw materials can be used as modifying admixtures to Portland cement. The effect consists in longer service life of concrete structures that operate in harsh climate conditions and aggressive sulphate environments. Besides, the cost of 1 ton of Portland cement with 15 % of the developed admixture is lower by 14 %.

Extension of this research can be focused on defining the influence of the developed admixture on durability of concrete products in natural conditions.

This research was supported by the Foundation for Assistance to Small Innovative Enterprises (FASIE), UMNIK program.

References

1. A.D. Egorova, A.E. Mestnikov, P.S. Abramova, Gornyi Zhurnal [Mining Journal], **9**, 65-68 (2016)
2. U.A. Gaziev, Othody promyshlennosti v proizvodstve stroitel'nyh materialov i izdelij [Industrial wastes in manufacture of building materials and products], 23-28 (2015) (in Russian)
3. L.I. Dvorkin, *Snizhenie rashoda cementa i topliva v proizvodstve sbornogo zhelezobetona* [Reducing of cement content and fuel during the production of prefabricated reinforced concrete], 53-54 (Vyshha shkola, Kiev, 1985) (in Russian)
4. AgroPraktik, Zasolenie pochv: problema i puti reshenija [Soil salinity: problem and solutions], available at: http://agropraktik.ru/blog/1097.html (accessed 07.07.2017) (in Russian)
5. K.A. Akmalaev, M.S. Saduakasov, E.K. Nurmaganbetov, Povyshenie sul'fatostojkosti betonov s primeneniem himicheskih dobavok [Increase of sulphate resistance of concrete by using chemical admixtures], available at: http://e-lib.kazntu.kz/sites/default/files/articles/akmalaev_2007_2.pdf (accessed 02.08.2017) (in Russian)
6. M.H. Shehata, G. Adhikari, Sh. Radomski, ACI Materials Journal, **6**, 594-602 (2008)
7. V.V. Turchin, L.V. Yudina, A.R. Ibatyllina, A.R. Sattarova, Intelligent Systems in Manufacturing, **2**(20), 173-179 (2012) (in Russian)
8. GOST 31108-2003. Common cements. Specifications (2016)

9. TU 5870-005-58042865-05. Specifications, Plasticizer C-3 (Poliplast CP-1)
10. GOST 10178-85. Portland cement and portland blastfurnace slag cement. Specifications (2008)
11. GOST 22266-94. Sulphate resistant cements. Specifications (2014)
12. GOST 8735-88. Sand for construction work. Testing methods (2006)
13. GOST 5802-86. Mortars. Test methods (2010)
14. GOST 10060.0-95. Concretes. Methods for the determination of frost-resistance. General requirements (1997)
15. V.A. Voznesenskiy, T.V. Lyashenko, B.L. Ogarkov, Chislennye metody reshenija stroitel'no-tehnologicheskih zadach na EVM [Numerical methods for the construction and technological tasks on computer] (Vyshha shkola, Kiev, 1989) (in Russian)
16. V.V. Kaymonov, D.V. Vasilyeva, E.I. Protodyakonov, V.A. Andreev, Current issues in construction and life support: safety, quality, energy and resources efficiency, Proceedings of the III All-Russian scientific and practical conference, 268-273 (March 3-4, 2014, Yakutsk, Russia) (in Russian)
17. V.S. Lesovik, I.L. Chulkova, Vestnik SibADI, **4**(44), 69-79 (2015) (in Russian)

Thermal insulation coating based on water-based polymer dispersion

Iuliia Panchenko[1,*], *Marina* Akulova[2], and *Dmitrii* Panchenko[1]

[1]Tyumen Industrial University, 625000 Tyumen, Russia
[2]Ivanovo State Polytechnic University, 153037 Ivanovo, Russia

Abstract. For Russia, due to its long winter period, improvement of thermal insulation properties of envelope structures by applying thermal insulation paint and varnish coating to its inner surface is considered perspective. Thermal insulation properties of such coatings are provided by adding aluminosilicate microspheres and aluminum pigment to their composition. This study was focused on defining the effect of hollow aluminosilicate microspheres and aluminum pigment on the paint thermal insulation coating based on water-based polymer dispersion and on its optimum filling ratio. The optimum filling ratio was determined using the method of critical pigment volume concentration (CPVC). The optimum filling ratio was found equal to 55%.

Introduction

Heat transmission through building envelope structures can occur in different directions depending on climate conditions of the region. The most relevant issue for countries with hot climate or long summer season is decrease of buildings heat gain caused by solar radiation. This problem is solved by covering the surface of building envelope structures with coatings that can reflect solar radiati0n [1-3].

For Russia, due to its long winter period, improvement of thermal insulation properties of envelope structures by applying thermal insulation paint and varnish coating to its inner surface is considered perspective [4-6]. Such materials represent viscous liquid compositions containing binder, hollow microspheres of various nature, such as polymers, glass, aluminosilicate, and titanium pigment [7-12].

It is economically reasonable to consider the possibility of using hollow alunimosilicate microspheres as filler for paint coatings, for their price is 8 to 10 times lower than that of polymer and glass microspheres. While producing paint pigment compositions, the color of microspheres has special relevance and depends on the ability of microspheres to disperse wave radiation in optical wavelengths. This ability is defined by the surface type, microsphere wall thickness and porosity. Shells of micropheres are light-gray and have a small number of closed pores due to which they provide sufficient transparency and do not change paint color and opacity, which allows saving high-priced pigments [13].

* Corresponding author: indyboot@list.ru

Optical values that characterize radiation properties of materials are absorption coefficient α and emissivity coefficient ε [14]. The most complex task is to obtain coatings with low emissivity coefficient values. Therefore, for a coating to be able to reflect thermal radiation it is required to add to its composition a component with high reflective capacity in IR region – aluminum pigment [15, 16].

Traditional aluminum pigments are powders PAP-1 and PAP-2 (aluminum pigment powder). However, such pigments have a number of drawbacks: poor wetting properties, oxidation of aluminum particles, fire and explosion hazards, inhalation toxicity, as well as presence of active aluminum able to interact with water [17]. German company Schlenk and American company Eckart developed the technology for obtaining spherical aluminum pigment particles [18]. Aluminum particles are covered with a transparent silicone dioxide layer, which allows using them for water-based paint compositions. Pigments are produced in the form of ready-to-use pigment pastes. In addition to aluminum pigment and water, such pastes contain admixtures for wetting and dispersion, antifoaming agents and special admixtures to optimize rheological properties.

Use of styrene acrylic dispersions as binder for highly filled compositions is considered reasonable due to their optimum price/quality ratio. The research is aimed to define the effect of hollow aluminosilicate microspheres and aluminum pigment on the paint thermal insulation coating based on water-based polymer dispersion and to find its optimum filling ratio.

Materials and Methods

Light-gray hollow aluminosilicate microspheres (Ural Ekibastuz class) produced by Inotek company were used during the study. Grain size distribution of microspheres was 75-300 microns, with the mean diameter of 152 microns. The image of a microsphere is given in Figure 1.

Fig. 1. Micrograph of a hollow aliminosilicate microsphere.

Pigment paste (series Hydroxal DC 5000) produced by Eckart company was used as aluminum pigment. The paste represents water-based dispersion containing 65% of aluminum pigment with the grain size of 50-70 microns (Fig. 2).

Fig. 2. Micrograph of aluminum pigment.

The task of making a paint formulation consists in selecting composition that provides the paint's stability, good technological properties and forms a film with as closed structure as possible [19]. Pigment concentration at which the filming agent completely covers pigment and filler particles and occupies all space between them is called critical pigment volume concentration (CPVC). As a rule, while increasing the filling ratio of a coating its properties smoothly vary, and at the point of CPVC either sharp variation or almost no further variation of properties occurs [20]. In other words, the best coating properties are obtained when the filling ratio is slightly less than the critical pigment volume concentration. Therefore, CPVC method was used for defining the optimum pigment filling ratio of a paint thermal insulation coating.

Results

The main characteristics for paint coatings meant for interiors are water absorption of the film, opacity and adhesion to surface. On the one hand, loosening of film-forming agents' structure can be observed in proximity to pigment and filler particles, which can cause degradation of coating properties. And, conversely, when tightened structures of film-formation agent are formed, the coating properties improve [20, 21]. It is known that CPVC of water-based styrene acrylic paints lies between 40 and 80% [22]. Hence, search of optimum filling ratio was performed in this range. The research results presented in Figures 3-5 show that degradation of coating properties occurs when the CPVC of the film reaches the value of 58%.

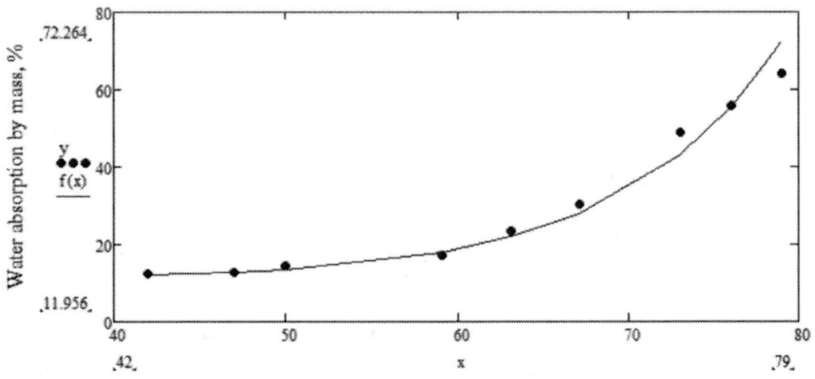

Fig. 3. Dependence of film water absorption on pigment and filler volume concentration. f(x)= $13.975 \cdot 10^{-3} \cdot \exp(0{,}1062 \cdot x) + 10.747$.

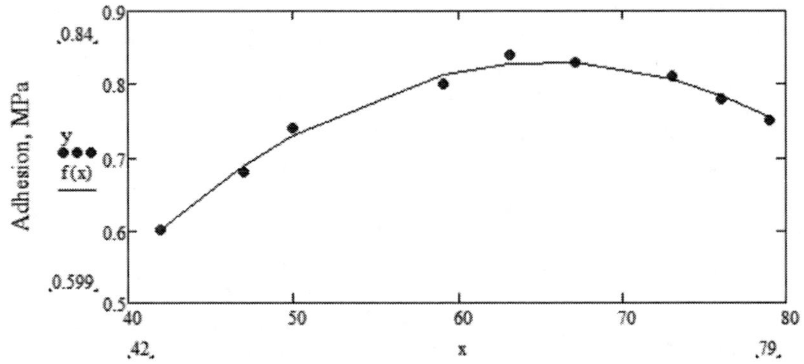

Fig. 4. Dependence of adhesion strength on pigment and filler volume concentration. f(x)= $-0.00042 \cdot x^2 + 0.055 \cdot x - 0.97$.

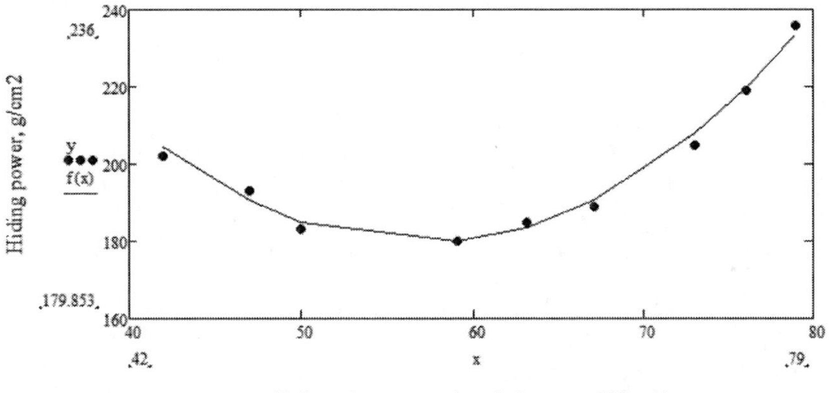

Fig. 5. Dependence of opacity on pigment and filler volume concentration. f(x)= $0.1114 \cdot x^2 + 0.055 \cdot x - 541.37$.

Conclusion

The conducted research proved the ability to obtain thermal insulation paint coating on the basis of styrene acrylic dispersion, aluminosilicate microspheres and aluminum pigment paste. It is found that CPVC of the thermal insulation coating film is 58%. The optimum filling ratio of the coating is proved to be always slightly lower than the CPVC and is 55%.

References

1. Y. Zhang, E. Long, Y. Li, P. Li, Energy Exploration & Exploitation, 1-19 (2017)
2. K. Banionis, E. Monstvilas, V. Stankevičius, R. Bliūdžius, K. Miškini, Journal of Civil Engineering and Management, **18**, 4 (2012)
3. M. Zinzi, G. Fasano, International Journal of Sustainable Energy, **28,** 1-3 (2009)
4. A.N. Trofimov, A.F. Kosolapov, Mezhotraslevoj al'manah, **48** (2014) (in Russian)
5. Y.Y. Kochetova, Bulletin USPTU. Science, Education, Economy. Series Economy, **1(7)** (2014) (in Russian)
6. N.N. Zhdanov, R.M. Garipov, A.I. Khasanov, Herald of Kazan Technological University, **16(17)** (2014) (in Russian)
7. L. Chen, Journal of Dispersion Science and Technology, **32**, 9 (2011)
8. V.S. Belyaev, Patent no. 2304156, Russian Federation (2007)
9. E.B. Bryantcev, Patent no. 2318782, Russian Federation (2008)
10. R.H. Fathutdinov, V.A. Maslov, S.A. Hafizova, Patent no. 2400506, Russian Federation (2010)
11. G.N. Jakunin, I.P. Prokop'ev, V.V. Burakov, Patent no. 102021, Russian Federation (2011)
12. V.A. Ryzhenkov, N.A. Loginova, A.I. Bychkov, Patent no. 133546, Russian Federation (2013)
13. A.D. Yakovlev, N.Z. Evtyukov, T.I. Gol'ts, Russian Coatings Journal, **9** (2011) (in Russian)
14. E.F. Itsko, Russian Coatings Journal, **5** (2004) (in Russian)
15. Y He, X Zhang, D Ph., and J Guan, Journal of Energy Engineering, **142**, 4 (2016)
16. L Yuan, X. L. Weng, J. L. Xie, and L. J. Deng, Materials Research Innovations, **19**, 1 (2015)
17. S.B. Prokhorov, L.F. Vagina, Stroitel'nye materialy [Construction materials], **6** (2006) (in Russian)
18. E.V. Grikshtas, Russian Coatings Journal, **9** (2005) (in Russian)
19. V.V. Verkholantsev, *Vodnye kraski na osnove sinteticheskih polimerov* [Water Paints Based on Synthetic Polymers] (Himija, Moscow, 1968) (in Russian)
20. E.A. Indeykin, L.N. Leibzon, I.A. Tolmachev, *Pigmentirovanie lakokrasochnyh materialov* [Pigmenting of paint and varnish coatings] (Himija, Moscow, 1986) (in Russian)
21. O.V. Orlova, T.N. Fomicheva, *Tehnologija lakov i krasok* [Paint and varnish technology] (Himija, Moscow, 1990) (in Russian)
22. E.E. Kazakova, O.N. Skorokhodova, *Vodno-dispersionnye akrilovye lakokrasochnye materialy stroitel'nogo naznachenija* [Water-dispersion acrylic paint and varnish coatings for construction purposes] (Paint-Media, Moscow, 2003) (in Russian)

Influence of cellulose fibers on structure and properties of fiber reinforced foam concrete

Valeriy Fedorov[1,*], and *Aleksey* Mestnikov[1]

[1]Ammosov North-Eastern Federal University, 677000 Yakutsk, Russia

Abstract. One of the promising means of foamed concrete quality improvement is micro-reinforcement by adding synthetic and mineral fibers to the base mix. This research is the first to investigate peculiarities of using recycled cellulose fiber extracted from waste paper for obtaining fiber reinforced foam concrete. The paper presents results of experimental research on the influence of cellulose fibers on structure and properties of fiber reinforced foam concrete by using methods of chemical analysis and scanning electron microscopy. The research determines peculiarities of new formations appearance and densification of binder hydration products in the contact zone between fiber and cement matrix, which boost mechanical strength of fiber reinforced foam concrete. Physico-mechanical properties of fiber reinforced foam concrete were defined depending on the amount of recycled cellulose fiber added to the base mix. It was found that the use of recycled cellulose fibers allows obtaining structural thermal insulating fiber reinforced foam concretes of non-autoclaved hardening of brand D600 with regard to mean density with the following improved properties: compressive strength increased by 35% compared to basic samples, higher stability of foamed concrete mix and decreased shrinkage deformation.

Introduction

The use of structural and thermal insulating wall materials allows producing lightweight envelope structures that meet modern requirements to thermal protection, environmental and fire safety. Among such materials are cellular concretes, particularly foamed concretes based on binders of hydraulic hardening [1].

In spite of obvious advantages cellular concretes, particularly foamed concrete, has a number of disadvantages which can be observed at all stages of production process. Here belong difficulties of producing uniform foam, foamed concrete structure management, high values of shrinkage deformations and relatively low strength characteristics.

Dispersed reinforcement with fibers is one of the promising methods to improve technological, construction and operational characteristics of cellular concretes. The use of this technology allows transmitting tensile force from composite matrix to fiber. This force transmission effect considerably reduces crack formation and opening and increases strength and crack resistance of fiber reinforced foam concrete [2, 3].

* Corresponding author: valeriyif.ykt@gmail.com

In practice, synthetic fibers (polypropylene, polyethylene, nylon, ect.) are the most widely used ones. [2]. Among their advantages is almost complete resistance to aggressive alkaline medium of set cement. Synthetic fibers contribute to formation of optimum mineral frame and decrease of shrinkage deformation in foamed concrete. According to foreign researchers [3, 4], apart from the mentioned advantages the use of discontinuous fibers enables to reduce mass of a structure, improve thermal protection characteristics and fire resistance of wall structures.

The paper [5] studies foamed concrete with dispersed reinforcement with coconut fiber from the Republic of Vietnam. The fibers were milled and added to the base mix in the amount of 0.5, 1, 1.5 and 2% of cement weight. The best parameters were observed for 1% of fiber content, which are 3.9 MPa with the mean density of 830 kg/m^3. Use of basalt fibers for foamed concrete production is studied in [6]. The research demonstrates that formation of crystal network on the surface of basalt fibers causes considerable increase of strength parameters of fiber reinforced foamed concrete. Polyamide fibers also intensify cluster formation processes and, as a consequence, increase aggregate stability of fiber reinforced foam concrete mix [2]. At the same time, fibers can control energy and geometrical parameters of spatial bonds in water in the material structure.

Thus, discontinuous fibers shall meet the following requirements: ultimate strain of fibers shall be higher than that of concrete, relation of elastic moduli of fiber and concrete matrix shall be equal to more that 1, ultimate tensile strength of fiber shall be higher than the same parameter for cement matrix, fibers shall be chemically stable in alkaline medium of set cement. With regard to the price/quality ratio, the so-called recycled cellulose fibers (RCF) extracted from paper waste completely meet the above requirements.

To date, both national and foreign research on the possibility of using RCF in fiber reinforced foam concrete production either lack or have the form of a review.

Due to this fact, investigation of structure formation processes in fiber reinforced foam concrete with recycled cellulose fibers, analysis of morphology and structural parameters of new formations in the contact zone between cement matrix and fibers are seen as relevant tasks in the field of obtaining foam concrete with improved properties.

The purpose of this study, therefore, is justification of the possibility to obtain fiber reinforced foam concrete of non-autoclaved hardening with improved properties based on Portland cement and the use of recycled cellulose fibers extracted from waste paper.

Materials and Methods

Technical requirements to non-autoclaved cellular concretes, including foamed concrete, are specified in the Russian Federation state standard – GOST 25485-89 "Cellular concretes. Specifications".

The subject of research is non-autoclaved structural and thermal insulating foamed concretes of the brand D600 with regard to the mean density.

Portland cement is the most often used binder for non-autoclaved concretes production - according to GOST 10178 (without adding tripoli, burnt clay, trasses, glinite, silica clay and ashes) it contains no more than 6% of tricalcium aluminate (C_3A).

Silica component acceptable for foamed concrete production is sand which according to GOST 8736 contains at least 90% of SiO_2, no more than 0.5% of mica and maximum 3% of silt and clay impurities.

Waste paper of brand MC-8B (newsprint) was used to extract RCF in compliance with GOST 10700-97 "Paper and cardboard waste. Specifications". This waste paper brand was selected due to its wide use and little amount of fillers and binders.

Fiber reinforced foam concrete samples were prepared by "traditional" method. Laboratory tests of the samples were performed with the use of certified equipment of the testing center of Yakutsk-Expert, LLC. Experimental data for qualitative and quantitative chemical analysis of the studied RCF and fiber reinforced foam concrete samples were obtained using JEOL JSM-7600F scanning electron microscope.

The experimental data and micropictures were processed in Microsoft office Excel 2007, MathCAD 2001i and ImageJ.

Experimental results

Cellulose fibers were obtained by repulping a waste paper sheet in aqueous medium. Cellulose fibers extraction from paper waste is done by combining physico-chemical and hydrodynamic processes which occur in waste paper when contacting with water. Prior to repulping waste paper was kept in aqueous medium for 12 hours. Disintegration of the sheet and its deflaking while mixing occurs under hydrodynamic and mechanical actions [7]. It shall be noted that strength of paper and cardboard fibers is much higher than that of wet sheet, therefore, mechanical actions allow splitting the sheet into fibers without their disintegration. Duration of deflaking process in a mixer was 1 hour.

Physico-mechanical parameters of fiber reinforced foam concrete samples were defined in compliance with the requirements of relevant standard technical documents [8, 9]. RCF belong to a big group of organic fibers. The fiber looks like a hollow tube with extended coarse surface (Fig. 1). According to the results of SEM analysis the obtained RCF have the following parameters: length 5100-5300 micron, width 20-35 micron, thickness 2-5 micron, bulk density 0.08-0.1 g/cm^3, temperature stability up to 235°C. RCF, like the majority of vegetable fibers, are hydrophilic, consequently, fibers are strongly glued with cement mortar.

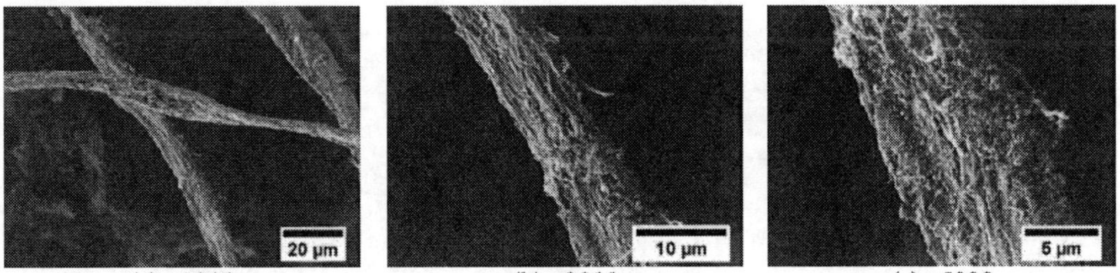

(a) x1000 (b) x3000 (c) x5000

Fig. 1. Microstructure of recycled cellulose fiber.

One of the key effects of using RCF in fiber reinforced foam concrete production is the ability to improve its physico-mechanical properties. The influence of RCF on strength parameters of fiber reinforced foam concrete samples was investigated in cubic specimens with the size of 100x100x100 mm. The specimens have been kept at the environment temperature of 20°C and relative air humidity of 100% for 28 days.

Fiber reinforced foam concrete brand with regard to mean density is D600. Content of RCF was 0.1, 0.2, 0.3, 0.4 and 0.5% of the binder mass on a dry matter basis. Ultimate compressive strength of fiber reinforced foam concrete under compressive load of 1.9 MPa after 28 days of natural hardening is reached with the fiber content of 0.3% wt. (compressive strength of a reference specimen is 1.4 MPa).

Strength decreases with further increase of fiber content. This correlation has one extremum point which, at the same time is the optimum one. Increase of strength up to the extremum value is explained by cumulative action of two factors. First, the load applied to fiber reinforced foam concrete is transmitted through cement matrix directly to fibers [2,

10, 11]. Second, the gain of strength is related to peculiarities of morphology of new cement formations around RCF.

Figure 2 demonstrates spectrograms of RCF and cement matrix at the age of 28 days. The following effect was determined after comparative analysis of spectrograms of RCF and products of cement hydration: higher content of calcium ions are observed in the zone of fibers location compared to the corresponding peak on the hardened cement paste spectrogram. Besides, not only the peak height increases, but its area, too. Appearance of silicon ions in RCF spectrogram shall be noted, the peak value of which is considerably lower than that of cement.

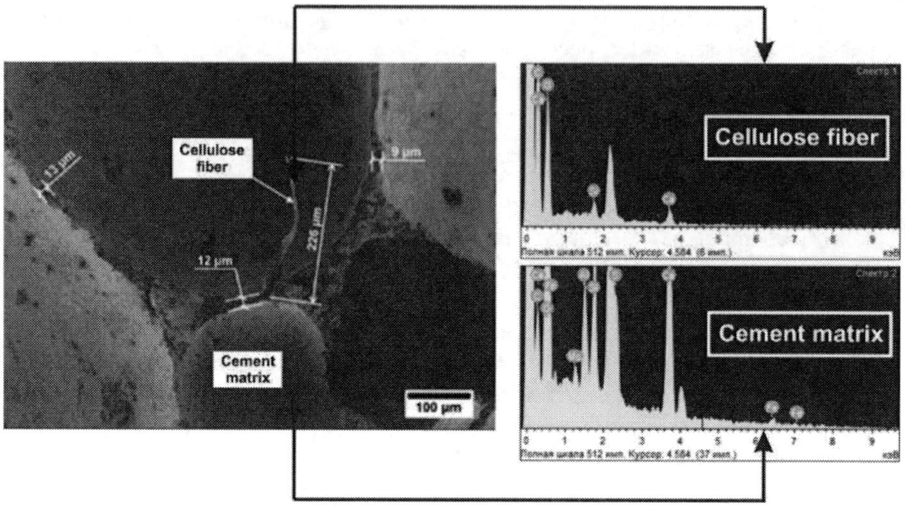

Fig. 2. Comparison of spectrograms for chemical composition of recycled cellulose fiber mixed with cement mortar and cement matrix of fiber reinforced foam concrete.

Based on the analysis of RCF and hardened cement paste spectrograms one can suggest that partial mixing of hydration products of cement clinker minerals occurred around the fiber surface layer [12, 13]. It shall be noted that higher content of calcium ions (Table 1) might be related to formation of crystallization centers in the contact zone between RCF and cement matrix, which is followed by setting of $Ca(OH)_2$ crystals [14, 15].

Table 1. Results of quantitative chemical analysis.

No.	Title	Element content (reference), %					
		C ($CaCO_3$)	O (SiO_2)	Si (SiO_2)	Ca (Wollastonite)	Mg (MgO)	Al (Al_2O_3)
1	RCF	31-58	33-43	1-4	6-19	1	0.6-1
2	Cement matrix	5-7	56-65	3-8	23-29	0.7-1	0.5-3

Microstructure of fiber reinforced foam concrete was studied for advance investigation of structure formation processes in deep layers of fiber reinforced foam concrete. Special attention was paid to studying appearance of new formations around RCF (Fig. 3).

Due to visual analysis of hydration products morphology it was found that microstructure of hardened cement paste generally consists of clusters which are composed of interlocked acicular crystals and hexagonal plates around spherical agglomerate.

First of all, new formations have high density which is proved by close packing of hydration products [12, 15]. Besides, a thin network of microfibrillar fibers was found all over the RCF surface, their diameter being between 50 and 300 nm (Fig. 3, on the left).

Presumably, microfibrils of cellulose fibers were formed because of mechanical defects during extraction from waste paper sheets. First, these fibers improve adhesion of RCF with cement matrix due to increase of fiber surface area and cause considerable gain of mechanical strength of fiber reinforced foam concrete. Second, due to ultrafine size microfibrils are sources of structure formation nuclei which increase density of interpore partition in fiber reinforced foam concrete. This leads to the increase of general strength of fiber reinforced foam concrete.

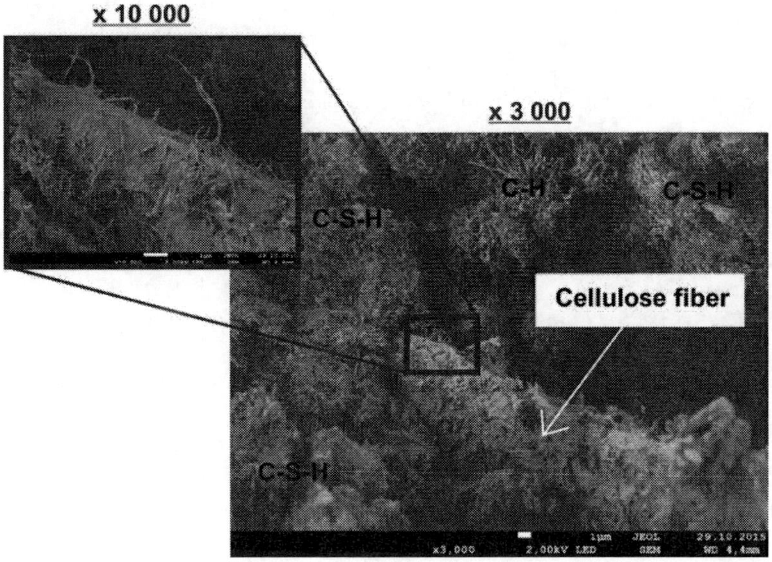

Fig. 3. Location of recycled cellulose fibers in cement matrix of fiber reinforced foam concrete. Zoom x3000 and x10000.

Conclusion

On the basis of the conducted research the conclusion can be made that the use of RCF allows increasing compressive strength of fiber reinforced foam concrete up to 35% (compared to foamed concrete without fiber addition). Besides, increase of stability of base foam concrete mixture and decrease of shrinkage deformations are observed in fiber reinforced foam concrete samples. Taking into account the fact that fibers were extracted from mere paper waste, the strength gain shall be considered quite high.

The research defines the possibility of obtaining structural thermal insulating fiber reinforced foam concrete of the brand D600 with regard to mean density and of the type B1,5 with regard to compressive strength with the use of RCF.

Positive effect of RCF on formation of optimum porous structure of fiber reinforced foam concrete shall be emphasized. The average diameter of macropores in samples of fiber reinforced foam concrete with RCF addition is 0.7 mm, whereas it is 1.1 mm for the reference sample. Considerable reduction of the average macropore diameter is supposedly connected with redistribution of internal stresses due to the effect of dispersed reinforcement of interpore partitions and as a consequence of coalescence effect.

The extension of this research will be focused on decreasing density of fiber reinforced foam concrete with the use of recycled cellulose fibers up to 500 kg/m^3 with B1.5 type of concrete with regard to compressive strength.

References

1. V.A. Perfilov, A.V. Atkina, News of higher educational institutions. Construction, **9**, 26 (2010) (in Russian)
2. F.N. Rabinovich, *Kompozity na osnove dispersno-armirovannyh betonov* [Composites based on dispersed-reinforced concrete], 23 (ACB, Moscow, 2004) (in Russian)
3. N. Yong, Natural fibre reinforced cement composites, doctoral thesis, 127 (Victoria University of Technology, Australia, July 1995)
4. R. Khiari, Z. Marrakchi, M. N. Belgacem, E. Mauret, F. Mhenni, Composites Science and Technology, **71**, 1867 (2011)
5. V.T. Nguen, Science and Technology, **9**, 10 (1981)
6. N.G. Vasilovskaya, I.G. Kalugin, News of higher educational institutions. Construction, **11**, 12 (2010) (in Russian)
7. D.M. Flyate, *Svojstva bumagi* [Paper properties], 35 (Lesnaja promyshlennost', Moscow, 1986) (in Russian)
8. GOST 10180-2012. Methods for strength determination using reference specimens (2013) (in Russian)
9. GOST 12730.1–78. Concretes. Methods of determination of density (1980) (in Russian)
10. V.I. Fedorov, A.E. Mestnikov, Industrial and civil engineering, **8**, 22 (2016) (in Russian)
11. Z. Li, X. Wang, L.Wang, Composites: Part A, **37**, 467 (2006)
12. Y. Chen, Evaluation of damage in cellulose fiber reinforced concrete using acoustic enission, doctoral thesis, 52 (University of Florida, 2010)
13. G.H.D. Tonoli, U.P. Rodrigues Filho, H. Savastano Jr., J. Bras, M.N. Belgacem, F.A. Rocco Lahr, Composites: Part A, **40**, 2046 (2009)
14. B.J. Mohr, J.J. Biernacki, K.E. Kurtis, Cement and Concrete Research, **36**, 1240 (2006)
15. R.D. Toledo Filho, K. Scrivener, G.L. England, K. Ghavami, Cement and Concrete Composites, **22**, 127 (2000)

Structure and properties of ceramic brick colored by manganese-containing wastes

Andrey Stolboushkin[1], *Danil* Akst[1,*], *Oksana* Fomina[1], and *Aleksandr* Ivanov[1]

[1]Siberian State Industrial University, 654007 Novokuznetsk, Russia

Abstract. The given paper investigates the influence of manganese-containing wastes on bulk coloring and physical and mechanical properties of wall ceramic materials with matrix structure. Conventional research methods to study physical and mechanical properties of ceramic specimens were used in the present study. The structure and phase state of decorative ceramic materials were investigated by means of physical-chemical analysis. Authors suggested the model to form coloring layer in ceramic matrix composites based on argillaceous raw materials and coloring pigment. Macrostructure was studied on ceramic samples manufactured according to the developed method from clay loam of moderate plasticity and wastes of manganese mining. The paper provides results of pilot testing on production of colored ceramic brick. It was found that incorporation of coloring component does not decrease strength properties of ceramic matrix materials. Authors defined the palette and color code for decorative ceramic specimens manufactured with different content of coloring component of manganese mining wastes according to RGB color codes chart.

1 Introduction

Rational selection of building materials plays an important role in construction. Building materials quality primarily defines strength, reliability and durability. Along with that, decorative properties are also of great importance as they define the outside look of a building.

Facing and decorative ceramic brick belongs to such materials. Various techniques have been used for its manufacture so far [1-5]. Currently, bulk coloring of ceramics has been considered to be one of the most popular techniques among them. Coloring of wall materials is conducted by incorporation of light-burning clays, various mineral additives or coloring metal oxides into furnace charge [6, 7]. Use of pure oxides for structural coloring of brick is retained due to their high cost, thus leading to appreciation of ceramic manufacture and reduced competitiveness of products on the market.

Efficient use of industrial wastes for obtaining ceramic brick with the required operational and decorative properties is one of the promising directions of wall ceramic materials development. Replacement of expensive coloring agents with industrial wastes containing coloring metal oxides [8] will enable to reduce expenses for brick manufacture

* Corresponding author: daniel_axt@mail.ru

and enhance environmental situation in industrial areas [9, 10].

Analysis of modern production of colored brick using high-performance color modifiers [11-16] shows that bulk coloring of wall ceramic materials with oxide pigments generally causes a number of problems. One of the main problems is disruption of sintering process of argillaceous minerals while burning which leads to degradation of physical and mechanical properties of ceramics [17, 18].

The present paper is focused on application of manganese-containing wastes for structural coloring of ceramics and development of the way to form coloring shells around the clay granules. Selective structural concentration of wastes in clay brick by their minimum content will provide high coloring effect and decrease of their negative effect while sintering of ceramic products.

The given work is aimed at studying the influence of manganese-containing wastes on bulk coloring and physical-mechanical properties of ceramic brick with matrix structure.

2 Materials and methods

2.1 Description of raw materials

Argillaceous raw material used in the given study was clay loam collected from Berdsk deposit (Novosibirsk Region) peculiar for Siberian Region. Clay loam properties were: semi-acid, moderate-plasticity, low-melting, containing coarse-grain particles (>0.01 mm) less than 3%. Tetravalent manganese oxide is one of the popular pigments used for ceramics bulk coloring. The wastes of manganese mining (WMM) of Selezen deposit (Kemerovo Region) were used as a coloring additive. Content of MnO_2 in wastes comprised more than 25%. Chemical and grain size composition of raw materials are given in Table 1 and Table 2 correspondingly.

Table 1. Chemical composition of raw materials.

Raw material	Mass content of components, % (per pure dry substance)											
	SiO_2	Al_2O_3	FeO	Fe_2O_3	CaO	MgO	TiO_2	MnO	R_2O	P_2O_5	SO_3	% of other impurities
Berdsk clay loam	65.74	14.71	5.71		5.23	1.79	0.95	0.17	3.48	-	0.01	2.21
Wastes of manganese mining	32.39	12.18	12.81		2.41		-	25.17	0.49	0.16	-	13.55

Table 2. Grain size composition of raw materials after blending.

Raw material	Fraction content, %, particles size, mm				
	>0.06	0.06-0.01	0.01-0.005	0.005-0.001	<0.001
Berdsk clay loam	-	2.6	65.05	4.18	28.17
Wastes of manganese mining	41.71	36.71	7.5	12.73	1.35

2.2 Research methods

The study of basic physical and mechanical properties of decorative ceramic materials with manganese-containing component was conducted on the laboratory cylinder specimens and factory-supplied bricks of dry compaction. Strength of bricks was defined according to the Russian National Standard GOST 8462–85 "Wall materials. Methods to define ultimate compressive and bending strength". Water adsorption and frost resistance were established in accordance with the Russian National Standard GOST 7025–91 "Bricks and ceramic and silicate stones. Methods to define water adsorption, density and frost resistance control".

The structure of ceramic samples was examined using the optical microscope Olympus GX-51 with polarizing accessory (Japan) and scanning electron microscope JSM-6460LV (JEOL, Japan).

X-ray investigation was performed with the help of diffractometer XRD-6000 (Shimadzu, Japan). The powders obtained from each of the layers were studied within the range of scanning 2θ from 4° to 80° at room temperature using FeKα radiation (Wavelength 1.936040 Å). Diffractometer settings were 40 kV, 30 mA, continuous scan, 2θ step width 0.02° and speed 2 deg/min.

3 Experimental

3.1 Investigation of decorative ceramic specimens

Previous works of authors were devoted to studying the influence of manganese ore production waste (**WMM**) on coloring of ceramic specimens produced from argillaceous raw materials [19]. The specimens were prepared in accordance with the developed methodology providing formation of matrix structure [19]. For that purpose fine-cut clay loam was aggregated in turbo-paddle type mixing grainer at simultaneous moisturization up to 11-13%. Angular spin velocity of paddled blender of grainer was 15-20 s^{-1}. Clay loam was granulated for 2-3 minutes until the granules with predominant size of 1-3 mm were formed. The final granulation stage included incorporation of additive from fine-grained **WMM** in order to form coloring layer along the granules surface. The actual moisture of granulated material "powdered with **WMM**" was 10-12%.

Cylinder specimens were compacted from granulated batch under the pressure of 15-17 MPa. Burning with isothermal curing for one hour was conducted at the temperature of 1000 °C.

Physical and mechanical properties of reference samples (without additive) and colored ones (with addition of 10 wt.% **WMM**) are given in Table 3.

Table 3. Physical and mechanical properties of ceramic samples.

Sample	Air shrinkage, %	Fire shrinkage, %	Average density, kg/m^3	Compressive strength, MPa	Water adsorption, %	Strength-density ratio
reference	1.6	1.4	1895	33.9	12.8	17.9
colored	1.0	0.9	1962	38.6	12.5	19.7

In order to conduct objective assessment of bulk coloring of ceramic specimens color codes were defined in accordance with RGB color code chart [20] and the resulting hex-code of color in hexadecimal number system formed by means of addition of basic color levels in the format #RRGGBB [21]. The results are given in Table 4.

Table 4. Color palette and color codes of ceramic specimens.

Specimen	Adapted specimen color	Color code according to RGB			Color hex-code
		R	G	B	
Reference		179	89	37	#B35925

Colored		56	42	29	#382A1D

The structure of ceramic specimens with **WMM** coloring additive was studied successively on a macro and micro level with different degree of increase of their internal surface area (Fig.1,2).

Macrostructure of specimens was investigated using binocular magnifier on ground surface of vertical saw cut (Fig.1,a). One could note distinctive matrix structure of ceramic fragment with a marked boundary between phases of different color. Matrix structure was formed at the stage of raw material by means of granulation of fine-grained clay loam and by coating of granules with **WMM** additive with further compaction. While burning the formed matrix structure of raw material was transformed into ceramic matrix composite; its macrostructure consisted of ceramic cores of brick-red color coated with shell from sintering products of brown-colored manganese-containing additive.

Observations of polished section in reflected light (Fig.1,b) showed even-grained microstructure with increments of relic minerals up to 0.2 mm in size. The boundaries of separation between the core and coloring matrix were clear, deep interpenetration between the dispersion phase and dispersion medium of ceramic composite was not observed.

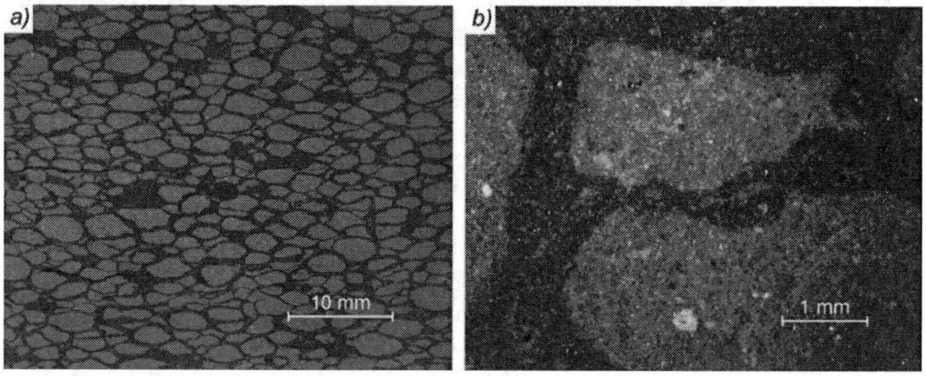

Fig. 1. Macrostructure of ceramic specimen based on Berdsk clay loam with coloring additive 10 wt.% **WMM**. Conditions: polished section, reflected light, binocular magnifier (*a*), optical microscope, nicol II (*b*).

Further detailed study of microstructure using electron microscope (Fig. 2) showed peculiarities of matrix and core composition. Microtexture inside the sintered granules was even, partly amorphized, with small closed pores of a round shape up to 20 μm in size. By ×1800 zooming one could observe hardened melt areas. The texture of matrix was presented by large-size aggregated stratum of irregular shape mostly 20-50 μm in size. Porosity of boundary coloring layer was more distinctive; pores were elongated, up to 100 μm long.

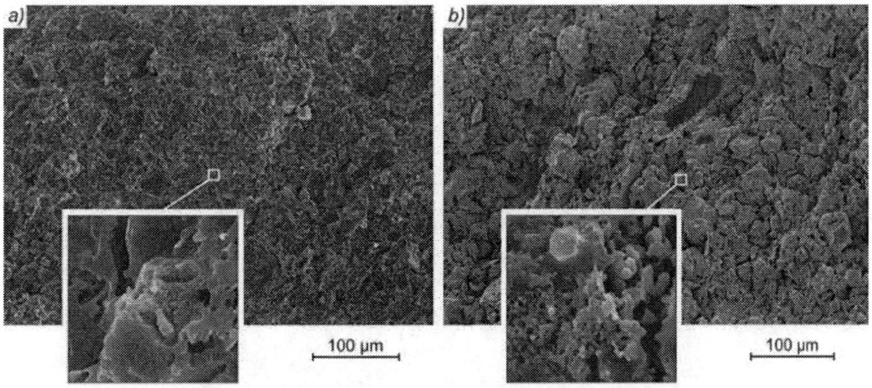

Fig. 2. SEM-micrographs of the structure of ceramic core (*a*) and coloring boundary layer (*b*) of ceramic matrix composite made from clay loam and **WMM** additive.

Investigation of phase state of ceramic specimens with **WMM** additive is presented in Figure 3.

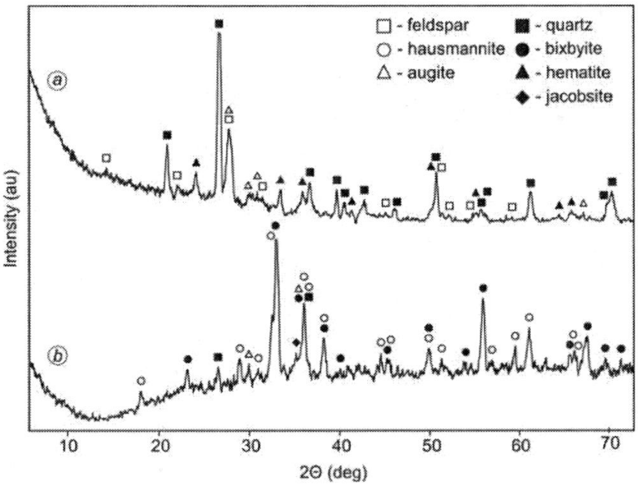

Fig. 3. X-ray diffractogram of the core (*a*) and coloring boundary layer (*b*) of ceramic matrix composite made from clay loam and **WMM** additive.

X-ray powder pattern showed that the basic mineral components of the core were quartz (4.255; 3.343; 2.457 Å etc.), feldspar (4.03; 3.247; 1.79 Å etc.), hematite (3.69; 2.69; 2.52; 1.697 Å) and augite (2.989; 2.917; 2.125 Å). The boundary layer was presented by bixbyite (3.84; 2.72; 1.664 Å etc.), hausmannite (4.92; 3.08; 2.76 Å etc.), jacobsite (2.563; 1.503 Å), quartz and augite.

3.2 Experimental production testing of decorative ceramic brick

Resulting from laboratory studies held at "Berdsk Brick Factory", LLC (Berdsk city, Novosibirsk Region) (factory uses semidry compaction) pilot batch of ceramic brick was manufactured (using bulk coloring with **WMM** additive). The appearance of reference brick (without additive) and colored one is given in Figure 4.

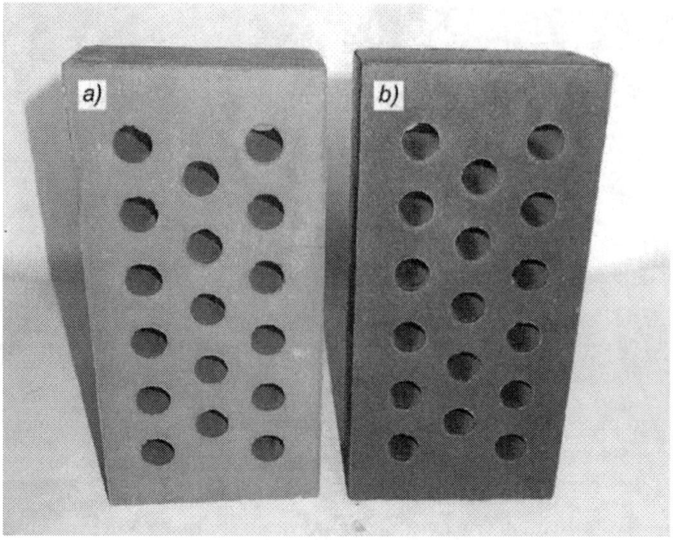

Fig. 4. Ceramic brick from red-burning Berdsk clay loam without additive (a) and with additive 10 wt.% **WMM** (b).

Berdsk clay loam was used in experimental production testing. It underwent factory preparation in accordance with drying-grinding technique. **WMM** was subject to two-stage milling in jaw breaker and rod mill up to fraction of -300 μm.

Molding material was prepared according to the developed methodology. Granules from Berdsk clay loam were aggregated up to 1-3 mm and powdered with fine-grained **WMM** additive incorporated in the amount of 10 wt.%. Molding-moisture content of powdered granulate was 11.6-12.4%.

Brick compacting was conducted using press-type machine SM-1085B. Raw products had clear boundaries and even surface without cavities and delamination cracks. Drying and burning of brick for 42 hours was held in the tunnel furnace at the temperature of 1000 °C, as specified in manufacturing instructions of the factory.

The brick with 10 wt.% WMM additive had brown color after burning (Fig.4,b) compared to the red color of the reference specimen (Fig.4,a).

Results of physical chemical tests of ceramic brick conducted in the factory laboratory are shown in Table 5.

Table 5. Physical and mechanical properties of ceramic brick of pilot batch.

Brick type	Ultimate strength		Average density, kg/m^3	Water adsorption, %	Frost resistance, cycles
	Compressive, MPa	Bending, MPa			
Reference	12.1	2.2	1770	13.6	35
Colored	15.3	2.4	1908	11.7	50

4 Results

The coloring effect of additive from powdered (-300 μm) wastes of manganese mining into red-burning clay raw during manufacture of ceramic brick of dry compaction was established.

The technique of obtaining decorative ceramic materials with matrix structure developed by the authors [19] provides bulk coloring of brick into brown color (RGB: 56;

42; 29) when 8-10 wt.% of WMM is incorporated into batch composition. Along with that the strength of such products is increased by 15-20% due to concentration of coloring component along the granules surface. Thus, it eliminates the negative effect on argillaceous raw material, which is generally observed in bulk coloring technique of wall ceramics using pigments.

Formation of ceramic matrix composite was defined, its macrostructure consisted of cores covered with shell from sintering products of manganese-containing additive of a brown color. After burning cores made from clay had even-grained structure with formation of glass phase and specific hematite coloring.

Investigations of phase composition showed that the core consisted of the following basic minerals: quarts, feldspar, hematite, and augite. The boundary layer was represented by quartz and augite, as well as minerals specific for manganese compounds: bixbyite, hausmannite, and jacobsite.

5 Conclusion

The developed method of raw materials preparation which implies fine grinding of raw material, its granulation and powdering of the granules surface with coloring agents, is considered promising among decorative coarse ceramics techniques. The technique of obtaining decorative ceramic products with matrix structure provides effective ceramics coloring by addition of 10% manganese-containing wastes. Concentration of the coloring pigment on the surface of clay granules does not reduce physical and mechanical properties of ceramic specimens. Further development of the mentioned technique will enable to solve the problem of disturbing processes of argillaceous materials sintering while burning.

Novel technique of bulk coloring will provide effective (high coloring effect) use of industrial wastes; it will reduce expenses for decorative brick manufacturing and improve ecological situation in industrial areas.

Further research in this field is aimed at broadening the range of coloring modifiers based on industrial wastes as well as development of process-dependent parameters and modes to obtain decorative ceramic materials.

The research was performed in the framework of State Assignment of the Ministry of Education and Science of the Russian Federation, project code No. 7.7285.2017/8.9 "Basic research on ceramic matrix composite building materials based on technogenic and natural raw materials".

References

1. M. Llusar, T. Bermejo, J.E. Primo, C. Gargori, V. Esteve, G. Monrós, Ceramics International, **43**(12), 9133-9144 (2017)
2. E. Adorni, Journal of the European Ceramic Society, **33**(13-14), 2801-2809 (2013)
3. A.M. Salakhov, V.P. Morozov, F.G. Vagizov, A.A. Eskin, A.R. Valimukhametova, A.L. Zinnatullin, Stroitel'nye Materialy [Construction Materials], **3**, 90-95 (2017) (in Russian)
4. C. Rathossi, Y. Pontikes, Journal of the European Ceramic Society, **30**(9), 1841-1851 (2010)
5. K.A. Ariskina, A.M. Salakhov, F.G. Vagizov, R.T. Akhmetova, Vestnik Tekhnologicheskogo Universiteta [Herald of Kazan Technological University], **19**(24), 25-28 (2016) (in Russian)
6. V.I. Reznik, Stroitel'nye Materialy [Construction Materials], **4**, 54-56 (2011) (in Russian)

7. G.P. Vasyanov, B.F. Gorbachev, E.V. Krasnikova, R.K. Sadykov, R.R. Kabirov, Georesursy [Georesources], **4**, 44-49 (2015) (in Russian)
8. A.Yu. Stolboushkin, D.V. Akst, O.A. Fomina, *Dolgovechnost stroitelnykh materialov, izdelii i konstruktsii* [Durability of building materials, products and structures], 154-160 (Publishing House of Mordovia State University, Saransk, 2016) (in Russian)
9. I.V. Petrov, D.Y. Savon, *Ecologiya. Prirodopolzovanie. Ekonomika [Ecology. Environmental Management. Economics]*, 43-56 (Publishing House of Moscow State University for Humanities, Moscow, 2013) (in Russian)
10. L.P. Schukina, E.V. Lyubova, I.V. Bilan, M.F. Kartavenko, Stroitel'nye Materialy [Construction Materials], **8**, 28-30 (2010) (in Russian)
11. A. De Bonis, G. Cultrone, C. Grifa, A. Langella, A.P. Leone, M. Mercurio, V. Morra, Ceramics International, **43**(11), 8065-8074 (2017)
12. M.B. Sedelnikova, V.M. Pogrebenkov, *Keramicheskie pigmenty na osnove prirodnogo i tekhnogennogo mineralnogo syr'ya* [Ceramic pigments based on natural and industrial mineral raw materials] (National Research Tomsk Polytechnic University, Tomsk, 2014) (in Russian)
13. S. Yeşilay, M. Çakı, H. Ergun, Ceramics International, **43**(12), 8912-8921 (2017)
14. V. Valanciene, R. Siauciunas, Z. Valancius, Applied Clay Science, **99**, 110-118 (2014)
15. S.R. Prim, A. Garcia, R. Galindo, S. Cerro, M. Llusar, M.V. Folgueras, G. Monros, Ceramics International, 39 (6), 6981-6989 (2013)
16. L. Maritan, L. Nodari, C. Mazzoli, A. Milano, U. Russo, Applied Clay Science, **31**(1-2), 1-15 (2006)
17. A.Yu. Stolboushkin, Stroitel'nye Materialy [Construction Materials], **8**, 24-32 (2013) (in Russian)
18. C. Gargori, S. Cerro, N. Fas, M. Llusar, G. Monros, Ceramics International, **43**(7), 5490-5497 (2017)
19. A.Yu. Stolboushkin, O.A. Fomina, D.V. Akst, A.I. Ivanov, M.S. Druzhinin, Stroitel'nye Materialy [Construction Materials], **12**, 38-44 (2016) (in Russian)
20. D. Pascale, *A Review of RGB Color Spaces ...from xyY to R'G'B* (The BabelColor Company, Montreal, 2003)
21. D.S. McFarland, *CSS: The Missing Manual* (O'Reilly Media, California, 2015)

Production of environmentally friendly aerated concrete with required construction and operational properties

Evgeniya Tkach[1,*], *Vladimir* Solovyov[2], and *Semen* Tkach[3]

[1]National Research Moscow State University of Civil Engineering, 129337 Moscow, Russia
[2]Euroasia MS LLP, 050060 Almaty, Republic of Kazakhstan
[3]Financial University under the Government of the Russian Federation, 125993 Moscow, Russia

Abstract. The purpose of these studies is to justify the feasibility of recycling different types of industrial waste instead of conventional expensive raw materials in production of environmentally friendly aerated concrete with required construction and operational properties. The impact of wastes from various industries on the environmental condition of affected areas, as well as the results of their environmental assessment were analyzed to determine whether these wastes could be used in production of high-performance building materials. The assessment of industrial wastes in aerated concrete production suggests that industrial wastes of hazard class IV can be recycled to produce aerated concrete. An environmentally friendly method for large-scale waste recycling, including a two-step environmentally sustainable mechanism, was developed. The basic quality indicators of the modified aerated concrete proved that the environmental safety could be enhanced by strengthening the structure, increasing its uniformity and improving thermal insulation properties. The modified non-autoclaved aerated concrete products with improved physical and operational properties were developed. They have the following properties: density – D700; class of concrete – B3.5; thermal transmittance coefficient – 0.143 W/(m·°C); frost resistance – F75.

Introduction

A critical review of scientific and technical literature shows that the disposal of wastes from various industries remains a challenge since their types and volumes are ever increasing [1–5]. This also applies to the disposal of waste from such industries as production of phosphate mineral fertilizers, alcohol and dairy products. One of the effective ways to dispose of these wastes is to recycle them for production of building materials [6–10], in particular, aerated concrete products. The purpose of these studies is to develop compositions and a production technology for modified non-autoclaved aerated concrete based on recycled industrial waste. The proposed technology is founded on an environmentally sustainable waste recycling mechanism. The mechanism is implemented

* Corresponding author: ev_tkach@mail.ru

through a two-step process: to prepare complex organic mineral modifiers and to produce non-autoclaved aerated concrete based on the developed modifiers that can reduce binder consumption by 20%. Thus, it can be concluded that the search for new ways of recycling industrial waste in the production of environmentally friendly and high-performance aerated concrete should be continued.

Materials and Methods

Portland cement of class B32.5H produced at Kaluga Cement Plant LLC (Kaluga, Russia) was used as a binder for producing modified non-autoclaved aerated concrete based on recycled industrial waste. This cement was found to comply with the requirements of GOST 31108-2003. Common Cements. Specifications [11].

Quartz sands were used as a silica component. In terms of fine clay impurities, particle-size distribution and radiation hygiene assessment, these sands comply with the requirements of GOST 8736-93. Sand for Construction Works. Specifications [12].

Aluminum powder of PAP-2 grade was used as a blowing agent. It complies with the requirements of GOST 5494-95. Powder Aluminum. Specifications [13]. The fineness of the aluminum powder is 5,000 cm^2/g. The active aluminum content is 82%. In terms of exposure to the human body, the dust of aluminum powder is assigned to hazard class III and belongs to the group of combustible substances as defined in GOST 12.1.005-88. General Hygiene Requirements for Working Zone Air [14]. It should be noted that aluminum is allowed to be used in paste form in aerated concrete production.

We analyzed the experience of using the common high-performance modifiers in aerated concrete production [15] and developed new compositions of complex organic mineral modifiers based on the industrial wastes listed below (Table 1).

Table 1. Compositions of complex organic mineral modifiers based on industrial waste.

Industrial waste	Ingredients, wt%			
	GM-SB*	GM-SBF*	GM-SB* plus STS	GM-SBF* plus STS
Phosphogypsum	-	40	-	25
Whey	40	20	15	15
Distillers grains	40	20	15	15
Hardener – sodium thiosulfate (STS)	-		45	20
Water	remaining			
*GM-SB – oil-in-water emulsion; GM-SBF – emulsion and suspension (paste); GM-SB plus STS and GM-SBF plus STS – dispersions with aqueous solutions of inorganic acid salts.				

In order to implement the industrial waste recycling process in aerated concrete production, we developed an environmentally sustainable mechanism that includes the following stages: environmental assessment of industrial waste and two-step recycling method. According to the environmental assessment, the test waste products are coded in the Federal Classificatory Catalog of Wastes (FCCW) and assigned to hazard class IV. (Table 2).

Table 2. Environmental assessment of industrial wastes.

FCCW code	Industrial waste	Hazard class
31224111395	Phosphogypsum	4
30121102395	Distillers grains from a federal state unitary enterprise	4
30115000000	Whey	4
94140152404	Sodium thiosulfate	4

The effective specific activity (A_{eff}) of natural radionuclides in the proposed phosphogypsum was found to be 110 Bq/kg. It is less than the limit value of 370 Bq/kg required by GOST 30108-94. In terms of radioactivity, the material is assigned to class I. It is non-toxic, fire- and explosion-proof.

To combine the selected components, we developed methods for preparing the proposed organic mineral modifiers. These methods are conventionally called the first step of the environmentally sustainable mechanism. The technology behind the proposed methods is based on the following processes: preparation of oil-in-water emulsion GM-SB; preparation of emulsion and suspension (paste) GM-SBF; combination of dispersions with aqueous solutions of inorganic acid salts GM-SB plus STS and GM-CBF plus STS.

The second step of the environmentally sustainable mechanism is to produce non-autoclaved aerated concrete using the developed complex organic mineral modifiers. To accelerate curing of non-autoclaved aerated concrete, we used a heat-insulating hood. This technology was found to reduce energy consumption by 25% compared to conventional autoclave treatment, while allowing aerated concrete to reach the required strength within 5–6 hours. Activation with a rotary pulser adds up to 20% to the content of "ordinary" phosphogypsum in the modified aerated concrete in the presence of the organic mineral modifier GM-CBF plus STS. This treatment increases the recycled volume of the test waste products.

Experiment

Experimental studies were based on concrete samples of reference and test compositions (Table 3). The tests were carried out by a laboratory of the Department of Building Materials, National Research Moscow State University of Civil Engineering. The strength of aerated concrete was tested according to the methods specified in GOST 10180-2012. Concretes. Methods for Determining the Strength of Reference Samples [16], using a Controls MCC 8 laboratory-scale plant.

Table 3. Compositions of non-autoclaved aerated concretes (Grade D700).

Type of composition	W/S ratio	Consumption of materials per m^3 of aerated concrete, kg					
		Binder	Sand	Aluminum powder	Caustic soda	Phosphogypsum	GM-SBF plus STS
Reference sample	0.65	300	320	0.3	0.55	-	-
Modified sample	0.55	240	300	0.3	-	55	5

Using a FEI Quanta 200 SEM scanning electron microscope, a microstructural analysis of the test samples was performed. In order to measure the specific activity of natural radionuclides in phosphogypsum, a MKSP-01 Radek portable spectrometer was used directly on the disposal site. The thermal transmittance coefficient of the aerated concrete was determined with a MG4 250 thermal transmittance meter. This meter is designed to determine thermal transmittance and thermal resistance of materials at mean temperature of a sample from + 15° to + 42.5° C.

The validity of the results was confirmed by conventional and standard research methods, as well as by statistical processing of the obtained data that provided sufficient accuracy at a confidence level of at least 95%.

Results

The analyzed effects of the organic mineral modifier on the properties of the concrete mixture and the concrete show that the proposed environmentally sustainable mechanism reduces the water-to-solid (W/S) ratio by 17% compared to the reference composition and increases the strength of the modified aerated concrete by 38% (Table 4). The increased strength of the test concrete can be explained by the fact that the matrix of interpore partitions is hardened by creating a modified structure. In terms of density, the proposed aerated concrete can be classified as a building thermal insulation material. The microstructural analysis of the modified aerated concrete samples (Fig. 1) confirms these findings.

Table 4. Properties of the modified aerated concrete.

Type of concrete	W/S ratio	Density grade	Ultimate tensile strength, MPa	Concrete strength class
Reference sample	0.70	D700	3.4	B2.5
Modified sample	0.58	D700	4.7	B3.5

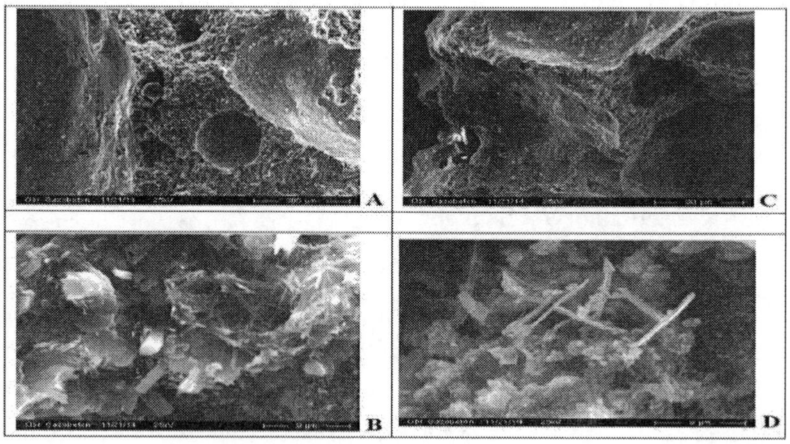

Fig. 1. Structure of the non-autoclaved aerated concrete: A, B – micro and macrostructure of the reference concrete sample; C, D – micro- and macrostructure of the modified concrete sample.

The results presented in Table 5 demonstrate that the organic mineral modifier also has a certain effect on the generation of the fine aerated concrete structure with a pore diameter of no more than 800 μm, i.e. the "crushing" of the interpore space is almost twice as fine as that in the reference composition. The quality at which a new hydrophobic structure is generated significantly contributes to the reduction of gel and capillary porosity by 1.5...2 times. This is due to the improved physical and operational properties of the modified non-autoclaved aerated concrete.

The further studies demonstrate that the proposed modifiers have a clear positive effect on the frost resistance of the produced aerated concrete (Table 6).

According to the test results, the frost resistance grade of the modified aerated concrete is two steps higher than the conventional one: 75 versus 35 cycles. The softening coefficient of the modified aerated concrete was found to be 22% higher than the reference sample. The ice formed in the modified concrete acquires a porphyritic structure that changes the ice formation pattern, reduces the strength of adhesion with cement membranes and increases its fluidity in the interpore space. As a result, when it crystallizes, the pressure in concrete pores decreases considerably. The results provide evidence that the aerated concrete destruction process is inhibited by the hydrophobic effect [16-18], which significantly reduces the penetration of water into the aerated concrete.

Table 5. Porosity of the modified aerated concrete.

Types of pores	Pore size, cm	Percentage of total porosity, %
Reference sample		
Gel	$1{\cdot}10^{-6} - 5{\cdot}10^{-6}$	15 – 20
Capillary	$8{\cdot}10^{-5} - 2{\cdot}10^{-4}$	20 – 30
Cellular	10^{-4} – (0.12 – 0.15)	60 – 75
Modified aerated concrete		
Gel	$1{\cdot}10^{-6} - 5{\cdot}10^{-6}$	6 – 10
Capillary	$8{\cdot}10^{-5} - 2{\cdot}10^{-4}$	10 – 20
Cellular	10^{-4} –(0.05 – 0.08)	75 – 85

Table 6. Frost resistance and water resistance of aerated concretes.

Type of concrete	Ultimate tensile strength, MPa,		Softening coefficient	Frost resistance grade
	dry	water saturated		
Reference sample	3.0	2.1	0.7	F35
Modified	4.7	4.0	0.85	F75

The organic mineral modifiers added to the composition improve the thermal transmittance of aerated concrete by improving the quality of its porosity and reducing the moisture transfer process. The graph presented in Figure 2 shows that the thermal transmittance coefficient of the modified aerated concrete is 0.143 W/(m·°C), which is 23% less than that of the conventional aerated concrete (0.177 W/m·°C) at equal mean density (700 kg/m^3).

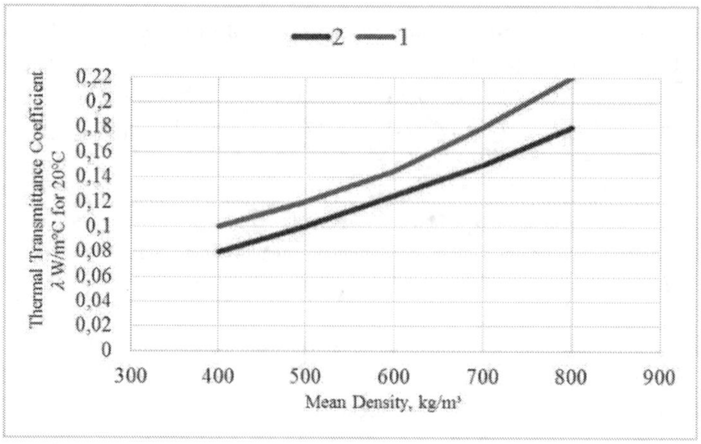

Fig. 2. Thermal transmittance coefficient: 1 – conventional aerated concrete; 2 – modified aerated concrete.

Thus, based on the experimental studies, it can be generally concluded that the proposed environmentally sustainable mechanism is appropriate not only for recycling industrial wastes from a variety of industries, but also for producing high-performance aerated concrete products with improved physical and operational properties. The produced concretes have improved quality indicators, including strength (by 38%), frost resistance (by two steps, up to F75) and softening coefficient (by 22%). This is accompanied with 23% decrease in the thermal transmittance coefficient, which reduces heat transfer through walls during the heating season as compared to the conventional aerated concrete and mitigates the greenhouse effect.

Conclusion

The results of the studies show that their purpose was achieved. The compositions and the production technology for modified non-autoclaved aerated concrete based on recycled industrial waste and ensuring the required level of environmental safety were developed. The application field of the results does not conflict with the applicable laws that regulate the environmental assessment for planned industrial or civil construction areas, and recycling of industrial waste in the production of aerated concrete with improved construction and operational properties.

References

1. V.I. Telichenko, D.V. Oreshkin, Ecology of urban areas **2**, 31-33 (2015) (in Russian)
2. M.V. Grafkina, A.D. Potapov, Vestnik MGSU **1**, 23-28 (2008) (in Russian)
3. V.I. Telichenko, M.Ju. Slesarev, Ecology of urban areas **4,** 6-9 (2014) (in Russian)
4. A.D. Potapov, E.V.Shubina, P.M. Zhuk, Vestnik MGSU **4**, 128-134 (2010) (in Russian)
5. L.A Suleymanova, V.S Lesovik, K.A. Kara, M.V. Malyukova, K.A. Suleymanov, Research Journal of Applied Sciences **9** (12), 1087-1090 (2014)
6. D.V. Oreshkin, A.N. Chebotaev, V.A. Perfilov, Procedia Engineering **111**, 607-611 (2015)
7. R. Duval, E.H. Kadri, Cement and Concrete **4,** 533-547 (1998)
8. C. Shi, J. Qian, Resource Conserve Recycle **29**, 195-207 (2000)

9. A. Grishina, E. Korolev, Proceedings of the XIII International Conference of Students and Young Scientists, 020003 (2015)
10. A. Grishina, E. Korolev, Proceedings of the XIII International Conference of Students and Young Scientists, 020016 (2016)
11. GOST 31108-2016. Common cements. Specifications (2016)
12. GOST 8736-2014. Sand for construction works. Specifications (2015)
13. GOST 5494-95. Powder aluminum. Specifications (1997)
14. GOST 12.1.005-88. General hygiene requirements for working zone air (1989)
15. E.V. Tkach, V.C. Semenov, S.A. Tkach, T.A. Rozovskaya, Procedia 24th Cep. XXIV R-S-P seminar – Theoretical Foundation of Civil Engineering (TFoCE), 763-769 (2015)
16. GOST 10180-2012. The concretes. Methods for determining the strength of reference samples (2013)
17. A.D. Zhukov, A.V. Chugunkov, A.O. Himich, A.A. Eremenko, N.A. Kopylov, Vestnik MGSU **7,** 167–175 (2013) (in Russian)
18. Yu.M. Bazhenov, E.M. Chernyshov, D.N. Korotkih, Stroitel`nye Materialy [Building materials] **3,** 6-14 (2014) (in Russian)

Hardened cement paste microstructure as the main factor of concrete durability

Kirill Shuldyakov[1,*], *Alena* Kirsanova[1], *Lyudmila* Kramar[1], and *Boris* Trofimov[1]

[1]South Ural State University, 454080 Chelyabinsk, Russia

Abstract. When studying durability of concrete the main emphasis is put on its resistance to frost, water and sulphate, which depends on type, dispersion and stability of cement hydration products. The paper focuses on interrelation between strength, open porosity and peculiarities of microstructure of hydrated phases and their crystallization tendency under cyclic and chemical actions. In order to find out these dependences, the following parameters were defined in compliance with standards: open porosity of concrete by water absorption, compressive and tensile strength in bending, frost resistance and sulphate resistance. Methods applied in the research are X-ray diffraction analysis and scanning electron microscopy. The most promising structure of hardened cement paste for obtaining modern paving concrete is the one formed by weakly crystallized gel phases of type C-S-H(I) which are resistant to crystallization and provide stability and durability of concrete. One of the most used methods to obtain such structure is addition of compounds containing activated pozzolanic materials and superplasticizers.

Introduction

Modern construction requires the use of durable and high-performance concretes with set performance properties [1-4]. Such concretes are used in construction of unique buildings, underground structures and are particularly demanded today in road construction. In practice, modern prefabricated and cast-in-place road and airfield pavements are not always durable for various reasons [1, 4-5].

Among the factors affecting their durability are, first of all, the used materials; they shall be of high quality and meet the required standards [3-4, 6-7].

Resistance of cement concretes to the operating environment is primarily dependent on the water-cement ratio, cement hydration degree and phase composition of hardened cement paste [2, 8, 9]. Water-cement ratio for high-performance concretes does not normally exceed 0.35, which is reached by using superplasticizers [3, 10]. However, decrease of water-cement ratio does not completely solve the issue of durability. Durable concrete can be obtained by adding modifying admixtures - highly reactive silica and others [11, 12]. When using pozzolanic admixtures cement hydration is activated, and hardened cement paste is preferably formed from low-basic amorphized calcium silicate hydrates. Such structure has low porosity and is resistant to crystallization, which considerably

* Corresponding author: kirill-shuld@ya.ru

improves strength of hardened cement paste in aggressive environments [3, 8, 13]. Consequently, concrete durability can be provided by managing its content, amount of hydrate formations and porosity.

According to the regulatory documents on road construction, concrete durability is defined by its resistance to water, frost and its capability to resist the action of aggressive environments [2, 14]. Such durability evaluation is carried out on the basis of secondary indications, without the account of structural characteristics of concrete hardened cement paste, which is a relevant issue according to many researchers [7, 15, 16].

This research is aimed at studying the influence of phase composition in hardened cement paste modified by various admixtures on microstructure and durability of concrete.

Materials and Methods

The following materials were used during the study:

- cement of brand CEM I 42..5 N, NG (ЦЕМ I 42,5 Н, НГ) 28% standard consistency corresponding to GOST 31108-2016;
- fine-grained filler - coarse sand, with the fineness modulus = 2.73 according to GOST 8736–2014;
- coarse-grained filler - granodiorite crushed stone, size 5-20 mm, brand with regard to strength - 1400, with regard to abrasion capacity - I1, frost resistance corresponds to the brand F400 according GOST 8267-93;
- superplasticizers: on the basis of polycarboxylate - Glenium ACE 430 (ACE), BASF, produced in Germany; on the basis of naphthalene-formaldehyde - SP-1, produced in Pervouralsk town, Russia;
- granulated silica fume (SF) with pozzolan activity of 1575 mg/g and S_{sp}=1500 m^2/kg, produced in Novokuznetsk city, Russia (TU 5743-048-02495332-96).

Specimens with the size of 10x10x10 cm and the same water-cement ratio = 0.33 were produced according to GOST 10180-2012 to define strength properties of concrete, its porosity and frost resistance [17]. Frost resistance was evaluated in compliance with GOST 10060-2012 by using the quick method No. 3 [18]. Sulphate resistance of hardened cement paste was defined according to GOST P 56687-2015 [19]. Concrete porosity was determined by water absorption according to GOST 12730.3-78 [20].

Phase composition of hardened cement paste in concrete was studied using X-ray diffraction analysis on the apparatus DRON-3M with PDWin attachment, as well as scanning electron microscopy using JSM-6460LA device.

Results

To conduct the research 5 series of samples were prepared - a reference one, without admixtures; those modified by plasticizers ACE and SP-1; and those with compound admixtures ACE+SF and SP-1+ SF.

Preliminary research was conducted on open porosity and strength of concretes with various admixtures. The results of studying water adsorption in concrete samples by weight (W_w, %) are given in Figure 1. According to the obtained data the highest open porosity which is directly related to water adsorption is observed in plain concrete samples. When adding only plastisizers the water absorption decreases from 6% to 4.2% and 3.8% for SP-1 and ACE respectively, which means performance of polycarboxylate plasticizer is higher than that of naphthalene-formaldehyde one. When compound admixture of ACE+ SF is used water adsorption decreases up to 3.2%.

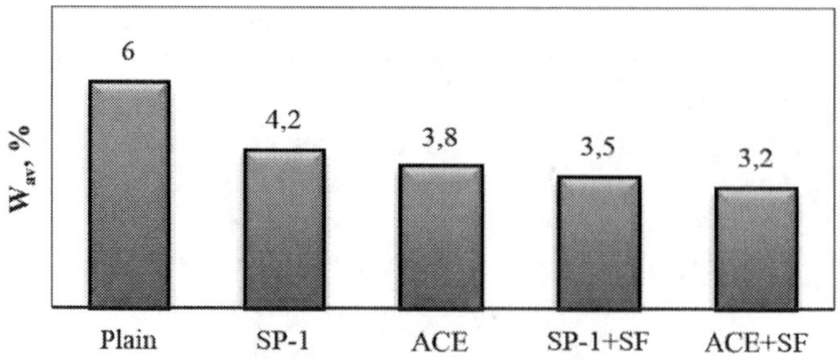

Fig. 1. Influence of admixtures on water adsorption of concrete.

Open porosity is interrelated with strength of concrete, the result of strength parameters variation is presented in Figures 2 and 3.

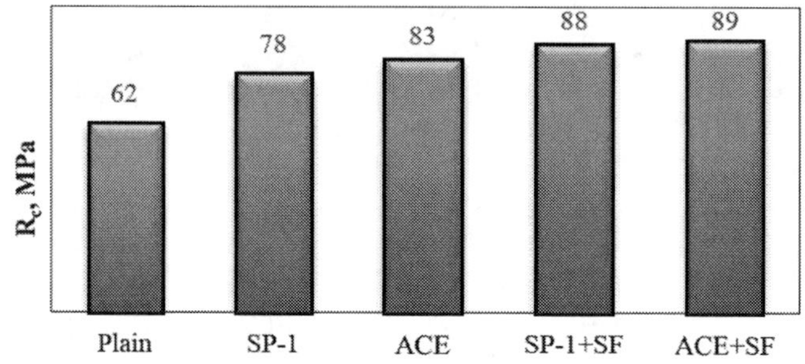

Fig. 2. Influence of admixtures on compressive strength (R_c).

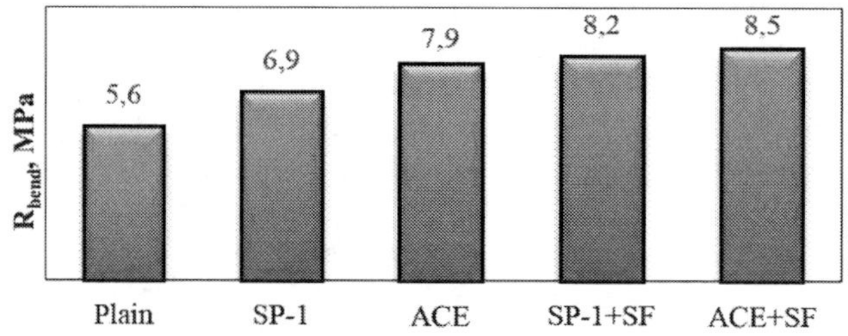

Fig. 3. Variation of tensile strength of concrete with admixtures in bending (R_{bend}).

The results of testing compressive strength and tensile strength in bending confirm the previously developed dependences for water adsorption and demonstrate that the highest values of the studied parameters are found in compositions modified by compound admixtures of SP-1+SF and ACE+SF. When comparing concretes with plasticizer modification only, concrete with ACE admixture demonstrates better mechanical properties, it is particularly visible for strength in bending with adding compound admixtures.

According to X-ray diffraction analysis, phase composition of hardened cement paste without admixtures and with plasticizers is preferably formed from high-basic hydrosilicate phases of type C-S-H(II) and has strong tendency to crystallization. This is confirmed by the data of scanning electron microscopy presented in Figure 4.

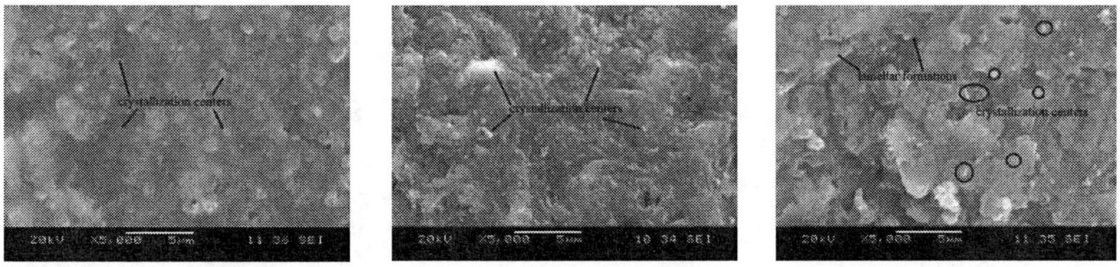

a) b) c)

Fig. 4. Microstructure of hardened cement paste: a) plain; b) with SP-1 admixture; c) with ACE admixture.

Hardened cement paste of plain concrete and concrete with SP-1 admixture has considerable amount of crystallization centers of hydroxide and calcium silicate hydrate all over the surface of the sample. Introduction of ACE admixture results in decrease of crystallization centers in the sample and increase of the number of laminate formations from high-basic phases.

According to the data of -ray diffraction analysis hardened cement paste with compound admixtures that include SF and SP-1 and ACE superplasticisers has higher content of hydrosilicate phases of type C-S-H(I). As can be seen in Figure 5, these phases form the structure from hydrosilicates of laminate type.

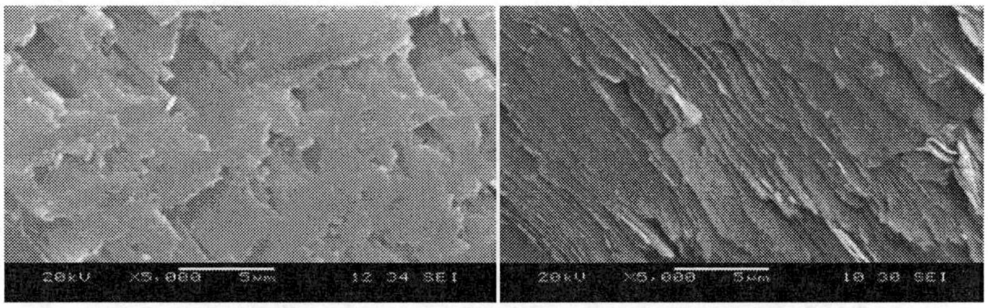

a) SP-1+ SF b) ACE+ SF

Fig. 5. Microstructure of hardened cement paste with compound admixtures.

Plates fit tight against each other, thus forming closed self-protecting structure resistant to moist and crystallization; such structure has improved compressive properties and particularly higher bending properties, which provides high durability of concrete. A part of crystallized calcium hydroxide is located between the plates, it is not subjected to leaching for a long time and causes no environmental pH variation under cyclic actions. Liquid phase stability in hardened cement paste under any external actions leads to slowdown of crystallization processes in calcium silicate hydrate.

When exploiting the roads in local conditions including the effects of low temperatures, anti-icing agents, aggressive ground waters, exhaust fumes and others, investigation of concrete's resistance to frost and sulphate is required. In this connection, durability of concrete was studied in this research with regard to its frost and sulphate resistance, the results are presented below. From Figure 6 it is seen that frost resistance of plain concrete corresponds to the maximum of brand F_2100.

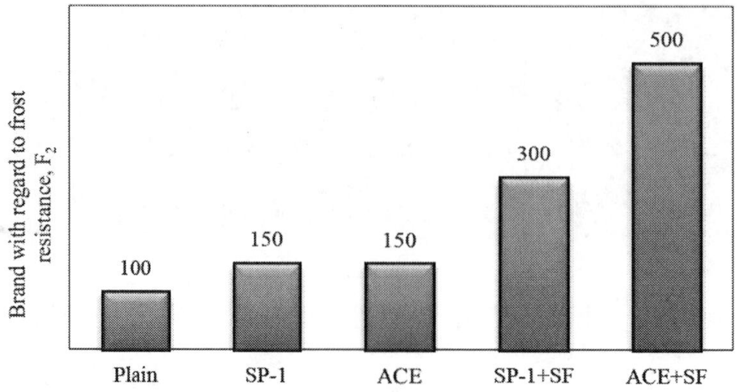

Fig. 6. Influence of admixtures on frost resistance of concrete.

Concretes with plasticizer admixtures sustained a higher, but still quite low number of cycles and thus were related to the brand F_2150. Whereas frost resistance of plain concrete and concrete with plasticizer admixtures has little difference, in concretes with silica fume admixture it increases up to F_2300 for SP-1+SF and up to F_2500 for ACE+SF. To better understand the processes occurring in the structure of concrete hardened cement paste scanning electron microscope investigations were carried out, the results thereof are given in Figure 7.

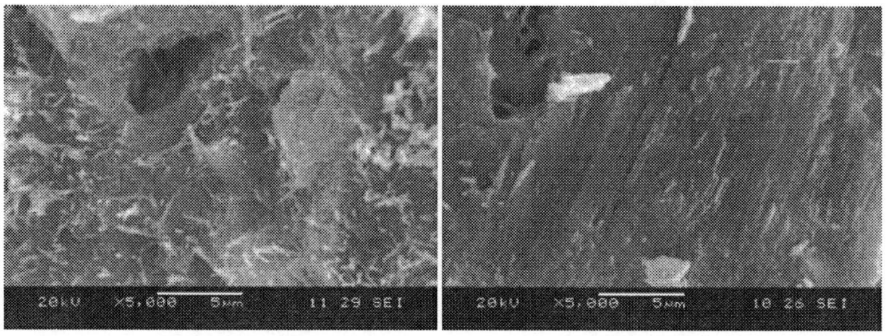

a) SP-1+ SF b) ACE+ SF

Fig. 7. Concrete hardened cement paste with a compound admixture after 55 freeze-thaw cycles.

Based on the obtained data supported by the X-ray diffraction analysis results, one can conclude that the structure of hardened cement paste of concrete with addition of ACE+SF is represented by low-basic amorphized calcium silicate hydrates of laminate structure and can sustain long-term cyclic actions without changing. Hardened cement paste of concrete with SP-1+SF admixtures has higher tendency to crystallization, which has negative effect on its performance properties.

The data of sulphate resistance tests confirmed higher properties of concretes with ACE+SF and SP-1+SF, which is demonstrated in Figure 8.

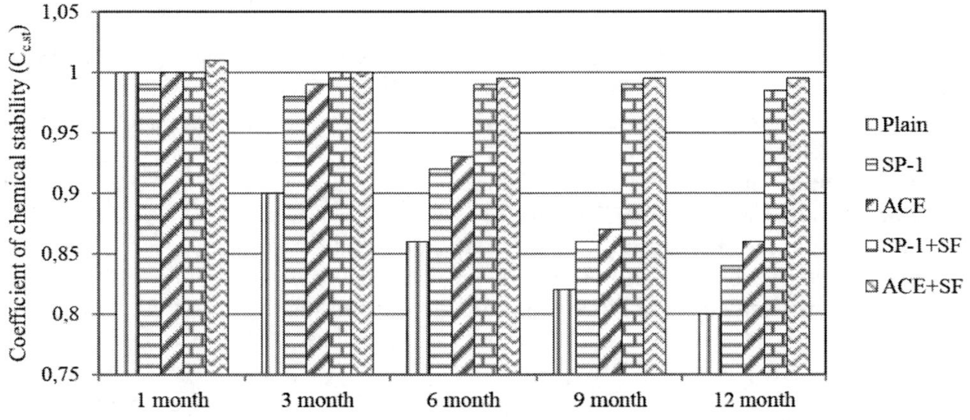

Fig. 8. Results of sulphate resistance tests of the studied concretes.

The obtained results demonstrate that practically all compositions aged 1 month have coefficients of chemical stability close to 1. First variations are observed at the age of 3 months, when strength of the reference composition drops significantly, which causes variation of $C_{c.st.}$ from 1 to 0.9 and the composition loses its resistance to sulphates. At the age of 6 months compositions with plasticizers have the value of $C_{c.st.}$=0.92 and 0.93 for SP-1 and ACE respectively. By the age of 9 months the value of $C_{c.st.}$ for concretes with plazticizers drops below the acceptable value of 0.9; whereas concretes modified by the compound of ACE+SF and SP-1+SF preserve their value of $C_{c.st.}$ close to 1 even at the age of 12 months. In that manner, the use of compounds of ACE+SF and SP-1+SF allows obtaining sulphate resistant concretes.

Conclusion

The most efficient means to increase strength and durability of high-performance concrete is the use of compound admixtures that modify the structure of hardened cement paste for obtaining durable road pavement. Use of only polycarboxylate superplasticizer as admixture causes formation of structure represented by high-basic calcium silicate hydrate which is irresistible to cyclic actions.

Compound admixture Glenium ACE 430+SF provides formation of dense structure from low-basic weakly crystallized hydrosilicate new formations of type C-S-H(I), such structure is resistant to recrystallization when freezing and thawing and has higher bend strength in sulphate attack. Further investigation of the influence of structure and phase composition of hardened cement paste on road concrete's resistance to cyclic mechanical actions are required.

The work was supported by Act 211 Government of the Russian Federation, contract № 02.A03.21.0011.

References

1. V.F. Stepanova, *Dolgovechnost' i zashhita konstrukcij ot korrozii* [Durability and protection of structures against corrosion], 32-37 (Moscow, 1999) (in Russian)

2. Ju.M. Bazhenov, *Tehnologija betona* [Concrete technology], 528 (ACB Publ., Moscow, 2011) (in Russian)
3. P.-C. Aitcin, *High-Performance Concrete*, 591 (E&FN Spon an imprint of Routledge, London and New York, 1998)
4. S. Caijun, W. Zemei, X. Jianfan, W. Dehui, H. Zhengyu, F. Zhi, Construction and Building Materials, **101**, part 1, 741-751 (2015)
5. W. Dehui, S. Caijun, W. Zemei, X. Jianfan, H. Zhengyu, F. Zhi, Construction and Building Materials, **96**, 368-377 (2015)
6. T.C. Powers, T.L. Brownyard, JACI, **77**(4), 264-268 (1980)
7. S.V. Shestopjorov, *Dolgovechnost' betona transportnyh sooruzhenij* [Durability of transport infrastructure facilities from concrete], 500 (Transport, Moscow, 1998) (in Russian)
8. L.Ya. Kramar, B.Ya. Trofimov, K.V. Shuldyakov, *Civil Engineering and Ecology: Theory, Practice, Innovation*, Proceedings of the I International Scientific Conference, 168-175 (Chelyabinsk, March 9, 2015) (in Russian)
9. I.M. Ivanov, D.V. Matveev, A.A. Orlov, L.Ya. Kramar, Bulletin of the South Ural State University, **17**(2), 42-49 (2017) (in Russian)
10. Z. Yanrong, K. Xiangming, Cement and Concrete Research, **69**, 1-9 (2014)
11. L.Ya. Kramar, B.Ya. Trofimov, E.A. Gamaliy, T.N. Chernykh, V.V. Zimich, *Modifikatory cementnyh betonov (Tehnicheskie harakteristiki i mehanizm dejstvija)*, 202 (Iskra-Profi, Chelyabinsk, 2012) (in Russian)
12. V.S. Izotov, *Himicheskie dobavki dlja modifikacii betona* [Chemical additives for concrete modification], 109 (Paleotip, Moscow, 2006) (in Russian)
13. A. Traetteberg, Cemento, **3**, 369-375 (1978)
14. S.W. Tang, Y. Yao, C. Andrade, Z.J. Li, Cement and Concrete Research, **78**, part A, 143-154 (2015)
15. B.Ya. Trofimov, L.Ya. Kramar, Stroitel'nye materialy [Building materials], **8**, 46-51 (2014) (in Russian)
16. H. Huang, C. Qian, F. Zhao, J. Qu, J. Guo, M. Danzinger, Construction and Building Materials, **110**, 293-299 (2016)
17. GOST 10180-2012 Concretes. Methods for strength determination using reference specimens (Moscow, 2012) (in Russian)
18. GOST 10060-2012 Concretes. Methods for determination of frost-resistance (Moscow, 2012) (in Russian)
19. GOST P 56687-2015 Protection of concrete and reinforced-concrete constructions from corrosion. Test method of sulfate resistance of concrete (Standartinform Publ., Moscow, 2015) (in Russian)
20. GOST 12730.3-78 Concretes. Method of determination of water absorption (Moscow, 2007) (in Russian)

MATEC Web of Conferences **143**, 02012 (2018) https://doi.org/10.1051/matecconf/201814302012
YSSIP-2017

Investigation of magnesium oxychloride cement at the initial hardening stage

Galina Averina[1,*], *Tamara* Chernykh[1], *Aleksandr* Orlov[1], and *Lyudmila* Kramar[1]

[1]South Ural State University (National Research University), 454000 Chelyabinsk, Russia

Abstract. The paper investigates the process of variation of magnesium oxychloride cement deformations at the initial hardening stage depending on the activity of magnesium oxide powder which is determined by the parameters of the source material burning. Investigation is focused on magnesium cements obtained from pure magnesium hydroxide. Source materials were burnt at various temperatures with the purpose to obtain magnesium oxide powder with different activity. Regular content of hydrated phases was determined in hardened magnesium cement prepared on the basis of binders with different activity. The study reveals the influence of magnesium oxide powder activity on the process of deformation occurrence in hardened magnesium cement and its tendency to crack formation.

Introduction

This research is aimed at finding the connection between the magnesium cement activity and its deformation behavior in the hardening process and active hydration period.

To date, the production of building materials based on magnesium cement is cut down because of insufficient awareness on the processes of its hydration and hardening, and because of connection of these processes with phase composition and properties of the obtained hardened cement.

It is known that up to date the effect of crystallite size in magnesium oxide (the activity of the cement) on the peculiarities of hydration and structure formation in hardened magnesium cement has not been determined, as well as the influence of the obtained hardened cement structure on its deformation behavior and crack formation tendency. Investigation of these processes will enable to determine the optimum conditions for obtaining magnesium cements in production environment and avoid failure of structures made from such cements because of uncontrolled expansion deformations.

Due to investigating hydration of magnesium cement the peculiarities of interaction between magnesium oxide and water solution of magnesium chloride and magnesium sulphate were determined [1-7].

Magnesium oxide powders mixed with solutions of magnesium chloride salts possess the highest strength properties and are most often used in production of construction products based on magnesium cement.

[*] Corresponding author: averinagf93@gmail.com

Many authors in their previous research showed that hardening of magnesium oxychloride cement results in formation of a hardened cement paste consisting of the mix of $Mg(OH)_2$ and magnesium chloride hydroxides of type $xMgO \times MgCl_2 \times yH_2O$. According to different authors, "x" varies between 1 and 10, and "y" - between 5 and 21 [1-4]. It was found that hardened magnesium chloride paste of natural hardening is usually formed by hydroxide, chloride pentahydroxide and chloride trihydroxide of magnesium, their different quantitative content having certain effect on the material properties [5-8].

Besides, the influence of activator density on the phase composition of hardened magnesium chloride paste was determined. Magnesium chloride pentahydroxide is synthesized with the average concentration of $MgCl_2$ and with the solution density of 1.20 g/cm^3 and higher. Mixtures of magnesium hydroxide and magnesium chloride pentahydroxide can be equilibrium products of magnesium chloride hardening in this environment. With higher concentration of activator the equilibrium phases are mostly represented by magnesium chloride pentahydroxide with insignificant content of hydroxide and chloride trihydroxide of magnesium. With low contents of $MgCl_2$ (lower than 1.80 g/cm^3) it is mostly magnesium hydroxide that is synthesized in the system. Increase in activator concentration leads to higher stability of chloride pentahydroxide phase, whereas magnesium hydroxide is either absent in such systems or is synthesized in small amount [9-12].

Burning of bulk magnesium raw material with non-uniform grain size distribution can result in instability of composition of the cement produced from it because of both overburn and underburn in the cement powder [13]. Underburn is represented by magnesium oxide crystals with high activity, which is characterized by their quick reaction with chlorides. By overburn are meant magnesium oxide crystals enlarged up to the size of 50 nm. Even small amount of overburn containing in the powder causes internal expansion deformations in hardened magnesium paste as a result of their hydration accompanied by increase of volume by 2.5 times at later hardening stages [14].

In case of using the waste of fire protection or mining industries as raw material for cement production, one of the reasons for such deformations in hardened magnesium cement is high content of calcium carbonate in the source rock [15]. According to GOST 1216-87, the calcium oxide admixture which is synthesized while decomposition of dolomite and calcite is considered adverse for magnesium cement and shall be limited up to 5% [16]. Negative effect of calcium oxide on the quality of magnesium cement is confirmed by many scientists, both national and foreign [17-20].

Materials and Methods of Research

Raw material for producing cements was pure magnesium hydroxide $Mg(OH)_2$. Pure magnesium chloride (bishofite MgCl26H2O) in the form of water solution with the density of 1.2 g/m^3 was used as activator and taken in the sufficient amount to provide standard consistency [16]. To find the dependences of deformations occurring in hardened magnesium oxychloride cement on the activity of magnesium oxide, five types of cements were prepared and burnt at the temperature of 400, 450, 700 and 900 °C for 1 hour in the laboratory batch furnace PKL-1,2.

Raw materials and the samples of hardened magnesium cement were investigated with the help of derivatography and other methods [16, 21-23]. Thermal analysis of minerals was performed using Luxx STA 409 derivatograph (Netsch, Germany).

Linear deformations in the material while hardening were studied using shrinkage measuring devices Schwindmessgerät Typ B in accordance with GOST 24544-81 and DIN 52450 [21, 22]. The samples were produced in stay-in-place forms. The moment of mixing the cement with activator was taken as a starting point for measuring deformations.

Soundness of cement while hardening was measured using the technique described in TU 5744-001-60779432-2009 [23].

The loss of activator while hardening was defined in percentage of the difference in masses of mixed cement paste and hardened magnesium cement.

Results

Preliminary soundness tests were performed using sample cement formed into pats. The samples produced with the use of magnesium oxide powders and burnt at the temperature of 400 and 450 °C developed cobweb-like through cracks within the first hour of hardening. The samples obtained on the basis of magnesium oxide and burnt at 700 and 900 °C preserved their integrity after all tests.

The final setting of the samples from magnesium oxide powder burnt at 400 and 450 °C was 20 and 27 minutes respectively; for samples burnt at 700 and 900 °C – 65 minutes.

The results of measuring activator loss while hardening are given in Table 1.

Table 1. Weight loss of hardened samples at the age of 1 day.

Burning temperature, °C	Weight loss, %
400	4.5
450	6
700	8
900	9

Deformations were measured every minute within the first hour after mixing, then measurements were taken once in two hours. The period of continuous measurements was 14 days. The graphs of deformation varied every two hours are presented in Figures 1-4. Measuring unit for deformation was accepted mm/m.

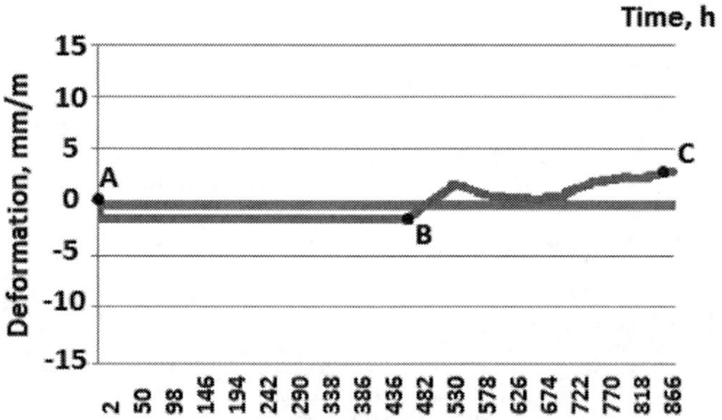

Fig. 1. Graph of deformation variations in samples produced from magnesium oxide powder and burnt at 400 °C.

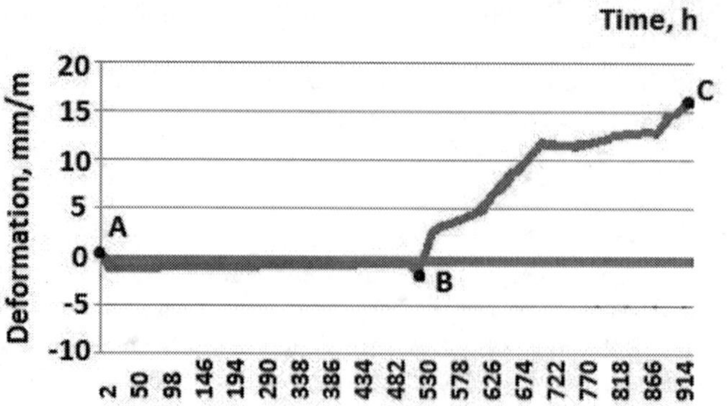

Fig. 2. Graph of deformation variations in samples produced from magnesium oxide powder and burnt at 450 °C.

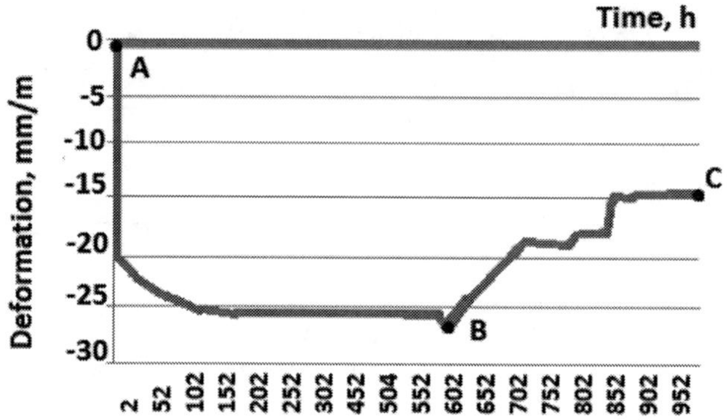

Fig. 3. Graph of deformation variations in samples produced from magnesium oxide powder and burnt at 700 °C.

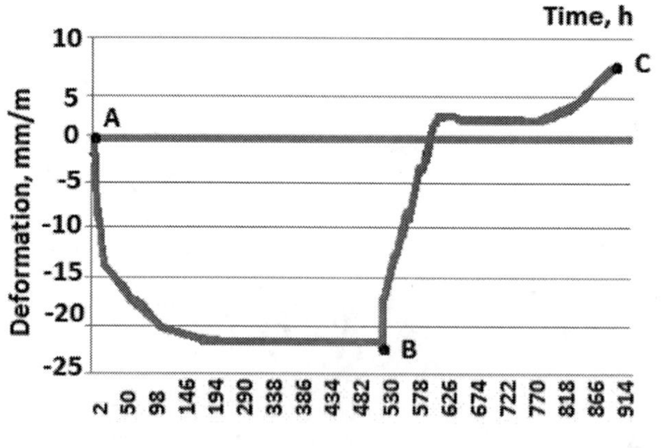

Fig. 4. Graph of deformation variations in samples produced from magnesium oxide powder and burnt at 900 °C.

The graphs demonstrate that deformations in all types of cements have similar behavior, but differ in value and time they originally occur. The highest activity is observed in magnesium oxide obtained at the lowest burning temperature (400 °C), and the magnesium oxide burnt at the highest temperature (900 °C) shows the lowest activity. Based on this fact one can conclude that the value and duration of initial setting (interval A-B) increase as the powder activity reduces. After initial setting, all samples experience two-stage expansion deformation (interval B-C). Duration of this process and the value of deformations also increase as the powder activity reduces.

Thermal analysis was performed to determine the main phases' composition and analyze their quantitative content in the samples. Using the data of derivatograms the summary tables were built with quantitative content of phases in samples measured in the first day (Table 2) and in the 14th day of hardening (Table 3).

Table 2. Phase composition of the samples in the first day of hardening.

Burning temperature, °C	$Mg(OH)_2$, %	$3MgO \times MgCl_2 \times 8H_2O$, %	$5MgO \times MgCl_2 \times 13H_2O$, %
400	26	25.7	41
900	26	12.5	52

Table 3. Phase composition of the samples in the 14t^{h} day of hardening.

Burning temperature, °C	$Mg(OH)_2$, %	$3MgO \times MgCl_2 \times 8H_2O$, %	$5MgO \times MgCl_2 \times 13H_2O$, %
400	25	22	53
450	20	21	59
700	14	18	68
900	13	15.7	70

The following conclusions can be made based on these data. Hardened magnesium cement paste obtained from magnesium oxide and burnt at low temperature has the highest content of magnesium hydroxide at the final hardening stages. With the increase of burning temperature the presence of magnesium hydroxide phase in the hardened magnesium cement paste decreases due to higher quantity of magnesium chloride pentahydroxide phase. Content of magnesium chloride trihydroxide is almost the same for all samples and shows slight decrease with the growth of burning temperature.

Hydration processes in cements produced from magnesium oxide powders with different activity have similar behavior and differ only in their rate. High activity magnesium oxide obtained by low temperature burning is characterized by quick reaction with chlorides, quick setting and hardening due to high growth rate of magnesium chloride hydroxide crystals. As the cement activity reduces with longer crystallization process, higher content of magnesium chloride pentahydroxide is synthesized.

The highest shrinkage deformations are observed in magnesium oxide with lower activity at early age. Table 1 shows that weight loss in samples produced from powders burnt at low temperature is lower than in those produced from powders burnt at high temperature. Thus, almost no deformations related to contraction shrinkage are observed in high activity cements due to their high setting rate.

With the increase of burning temperature, the percentage of expansion deformations in the obtained cement grows, since a higher amount of free magnesium oxide is synthesized and reacts with chlorides. The cement powder burnt at 400 °C is characterized by greater amount of underburnt raw magnesium hydroxide which didn't react with chlorides.

Consequently, the lowest percentage of magnesium chloride hydroxides was synthesized while hardening, which led to the minimum total deformations.

However, in spite of low level of deformation variation, cement powders with excessively high activity are not suitable for building materials production because of their tendency to cracking.

The optimum region for obtaining cements with controlled deformations is within the temperature range between 700 and 900 °C. In hardened magnesium cement paste based on such powders, the processes of magnesium chloride hydroxide crystallization compensate for the initial shrinkage and stabilize within two weeks. The final phase composition of such hardened cement paste is represented by the greatest amount of magnesium chloride pentahydroxide which has the highest strength.

Conclusion

1. Based on the outcomes of the research the connection was determined between the activity of magnesium oxide powder and the behavior of deformations occurring in magnesium oxychloride cement at early age.
2. The results demonstrate the capability of producing cements with required deformations while hardening and describe the method to avoid degradation of products and structures by managing the activity of magnesium oxide powder at the stage of burning source magnesium material.
3. In order to develop mathematical models for interrelation between magnesium oxide powder activity and deformations occurring in magnesium oxychloride cement paste while hardening, it is required to perform a detailed investigation of phase composition of hardened magnesium cement paste at each of the specified hardening stages and to define the size of magnesium oxide crystals typical for every stage of powders activity.

The work was supported by Act 211 Government of the Russian Federation, contract № 02.A03.21.0011.

References

1. H. Biliuski, B. Matcovic, C. Mazuranic, Zumic Amer. Ceram. Soc. **67,** 266-9 (1984)
2. I. Nanazashvili, *Stroitel'nye materialy iz drevesno-cementnoj kompozicii* [Building materials from wood-cement compositions] (Stroiizdat, Moscow, 1990) (in Russian)
3. A. Yanshina, Ogneupory [Refractory Materials] **11,** 505-15 (1960) (in Russian)
4. B. Matcovic, S. Popovic, V. Rogic, Zunic Amer. Ceram. Soc. **60,** 504-7 (1977)
5. V. Chumak, Stroitel'nye materialy [Building Materials] **9,** 10-11 (2003) (in Russian)
6. V. Rogic, B. Matkovic, Cement (Zagreb) **16,** 61-9 (1972)
7. C. Sims, Industrial minerals **2,** 43-8 (1987)
8. X. Zhang, H. Wang, R. Chen, Construction and Building Materials **150**, 409-17 (2017)
9. G. Bergman, I. Vyrodov, RJAC **3,** 119-24 (1958) (in Russian)
10. G. Bergman, I. Vyrodov, RJAC **32,** 504-9 (1959) (in Russian)
11. V. Vorobiev, V. Kolokolnikov, *Proizvodstvo mineral'nyh vjazhushhih* [Manufacture of mineral binders] (Gosstroizdat, Moscow, 1960) (in Russian)
12. I. Vyrodov, RJAC **33**, 2399-404 (1960)
13. M. Mayska, D. Jindrich, Ceramics–Silikaty **41**, 121-23 (1997)
14. G. Averina, T. Chernykh, L. Kramar, Proceedings of the XIII International Conference "Trends in the Development of Science and Education **4**, 5-7 (2016) (in Russian)

15. G. Averina, T. Chernykh, L. Kramar, B. Trofimov, AIP Conference Proceedings **1800**, 020003 (2017)
16. GOST 1216-87. Poroshki magnezitovye kausticheskie. Tehnicheskie uslovija [Magnesite caustic powders. Specifications] (2004) (in Russian)
17. D. Beruto , R. Vecchiattini, M. Giordani, Thermochimica Acta **10**, 25-33 (2003)
18. W. Noll, Angew. Chem. **62**, 567-72 (1950)
19. R. Haul, H. Heystek, Am. Mineralogist **1**, 166-179 (1952)
20. J. Hedvall, Anorg .allg. Chem **4**, 22-25(1953)
21. GOST 24544-81. Betony. Metody opredelenija deformacij usadki i polzuchesti [Concretes. Methods of shrinkage and creep flow determination] (1987) (in Russian)
22. DIN 52450. Testing of inorganic non-metallic building materials; determination of shrinkage and expansion on small specimens (German Institute for Standardization, 1985)
23. TU 5744-001-60779432-2009 Magnezial'noe vjazhushhee stroitel'nogo naznachenija. Tehnicheskie uslovija [Magnesium oxychloride cements for construction purposes. Specifications] (2009) (in Russian)

Mortars for 3D printing

Olga Demyanenko[1,*], *Ekaterina* Sorokina[1], *Natalya* Kopanitsa[1], and *Yurij* Sarkisov[1]

[1]Tomsk State University of Architecture and Building, 634003 Tomsk, Russia

Abstract. The paper is aimed at developing scientifically proven compositions of mortars for 3D printing modified by a peat-based admixture with improved operational characteristics. The paper outlines the results of experimental research on hardened cement paste and concrete mixture with the use of modifying admixture MT-600 (thermally modified peat). It is found that strength of hardened cement paste increases at early age when using finely dispersed admixtures, which is the key factor for formation of construction and technical specifications of concrete for 3D printing technologies. The composition of new formations of hardened cement paste modified by MT-600 admixture were obtained, which enabled to suggest the possibility of their physico-chemical interaction while hardening.

Introduction

Today, the construction industry faces a number of serious problems, such as: low labor efficiency, high statistical rate of emergencies on construction sites, complications of performing control of construction processes. Besides, construction industry is one of the main consumers of non-renewable natural resources [1].

Adaptive technologies opened new opportunities to many dynamically developing branches of industrial production. 3D printing brought technologies to the new level due to their versatility; they are also widely used in construction [2-10]. Compared to traditional ones, 3D technologies enable complete automation of the process, reduce the amount of waste, improve sustainability of construction and architectural expressiveness of structures, as well as provide solidity of structures due to continuous layer-by-layer application. The advantage of such technologies is high speed of construction, for example, Shanghai company WinSun constructed 10 single-floor houses 200 m^2 each using a 3D printer in one day [11, 12].

In order to apply adaptive technology in construction the recent achievements in the field of material science shall be used. For construction of real buildings and structures using 3D printing special fine-grained concretes (mortars) are required, which can provide safe exploitation of buildings, their robustness and durability. These issues today are not sufficiently studied. Traditional concrete based on cement does not meet the requirements to the materials for constructive 3D printing [13].

To optimize the 3D printing process 2 conditions shall be accounted. First, the adhesion between layers decreases with the increase of time interval between deposing layers [14].

* Corresponding author: demyanenko.olga.v@gmail.com

Second, concrete mixture shall develop sufficient initial strength to bear the following layer without deforming. The need in setting of previously printed layers can slow down the construction. Comparison of these two limitations leads to the paradox of 3D printing speed optimization. The time interval between the two set layers shall be, on the one hand, long enough for providing the required strength, and on the other hand, short enough for providing good adhesion between the layers [15].

In this regard, a conclusion can be made that mortar for 3D printing shall have the following characteristics: high strength at early age, fast setting, cohesion, sufficient yielding stress (to bear the load with layer-by-layer construction). In order to provide these characteristics, scientifically proven compositions of concrete mixtures shall be developed with the use of complex modifying admixtures. The research results on the impact of effective modifying admixtures based on local peat materials can be relevant [16].

The purpose of this study is development of scientifically proven compositions of mortars for 3D printing modified by a peat-based admixture with improved operational characteristics.

Selection of admixture

The wide range of application properties of concrete mortars for 3D printing is caused by the need to use a complex of modifying admixtures with different purposes. Production of import-substituting admixtures to building mortars competitive in terms of their properties is relevant for Russia; it can be provided by manufacturing of modifying admixtures based on local raw materials. The resources of Siberian region (Russia) enable producing admixtures based on peat, which also allows managing properties of building mixtures and mortars for 3D printing. Stock of this partially renewable resource in Tomsk region is 29.3 bn tons with regard to 40% humidity. Tomsk region occupies the second position in Russia in terms of peat stock.

Material composition of peat is represented by various organic and mineral compounds. Due to the presence of chemically active organo-mineral functional groups it can be used for various techniques of raw material modification: thermal, chemical, mechanical or a combination of several types for obtaining products for various purposes, including admixtures for building mixtures.

To perform this research a modifying admixture (MT-600) was prepared in a laboratory by thermal anaerobic activation of peat, which corresponds to the technique of low-temperature carbonization. Pyrolysis of the source materials was performed with the use of specially developed apparatus at the temperature of 600°C [17]. The obtained admixture is a hydrophobic finely dispersed consistent powder of a dark-gray color. Physico-chemical studies of this admixture demonstrate that the element composition of thermal modified peat is represented mainly be silicone, aluminum, carbon and calcium, which corresponds to the data on the composition of mineral part of peat.

The results of electron microscopy analysis showed that the MT-600 admixture contains also nano-size elements which are nano-carbons of various shapes, such as fullerenes sized 5 nm and bigger, nanotubes and nanofibers sized up to 200 nm and more with the crosswise size of 20 nm and less [17].

According to the experimental data in [18-23], the nano-dispersed particles are most efficient in compound materials when added in microdoses.

It was found that the average size of a peat admixture particle is (98±5) microns. Specific surface area of the obtained admixture is 0.6 m^2/g. It is known that the most part of modifying admixtures in compound materials based on cement react with binder by controlling their surface properties or densifying the structure of cement systems. To define the reasonable cement-admixture ratio the finely dispersed substance obtained after thermal treatment of peat was mixed with cement in the amount of 0.25; 0.5; 0.75 % of cement weight. Dry components were mixed first, after which the mixture was activated by water. Water-cement ratio was correspondent to the standard concentration and equaled 0.28. Compressive strength of the hardened cement paste was defined in cubic samples 20x20x20 mm in size, at the age of 7, 14 and 28 days. Each series of experiments included at least 5 samples. Coefficient of strength variation is no more than 5.2%.

Figure 1 demonstrates the research results on the impact of the MT-600 admixture on strength of hardened cement paste. It is showed that the reasonable content of modifying admixture shall be 0.5% of cement weight. Adding the admixture in the amount less than 0.5% of cement weight does not provide the required result.

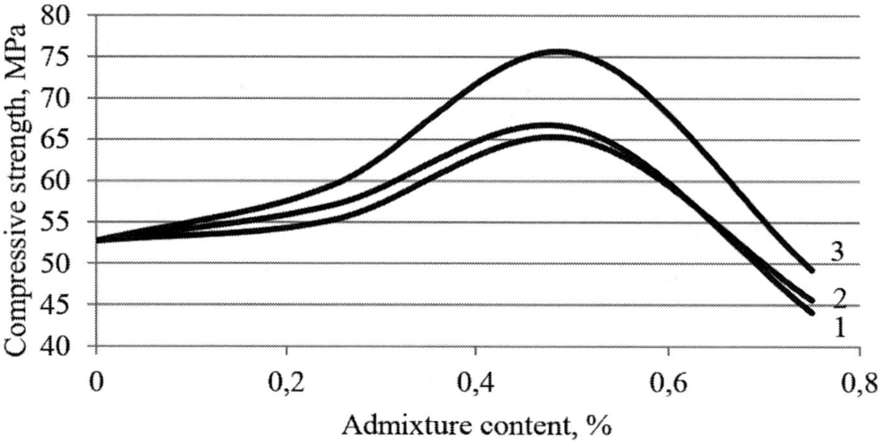

Fig. 1. Impact of MT-600 admixture on compressive strength of hardened cement paste:
1 – hardened cement paste modified by admixture based on peat of Kandinskoe deposit;
2 – hardened cement paste modified by admixture based on peat of Chelbak deposit;
3 – hardened cement paste modified by admixture based on peat of Gusevskoe deposit;

Properties of the end product were found to be greatly dependent on the properties of source materials. Figure 1 demonstrates that the admixture based on peat of Gusevskoe deposit (Tomsk region) provides the best compressive strength properties of the hardened cement paste. It is known that peat obtained from this deposit has higher ash content and the average decomposition grade compared to peats from other deposits. Higher ash content causes high probability of formation of mineral-organic compounds which are the main reason of increase in strength of hardened cement paste. Further investigation was performed with the use of peat of Gusevskoe deposit [17].

Materials and Methods

The following materials were used during the study: Portland cement of grade CEM I 42.5H (ЦЕМ I 42,5H) corresponding to GOST 31108-2016 and construction sand corresponding to GOST 8736-2014. To investigate the properties of cement-sand mortars modified by MT-600 admixture the cubic samples sized 70.7×70.7×70.7 mm were prepared, the sand:cement ratio was 1:2, samples were hardened in air-humid conditions. Strength of samples was evaluated at the age of 3, 7 and 28 days. The following rheological parameters of freshly mixed mortars were investigated: mobility, water retention capacity according to GOST 31356-2007 (Fig. 5).

Results and discussion

The impact of polyfraction admixture MT-600 on the properties and concrete mixtures was previously investigated in [18]. However, there has been no research on monofraction admixtures so far. The admixture obtained as a result of pyrolysis and milling were divided by classification into the following fractions: less than 0.08 mm; 0.08-0125 mm; 0.125-0.18 mm in the 0.5% ratio of cement weight; then, they were mixed with dry cement and activated with water. Compressive strength of the modified samples of hardened cement paste was evaluated at the age of 3, 7 and 28 days. The experimental results are given in Table 1.

Table 1. Strength parameters of samples.

Samples series	MT-600, % wt. of cement	water-cement ratio	Compressive strength, MPa		
			3 days	7 days	28 days
1	-	0.258	40.2	50.8	57.65
2	0.5	0.25	49.6 (+23%)	65.3 (+20%)	71.9 (+24%)
3	0.5	0.258	60 (+49%)	63.5 (+25%)	78.4 (+35%)
4	0.5	0.26	56.7 (+41%)	75.5 (+49%)	76.9 (+33%)
5	0.5	0.268	58.5 (+45%)	76 (+50%)	77.3 (+34%)

Series: 1-reference (Cement+Water); 2-Cement+MT-600 (fraction less than 0.08 mm); 3-Cement+MT-600 (fraction 0.08-0.125 mm); 4-Cement+MT-600 (fraction 0.125-0.180 mm); 5-Cement+MT-600 (polyfraction).

According to the obtained data, in samples of hardened cement paste modified with a polyfraction composition strength increase by 34% is observed on the 28th day of hardening, whereas in hardened cement paste with MT-600 with the fraction of 0.08-0.125 mm strength increased by 35% compared to the reference sample. At early age (3 days), strength of hardened cement paste with MT-600 with the fraction of 0.08-0.125 mm is 49% higher than that of the reference sample. Thus, the use of MT-600 admixture gives cement compositions the ability to develop strength at early age, which is an important feature for 3D printing technologies.

X-ray diffraction analysis was performed to define the composition of new formation in the modified hardened cement paste. Diffractograms were obtained using XRD-6000 diffractometer with CuKα radiation (Fig. 2, 3). As a result of analyzing hardened cement paste composition performed with the use of the databases PCPDFWIN and PDF-4+, as well as POWDERCELL 2.5 program of full-profile analysis, the peculiarities of structure and composition formation in hardened cement paste modified with MT-600 admixture were determined.

X-ray diffraction analysis of the hydration products of the reference sample and the one modified with MT-600 admixture confirms presence of new crystalline phases in the modified hardened cement paste. When introducing admixtures based on thermally treated peat in hydrated cement, one can observe occurrence of additional peaks of stable hydrated calcium silicate (d/n= 7,40; 3,04 Å) and low-basic hydrated calcium silicate of type C-S-H (d/n=11, 4,94; 2,92; 2,19; 2,09 Å); besides, intensity of peaks in crystalline phases $Ca(OH)_2$ (d/n=4,9; 2,63; 1,68; 1,48 Å) decreases, which causes higher strength of hardened cement paste.

The paper provides the result of investigation on physico-mechanical properties of building mortar based on cement modified with MT-600 admixture with various fractions, such as water retention capacity, mobility of mortar mixtures and mortar strength (Fig. 4, 5). The investigation demonstrated that the use of MT-600 admixture reduces water retention capacity of mixtures by 0.2-0.7% compared to the reference compositions. The obtained results shall be accounted in further research.

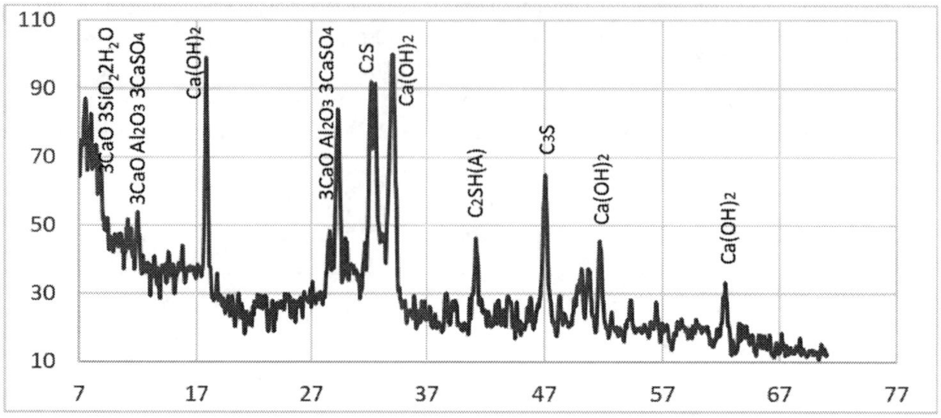

Fig. 2 Diffractogram of hardened cement paste at the age of 28 days: reference sample.

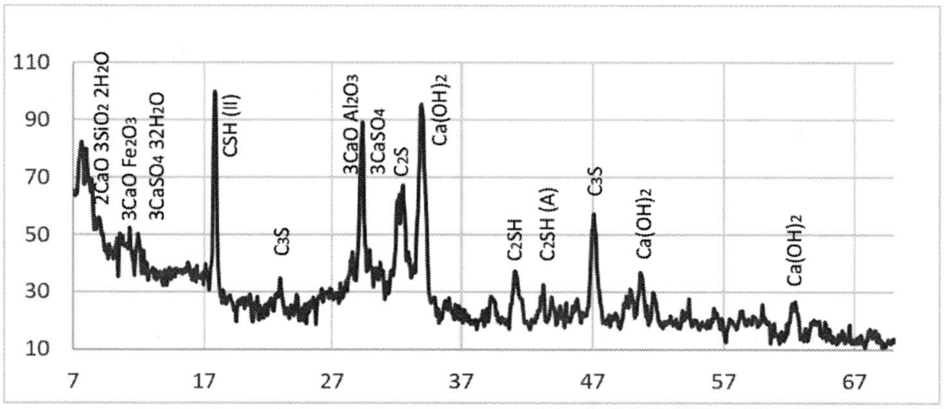

Fig. 3. Diffractogram of hardened cement paste at the age of 28 days: Cement + MT-600.

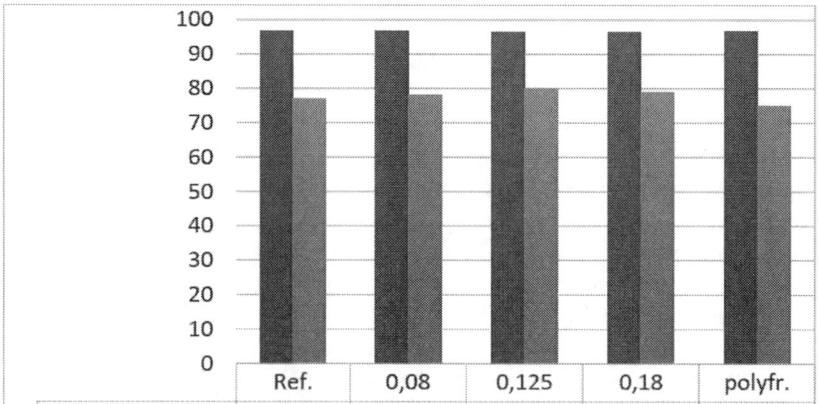

	Ref.	0,08	0,125	0,18	polyfr.
Water retention capacity, %	96,86	96,65	96,56	96,37	96,69
Mobility, mm	77	78	80	79	75

Fig. 4. Water retention capacity and mobility of cement-sand mortars with MT-600 admixture .

Mobility of mixture increases by 1-4% compared to the reference sample. The admixture with hydrophobic properties reduces friction between solid particles and by doing so increases the mixture mobility.

The study of impact of peat-based admixtures on strength of cement-sand mixture revealed that even at early age the samples modified with MT-600 admixture with the fraction of 0.125-0.180 mm showed 20% higher strength compared to the reference samples. By the age of 28 days strength of samples modified with MT-600 admixture with various fractions increased by 18-25% in comparison with the reference samples (Fig.5). Maximum strength was observed in samples modified with MT-600 admixtures with the fraction of 0.125-0.180 mm; the addition of strength is 25%. Coefficient of compressive strength variation is no more than 4.6 %. Samples modified with MT-600 admixture with polyfraction composition and with the fraction of 0.125-0.180 mm showed the maximum strength, which enables to conclude that particles larger than 0.125 mm have strong effect on the structure formation process in hardened cement paste.

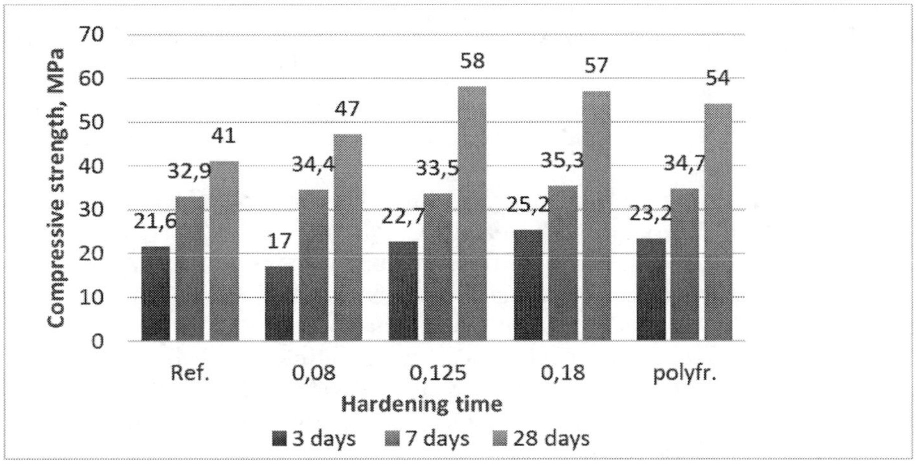

Fig. 5. Strength of concrete with MT-600 admixture at the age of 3, 7 and 28 days.

Strengthening of hardened cement paste when introducing peat-based admixtures can be explained by formation of a denser spatial and organizational hierarchical structure due to directed variation of adsorption and hydration processes, as well as due to kinetics of nucleation. At the same time, development of microcracks in the material slows down, which is connected with the effect of micro-reinforcement and densification of hardened cement paste and stress distribution [24]. Besides, the effectiveness of densifying admixture is caused by formation of additional crystallization centers and high surface energy, which brings about higher hardening rate and higher strength of hardened cement paste [25].

Conclusion

The paper studies strength parameters of hardened cement paste and hardened mortars for 3D printing. It is found that the use of MT-600 admixture provides higher strength of hardened cement paste and mortar both at early age and at the age of 28 days. Moreover, the effect of admixture dispersion on the main technological parameters of a composite material was determined. The study demonstrates that it is reasonable to use the MT-600 admixture as a component for obtaining polyfractional admixtures to building mixtures for 3 printing. The obtained results satisfy the initial purpose of the research which is development of building mortars for 3D printing technologies.

Further research on the impact of peat-based admixtures on the properties of building mortars for 3D printing are relevant, since the use of such admixtures allows reducing costs by decreasing the amount of imported foreign modifiers in the mortars composition. The use of peat as a basis for obtaining modifying admixtures improves the environmental situation of peat producing regions.

References

1. M.I. Ibrahim, Sustainable Cities and Society, **23**, 78-93 (2016)
2. L. Chen, *The International Journal of Advanced Manufacturing Technology,* **89**, 3651–3660 (2017)
3. A.V. Valter, Mining Informational and Analytical Bulletin, **12**, (2011), (in Russian)
4. E.N. Kablov, Aviation Materials and Technologies, **5**, 7-17 (2012), (in Russian)
5. J.A. Gálvez, 3D Printing and Additive Manufacturing, **3**, 160-165 (2016)
6. A. Barazanchi, Journal of Prosthodontics, **26**, 156-163 (2016)
7. L.J. Kumar, *Advances in 3D Printing & Additive Manufacturing Technologies,* 39-54 (Springer, 2016)
8. M.A. Kreiger, B.A. MacAllister, J.M. Wilhoit, M.P. Case, *The Proceedings of the 2015 Conference on Autonomous and Robotic Construction of* Infrastructure, 149-158, (2015)
9. S.H. Huang, P. Liu, A. Mokasdar, L. Hou, The International Journal of Advanced Manufacturing Technology, **67**, 1191-1203 (2013)
10. S.A. Udodov, F.A. Belov, A.E. Zolotukhina, *Sbornik statej pobeditelej VI Mezhdunarodnoj nauchno-prakticheskoj konferencii* [Selected papers of the VI International Scientific Conference] (2017) (in Russian)
11. S.V. Litovkin, Yu.R. Pet'kova, Proceedings of the International Scientific Conference «Relevant issues of modern mechanical engineering», 433-435 (2015) (in Russian)
12. V.P. Grakhov, S.A. Mokhnachev, O.V. Borozdov, Fundamental research, **11-12**, 2673-2676 (2014) (in Russian)
13. I. Hager, A. Golonka, R. Putanowicz, Procedia Engineering, **151**, 292-299 (2016).
14. E. Lloret, A.R. Shahab, M. Linus, R.J. Flatt, F. Gramazio, M. Kohler, S. Langenberg, Computer-Aided Design, **60**, 40-49 (2015)
15. T.T. Le, Cement and Concrete Research, **3,** 558-566 (2012)
16. O.V. Demyanenko, Vestnik of TSUAB, **4 (63),** 101-106 (2017) (in Russian)
17. A.V. Gorshkova, Suhie stroitel'nye smesi s modificirujushhej dobavkoj na osnove torfa [Dry construction mixes with a modifiing additive on the basis of peat], Candidate thesis, Tomsk State University of Architecture and Building, Tomsk (2015) (in Russian)
18. Yu.S. Sarkisov, Vestnik of TSUAB, **4**, 226 – 234 (2012) (in Russian)
19. L.A. Urkhanova, Stroitel'nye materialy [Builind Materials], **8**, 52-55 (2014) (in Russian)
20. E.V. Korolev, Stroitel'nye materialy – Nauka [Builind Materials – Science], **8**, 2-4 (2006) (in Russian)
21. V.G. Hozin, Stroitel'nye materialy [Builind Materials], **2**, 25-33 (2015) (in Russian)
22. E.E. Ezupenok, Soderzhanie himicheskih jelementov v torfah i torfjanyh pochvah juzhno-taezhnoj podzony Zapadnoj Sibiri [The content of chemical elements in peat and peat soils of the southern taiga subzone of Western Siberia], Candidate thesis, Tomsk State Pedagogical University, Tomsk 2005, (in Russian)
23. A.A. Guvalov, Stroitel'nye materialy [Builind Materials], **9,** 94-95 (2013) (in Russian)

24. G.I Berdov, Mezhdunarodnyj zhurnal prikladnyh i fundamental'nyh issledovanij [International Journal of Applied and Fundamental Research], **9,** 87-91 (2010) (in Russian)
25. P.N. Kurochka, Engineering journal of Don, **1** (24) (2013) (in Russian)

Building ceramics based on composition of low-quality clay and slurry

Viktoriya Guryeva[1,*], *Alexander* Doroshin[1], and *Yuliya* Andreeva[1]

[1]Orenburg State University, 460018 Orenburg, Russia

Abstract. At present, there is a trend towards increasing demand for ceramic products with enhanced physical and mechanical properties. This study is aimed at developing scientifically grounded compositions of ceramic products that maximize the use of local low-quality aluminosilicate raw materials and secondary raw materials in brick-making technology. Characteristics of local low-melting clays and industrial waste in the form of drilling slurry with minimum percentage of petroleum product were studied and presented in this paper. Standard methods for determining physical and mechanical properties of bricks, physicochemical and statistical methods of research are used as a methodological framework. The paper shows effects of drilling slurry with an increased content of calcium oxide on the composition, physical and mechanical properties of wall ceramics, as well as effects of composition of low-quality local clays and slurries on the structure and properties of ceramic materials. The modifying role of container glass and its effects on properties of a synthesized ceramic piece were studied and presented.

Introduction

Priority objectives for the building materials industry, including manufacturers of building ceramics, are to improve energy efficiency of production and to reduce the ecological footprint of industrial products. Currently, clays are one of the main components of ceramic products around the world. However, despite the presence of basic minerals in the composition of aluminosilicate raw materials, each clay deposit has its own features. These differences arise from conditions of deposit formation in different geological periods, mineralogical and chemical compositions, thickness of accumulations of useful raw materials and overburdens [1]. Most of local argillaceous rocks around the world are low-melting. Rock forming minerals are represented by mixed layered hydromica-montmorillonite formations. Disadvantages of these clays are a narrow sintering range and low fire resistance. A ceramic piece is characterized by high water absorption capacity and loose structure due to high content of carbonate inclusions, silt and sand.

Use of industrial waste in production seems to be a promising solution for improving the quality of building ceramics based on low-melting clays [2]. In developed countries such as the United States and Germany, the percentage of recycled waste has exceeded 30%

*Corresponding author: victoria-gurieva@mail.ru

[3]. According to a literature review, ceramic products with a spatially organized structure are found to have been produced on the basis of industrial waste [4]. A promising solution is to produce composite ceramic materials on the basis of their modeled structure by optimizing material and fractional compositions of a batch and industrial filler, as well as by achieving the maximum packing density in a semi-finished product [4]. Russian and foreign scientists, such as A. Yu. Stolboushkin, V. N. Makarova, R. Socolar, P. Raut, et al. [2–8] show that various kinds of industrial products can be used as raw materials in production of high-quality wall ceramics.

However, a diverse range of mineralogical and chemical compositions of industrial products, poorly studied raw materials and batch compositions during heat treatment limits their use in the production of building materials. The literature review shows that the use of drilling slurries as raw materials in ceramic products has not been studied yet. Therefore, the authors carried out this study to develop building ceramics products based on scientifically grounded compositions in low-temperature firing conditions with the aim of improving the quality of ceramic products, maximizing the use of locally extracted raw materials, and introducing secondary raw materials in industrial production. The object of the study is a ceramic brick based on low-quality clay and drilling slurry with high calcium content. The scope of the study is phase and structure formation processes, as well as properties of a synthesized composite ceramic material.

Materials and Methods

The initial aluminosilicate raw materials are represented by low-melting clays from the Buzuluk deposit in the northwest of the Orenburg Region and the Alimsay deposit located on the southeastern margin of Orenburg (capital city of the Orenburg Region), Russia. A previously unexplored drilling slurry of Orenburgneft PJSC was chosen as an additive due to the proximity and availability of this source [7].

The methodological framework of the study includes standard methods for determining chemical, particle-size and mineralogical compositions, physicochemical methods for determining the phase composition of samples (X-ray diffraction analysis, differential thermal analysis), optical microscopy methods, methods for determining physical and mechanical properties of ceramics, and statistical methods for planning and processing experimental data.

Raw materials, molding compounds with moisture content of 8–10% and samples were prepared according to the standard procedure adopted in ceramic production. The samples were molded in the form of cylinders with a diameter of 50 mm. They were dried at 100 °C for 12 hours to constant weight. Then they were fired in the range of 900 to 1,100 °C at a heating rate of 5 °C/min and held at the maximum temperature for 60 minutes.

Results

At the first stage of the study, chemical and particle-size compositions of the test clays were determined (Tables 1, 2).

Table 1. Chemical composition of test clays.

Oxide content, % on a dry matter basis											
Deposit	SiO_2	TiO_2	Al_2O_3	Fe_2O_3	MnO	CaO	MgO	Na_2O	K_2O	P_2O_5	loss on ignition
Buzuluk	41.71	0.23	3.92	2.10	0.01	25.6	0.37	0.18	0.94	0.12	23.05
Alimsay	56.50	0.67	12.75	4.78	0.01	9.27	2.75	1.1	1.84	0.6	10.47

Table 2. Particle-size compositions.

Clay deposit	Particle size, mm / Size frequency, %,				
	1.0-0.063	0.063-0.01	0.01-0.005	0.005-0.001	<0.001
Buzuluk	14	10	1	6	42
Alimsay	4	32	12	14	38

The mineralogical composition of the clays was determined by petrographic and X-ray diffraction analyses. The data are given in Table 3.

Table 3. Mineralogical composition of the clay from the Buzuluk deposit.

Quartz	Feldspars		Mica and hydrous mica		Carbonate minerals	
	Albite	Microcline	Muscovite	Chlorite	Calcite	Dolomite
32.8	17.5	5.3	8.1	7.5	25.4	4.0

The obtained data indicate that both clays are sandy, which agrees with the data in Table 2. According to the X-ray diffraction analysis, particles larger than 10 μm are represented by primary minerals such as quartz, feldspar and mica. Particles less than 5 μm are composed of argillaceous materials, mostly of hydromica-montmorillonite formations. The clay from the Buzuluk deposit is characterized by a high content of carbonate compounds represented by limestone and dolomite, with a particle size of 5–10 μm and more. In terms of clay particles (less than 5 μm) and dust / sand particles (more than 5 μm), the tested aluminosilicate raw material belongs to the group of clays. The low alumina content, especially in the clay from the Buzuluk deposit, indicates low fire resistance of clays and sintering temperature.

In order to determine properties of the aluminosilicate raw materials, we tested laboratory samples using standard techniques subject to the requirements of [8]. At the first stage, the main pre-firing properties of clays (Table 4) were studied. These properties determine the choice of the most appropriate method for molding products and preparing molding compounds.

Table 4. Pre-firing properties of raw materials for ceramic products.

Deposit	Plasticity index (PI, %)	Molding-moisture content %	Air shrinkage, %	Coefficient of drying sensitivity
Buzuluk	24.01	24.01	4.4	0.82
Alimsay	24.4	24.9	10.8	1.93

The obtained results show that the test clays are classified, respectively, as clays of moderate and medium plasticity in terms of the plasticity index; as clays of low and high sensitivity in terms of the coefficient of drying sensitivity. Firing properties of the clays were determined within a range of 900 to 1150 °C. (Fig. 1, 2)

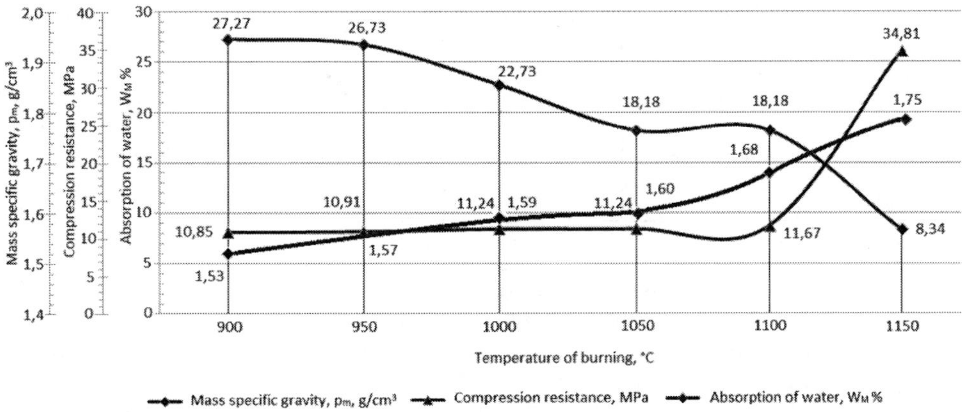

Fig. 1. Firing properties of the clay from the Buzuluk deposit, Orenburg Region.

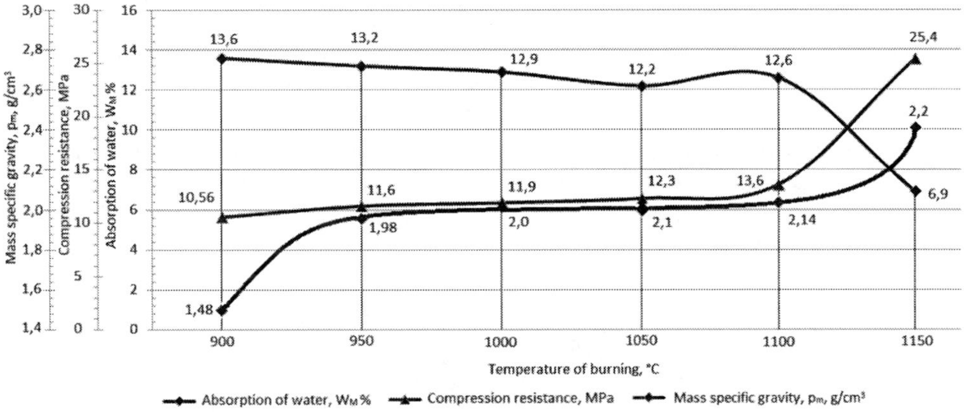

Fig. 2. Firing properties of the clay from the Alimsay deposit, Orenburg Region.

As it can be seen from the graphs, both clays demonstrate a decrease in water absorption to the limit specified in [9] and, at the same time, an increase in strength in a temperature range of 1,100 to 1,150 °C. Since the water absorption of the clays in the tested temperature range exceeds 5%, it can be concluded that the fired clays form a piece which is difficult to sinter. The obtained data suggest that the test low-quality clays can be used in the production of ceramic products in combination with corrective additives that affect physical and mechanical properties of products in accordance with applicable standards.

In further research, the clay of the Buzuluk deposit was used as the basic low-quality aluminosilicate raw material.

The material composition of the disposed slurry was analyzed according to the data of Orenburgneft PJSC. The results show that the content of petroleum products in the slurry ranges from 800 to 9,870 mg/kg. In samples of asphalt, resin, and paraffin deposits taken from pits of oilfields in the Southern Urals, the content of paraffin and ceresin components with melting points of 66–84 °C is 40–70% by weight. The content of the organic part is 72–90% by weight. [11]. In the experimental part, drilling slurry with a minimum percentage of petroleum product was used. The chemical composition is shown in Table 5.

Table 5. Chemical composition of the samples taken from the mineral constituent of slurry.

Name	Chemical composition, % on a dry matter basis								
	SiO_2	Fe_2O_3	CaO	MgO	SO_3	R_2O	Al_2O_3	loss on ignition	Σ
Slurry from fields of Orenburgneft PJSC	23.84	10.8	21.28	2.28	1.81	10.83	3.72	29.24	100.0

The comparison between chemical compositions of drilling slurries from various slurry pits in the Russian Federation shows that they are similar to each other. At the same time, secondary raw materials in all regions have significant fluctuations in chemical and mineralogical compositions, even if they come from the same slurry pit. The test slurry is featured by increased carbonate content. Therefore, theoretical and experimental studies of raw materials are required to determine the possibility of using it in ceramics engineering.

In order to use drilling slurry as a component of ceramic compound, its sinterability should be taken into account as an important parameter. For this reason, sinterability of the slurry was assessed at the first stage on the basis of firing properties. The average density of the slurry increases with temperature and, at t = 1,150 °C, reach 1.22 g/cm^3 (Fig. 3). Water absorption in the firing range of 900 °C to 1,150 °C decreases by 57.7% and reaches 19.2% (Fig. 4).

Shrinkage was not determined due to lack of plasticity. Since the fire resistance of slurry particles exceeds 1,200 °C, the experimental temperature range did not allow us to reach the required sinterability of the material. The fired slurry samples were characterized by low compressive strength.

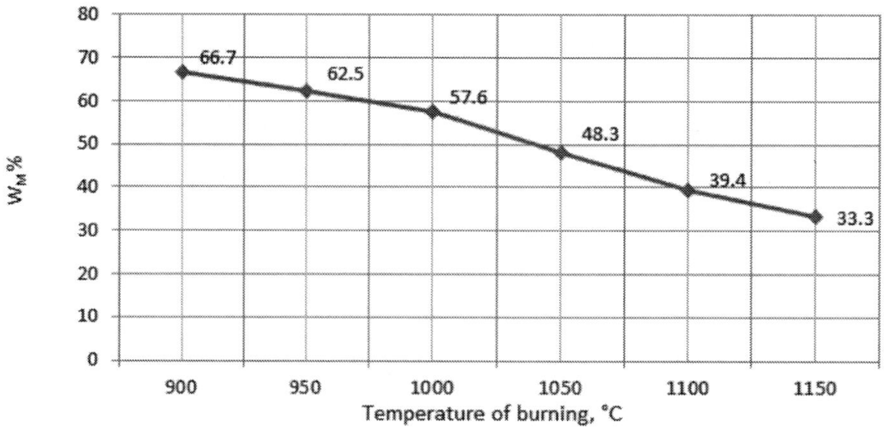

Fig. 3. Average density of the slurry.

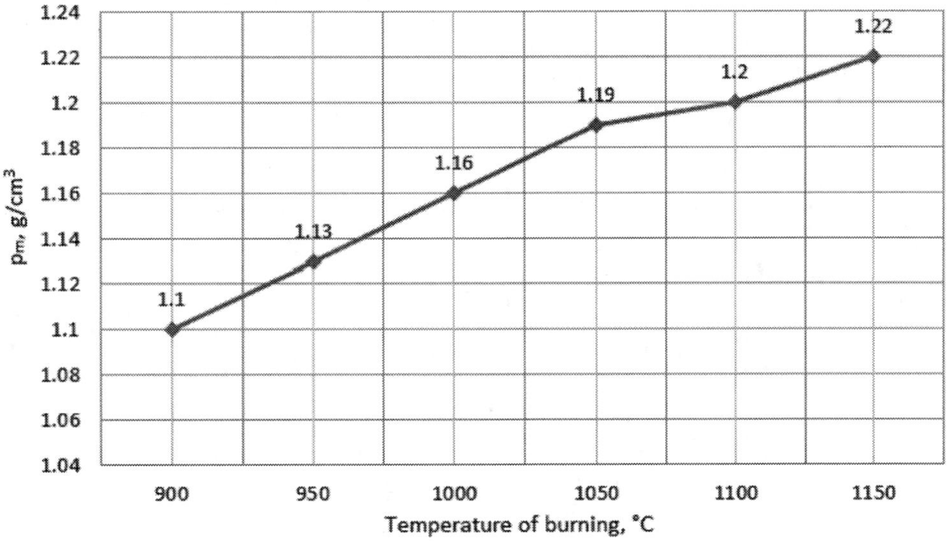

Fig. 4. Water absorption of the slurry.

Due to the fact that the previous experiments indicate low sinterability of the drilling slurry taken in bulk from a landfill, it is necessary to increase the temperature of burning as in the case of low-melting raw materials — within 1,050–1,200 °C, in order to achieve satisfactory mechanical results [8]. However, this technological solution will inevitably lead to higher energy costs and higher cost of production. To solve this problem, and increase the density and strength of the synthesized artificial composite stone, we decided to add an additional additive, sintering aid (flux), to a batch [10].

The most common type of a flux is cullet, which has a low melting eutectic and initiates the formation of a liquid phase in the ceramic structure. Part of the melt formed during firing is used to fill large pores and coat grains of the material, acting as a binder between ceramic particles [14]. In a cullet production experiment, container glass was used according to GOST 54170-2010. Its chemical composition is presented in Table 6.

Table 6. Chemical position of container glass.

Name	Chemical composition, %								
	SiO_2	Al_2O_3	Fe_2O_3	CaO	MgO	Na_2O	K_2O	loss on ignition	Σ
Container glass	67.40	5.81	1.76	7.21	3.38	12.73	2.0	-	100.02

In order to optimize the batch composition depending on the compound and firing conditions [10], we developed compounds of the three-component system "clay – slurry – cullet" (Table 7). Prior to batching, the container glass was coarsely ground to a particle size of 1.0 mm and then finely ground to a particle size of 0.3 mm in a laboratory ball mill with uralite balls. The experimental data are presented in Table 7.

Table 7. The matrix and results of the three-factor experiment.

No.	Temperature of burning °C	Batch composition, %			Compressive strength R_c, MPa	Water absorption W, %	Density P, g/cm^3
		Clay	Slurry	Cullet			
1	900	85	10	5	28.4	18.5	1.54
2	900	70	20	10	26.3	17.8	1.49
3	900	55	30	15	24.6	17.5	1.45
4	1,000	85	10	5	30.1	16.4	1.62
5	1,000	70	20	10	28.9	15.7	1.57
6	1,000	55	30	15	26.5	15.5	1.55
7	1,100	85	10	5	31.3	13.1	1.76
8	1,100	70	20	10	29.9	12.8	1.73
9	1,100	55	30	15	29.2	12.3	1.69

It can be seen from the table that the most optimal compounds are compounds No. 3, 6 and 9 that have the highest content of the industrial product — 30%. The required sintering level of the raw material according to GOST 530-2012 is achieved with 15% cullet content.

By varying the maximum temperature of burning from 900 to 1,100 °C, we produced a ceramic piece with different volumes of the pore structure, average density, g/cm^3, (1.45, 1.55, 1.69) and water absorption, %, (17.5, 15.5, 12.3).

Conclusion

As a result of the research, new knowledge of effects of slurries with high calcium content on physicochemical processes and the structure formation mechanism of a ceramic piece, properties of molding compounds, as well as physical and mechanical properties of ceramic products was gained.

Two-component batch compositions for the production of ceramic bricks based on a composition with low-quality local clay and drilling slurry using the semi-dry pressing method were developed. These raw materials ensure a high compaction degree of the molding compound, a homogeneous structure of the material throughout the production cycle, and making products with a strength grade M 100-150 that meet the requirements of [8].

The feasibility of introducing cullet additive (15% by weight) to the two-component batch and the possibility of producing a wide range of ceramics (porous wall ceramics, floor and roof tiles) on the basis of local low-quality clays in combination with a drilling slurry and flux were proven.

Thus, the obtained results will make it possible to fully adapt the production of ceramic bricks to the local raw material base, reduce the cost of bricks, to react quickly to market needs, and to improve the environmental situation in regions.

References

1. T.V. Vakalova, V.M. Pogrebenkov, Building materials, **5**, 35 (2009) (in Russian)
2. S.P. Raut, R.V. Ralegaonkar, Construction and Building Materials, **5**, 28 (2013) (in Russian)

3. A. Benlalla, M. Elmoussaouiti, M. Dahhou M, Applied Clay Science, (2015)
4. V. Makarov, O. Suvorova, V. Kumarova, Proceedings of the XVI Balkan Mineral Processing Congress (Belgrade, Serbia, June 17–19, 2015)
5. A.Y. Stolboushkin, News of higher educational institutions. Construction (2015) (in Russian)
6. V. A. Mymrin, K. P. Alekseev., Construction and Building Materials, **5**, 42 (2014) (in Russian)
7. V.V. Dubineckij, V.A. Guryeva, Building materials, **5**, 40 (2015) (in Russian)
8. GOST 21216-2014 *Raw clay. Test methods* (2015) (in Russian)
9. GOST 530-2012 *Ceramic brick and stone. General specifications* (2013) (in Russian)
10. V.V. Dubineckij, V.A. Guryeva, K.M. Vdovin, Proceedings of the All-Russian Scientific and Practical Conference OSU,145, (2014) (in Russian)
11. A.L. Ukwatta, A. Mohajerani, Construction and Building Materials (2015)
12. V.A. Mymrin, R.E. Catai, A. Nagall, R.L. dos Santos Izzo, Applied Clay Science (2015)
13. R. Kusiorowski, T. Zaremba, J Construction and Building Materials (2016)

MATEC Web of Conferences **143**, 03001 (2018) https://doi.org/10.1051/matecconf/201814303001
YSSIP-2017

Stress-strain dependence and the evolution of dislocation structure in Cu-Mn polycrystalline concentrated solid solutions

Nina Koneva[1,*], *Yuliya* Sentsova[2], *Lyudmila* Trishkina[1], and *Tatiana* Cherkasova[1,2]

[1]Tomsk State University of Architecture and Building, 634003 Tomsk, Russia
[2]National Research Tomsk Polytechnic University, 634034 Tomsk, Russia

Abstract Deformation of metallic materials results in dislocations occurrence and accumulation, with establishment of their definite distribution (dislocation substructure). The type of dislocation substructure (DSS) defines largely resistance to deformation and fracture of materials. Quantitatively the interrelation of DSSs and resistance to deformation are studied on a limited number of alloys. Foreign literature lacks studies referring to those issues. The given paper investigates the stress-strain dependences in Cu-Mn polycrystalline solid solutions. Mn concentration in the alloys varied within the range of 10…25 at.%. Polycrystals with 10 and 240 μm mean grain size were investigated. Deformation was applied by means of tension at the velocity of $2 \cdot 10^{-2}$ s^{-1} and the temperature of 293 K. Using transmission electron microscope at the accelerating voltage of 125 kV, microstructure of the samples was investigated when they were deformed up to various deformation degrees. The types of DSSs were defined. The connection of the deformation stage under tension with the formed types of DSS was discussed. The sequence of transition of DSSs during the process of alloys deformation was defined. Appearance of the new stage of deformation hardening is attributed to the occurrence of the new type of substructure. The occurring "new" DSS develops during the deformation process, while "the old" DSS gradually disappears. Each of the stage of plastic deformation generally has two types of DSSs. The connection of the deformation stage and the strain hardening coefficient with the DSS was defined and it was shown to possess the definite dislocation density.

Introduction

The processes of plastic deformation are usually characterized by the clearly defined stages of strain hardening [1–4]. The staging of the dependency $\sigma = f(\varepsilon_{true})$ is identified from the mechanical testing. The number of stages of strain hardening and their duration depends on the type of polycrystalline aggregate, testing temperature and the deformation mechanisms.

[1] Corresponding author: koneva@tsuab.ru

The stages vary by the value of strain hardening coefficient $\theta = \frac{d\sigma}{d\varepsilon_{true}}$ (σ – stress, ε_{true} – strain), and by its dependence from ε_{true} [1]. A number of alloys enabled to establish that initiation of each of the stages was conditioned by appearance of the new substructure in local areas with the increased dislocation density and presence of the prior substructure in the material bulk [5]. The critical dislocation density [6] in the grain bulk cannot be reached yet, however it is realized in the grain boundary zone. The new substructure starts to form in that area, and the transition from one stage to another is dissolved at the dependence $\sigma = f(\varepsilon_{true})$. Broadening the number of alloys where the mentioned peculiarities are observed is the issue of the great relevance.

The present research is aimed to define the types of the formed substructures and their connection to the plastic deformation stages considering various grain sizes of Cu-Mn polycrystalline FCC alloys.

Materials and methods

Cu-Mn polycrystalline FCC solid solutions with the mean grain size <d>= 10 and 240 μm were investigated. Mn concentration in the alloys was varied within the range of 10…25 at.%. Flat samples (sized 100×12×2 mm^3) were deformed by tension with the aid of Instron testing machine at the velocity of $2\cdot10^{-2}$ s^{-1} at the temperature of 293 K. Further, the dependences "σ – ε_{true}" were built and the value of the strain hardening coefficient $\Theta = d\sigma/d\varepsilon_{true}$ was defined. Samples microstructure deformed up to the various degree of deformation was investigated using Tesla–BS-540 electron microscope, equipped with the goniometer at the accelerating voltage of 125 kV. The obtained experimental data of polycrystalline alloys with the grain size 10 and 240 μm was used for establishing the connection between the strain hardening stages and the dislocation structure.

Results

The dependences σ – ε_{true} and θ – ε_{true} were built for the purpose of defining the deformation stages. For the dependencies $\sigma = f(\varepsilon_{true})$ the connection Θ was found with the type of DSS with various grain size.

Let us consider the dependences of the flow stress and the strain hardening coefficient on the deformation degree for Cu+10 at.% Mn samples with the grain size <*d*> = 10 and 240 μm. The dependences $\sigma - \varepsilon_{true}$ and $\theta - \varepsilon_{true}$ for these grain sizes are given in Figure 1*a*. The analysis of dependence $\theta - \varepsilon_{true}$ demonstrated the presence of short transition stage π, stages II and IV with almost constant strain hardening coefficient and stage III with the decreasing strain hardening coefficient. Stage II is characterized by the higher value of θ. The values of θ and σ coefficients are higher in the alloy with the grain size of 10 μm. Figure 1*b* illustrates the dependences $\sigma - \varepsilon_{true}$ and $\theta - \varepsilon_{true}$ for the alloys Cu+13 at.% Mn and Cu+25 at.% Mn with the grain size of 10 μm. It can be seen from the figure that the stress and strain hardening coefficient are higher in Cu+25 at.% Mn.

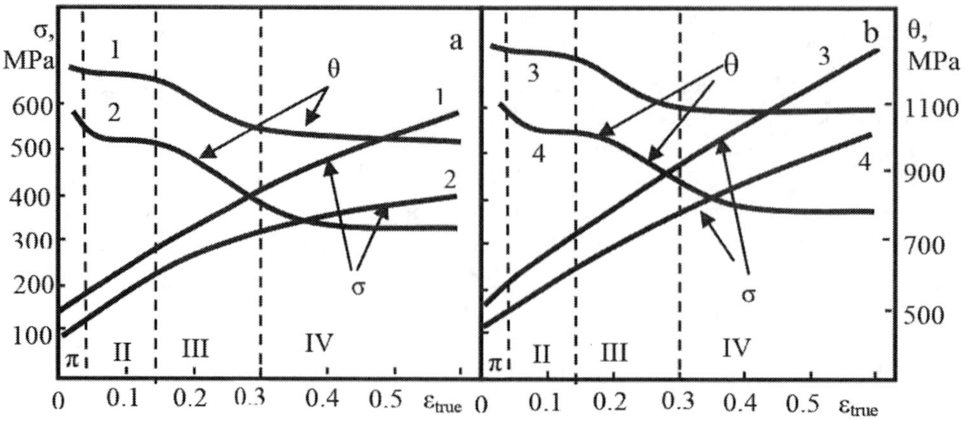

Fig. 1. The dependences of flow stress (σ) and the strain hardening coefficient (θ) on the deformation degree (ε_{true}): a – for Cu+10 at.% Mn alloy with the grain size of 10 μm (1) and 240 μm (2) and b – for Cu+13 at.% Mn (4) and Cu+25 at.% Mn (3) alloys with the grain size of 10 μm. Dashed lines and Roman numbers define the stage of the strain hardening.

Let us consider the evolution of dislocation structure with deformation for the alloys in the concentration range 10…25 at.% Mn. Figure 2 presents the sequence of transition of the DSSs types depending on their deformation degree.

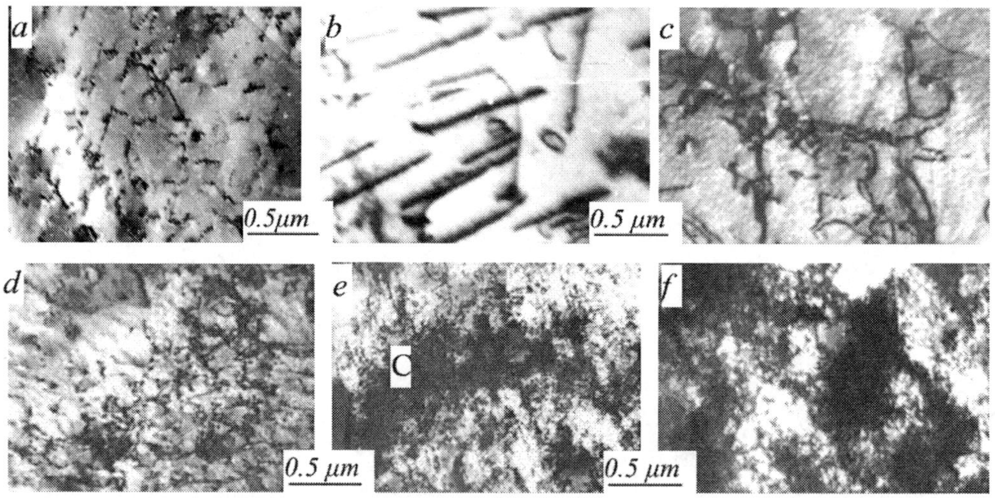

Fig. 2. The evolution of DSS in the alloys Cu+10 at.% Mn with deformation: a – dislocation accumulation, b – dislocation loops, c – cellular DSS, d – non-disoriented cellular-network DSS, e – disoriented cellular-network DSS, f – fragmented DSS. The grain size is 10 μm. C – bending extinction contour.

When the deformation degree is not high ($\varepsilon_{true} \approx 0.05$) chaotic distribution of dislocations and dislocation accumulation is observed (Fig. 2*a*). Dislocation loops given in Figure 2*b* can seldom be found. Increase in the deformation degree ($\varepsilon_{true} \approx 0.15$) is followed by occurrence of the rectilinear dislocations where thresholds and responses are observed. This will further lead to the formation of network DSS (Fig. 2*c*), and then cellular-network DSS (Fig. 2*d*). The increase in the deformation degree is followed by occurrence of disorientation in the substructure, which is characterized by the presence of bending extinction contours [7-10] (see e.g. Fig. 2*e*). Disoriented cellular-network DSS (Fig. 2*e*) and fragmented DSS (Fig. 2*f*) are observed.

It was established that the sequence of transitions of DSSs during the deformation process of Cu+10 at.% Mn – 25 at.% Mn alloys is as follows: chaotic dislocations distribution and dislocation accumulations → homogeneous network DSS→ non-disoriented cellular-network DSS → cellular-network DSS with disorientations → fragmented DSS. Result analysis revealed that occurrence of the new type of DSS corresponds to the occurrence of the new stage of strain hardening. The "new" DSS in the areas with increased dislocation density is evolving, and the previous "old one" disappears by the end of the stage. Redistribution of dislocations is followed by the reduction of the stored deformation energy [11, 12] and that is why it is energetically profitable. Thus, each of the stages generally has two types of dislocation substructure: "the new one" and "the old one". Stage II of the deformation is characterized by non-disoriented substructures: homogeneous network and cellular-network DSS. Disorientations start to form only at the end of stage II. The large bulk of the material at the stage III corresponds to the disoriented DSS. Stage IV is characterized by only disoriented DSS: cellular-network with disorientations and fragmented DSS are observed.

Conclusion

The stress-strain dependences and evolution of dislocation substructures were investigated in Cu-Mn polycrystalline solid solutions with the mean grain size <d>= 10 and 240 μm. Mn concentration in the alloys varied within the range of 10…25 at.%. On the dependences $\sigma = f(\varepsilon_{true})$ four stages of strain hardening were observed, which varied by the value of the strain hardening coefficient: transition stage (π), stages II, III and the final longstanding stage IV.

The grain size does not influence the amount of the formed stages of plastic deformation; however the values σ and Θ depend on the grain size. Occurrence of the new stage of deformation hardening is connected to the formation of the new substructure type, where the dislocation density is higher than in the prior DSS. Dislocations redistribution resulting in the formation of the new type of DSS is attributed to the dislocation subsystem tendency to reduction of the stored deformation energy.

The studies were carried out within the support of the Government Task for research, project No. 3.8320.2017/БЧ.

References

1. N.A. Koneva, E.V. Kozlov, Russian Physics Journal, **8**, 90-98 (2004)
2. S.A. Firstov, G.F. Sarzhan, Russian Physics Journal, **34**(3), 23-34 (1991)
3. E. Nes, Progr. Mater. Sci., **41**, 129–193 (1998)
4. X.F. Fang, W. Dahl, Mater. Sci. Eng, **A203**, 14–25 (1995)
5. N.A. Koneva, E.V. Kozlov, Russian Physics Journal, **2**, 89-106 (1990)
6. N.A. Koneva, E.V. Kozlov, L.I. Trishkina, T.B. Cherkasova, Fundamental'nye problemy sovremennogo materialovedenia [Basic Problems of Material Science (BPMS)], **13**(2), 162-167 (2016) (in Russian)
7. P.B. Hirsch, *Electron Microscopy of Thin Crystals* (Mir, Moscow, 1968) (in Russian)
8. N.A. Koneva, L.I. Trishkina, E.V. Kozlov, Mater. Sci. and Eng. A, **319-321**, 156-159 (2001)

9. T.E. Konstantinova, *Mezostruktura deformirovannyh splavov* [Mesostructure of deformed alloys] (Don FTI NAN Ukrainy, Donetsk, 1997) (in Russian)
10. A.N. Tyumentcev, A.I. Ditenberg, A.D. Korotaev, K.I. Denisov, Physical Mesomechanics, **16**, 3, 63-79 (2013) (in Russian)
11. V.L. Indenbom, *Nekotorye voprosy fiziki plastichnosti kristallov* [Issues of the physics of the crystals plasticity] (Izd-vo AN SSSR, Mosscow, 1960)
12. W.A. Backofen, *Deformation processing* (Mir, Moscow, 1977)

Primary gas discharge transducer and its application

Gennadii Smirnov[1,*], and *Aleksey* Kovalsky[1]

[1]Tomsk State University of Control Systems and Radioelectronics, 634050 Tomsk, Russia

Abstract. The aim of this study is to develop a high-speed fault transducer and, based on it, to create an automated system for monitoring and repairing faulty insulation of winding wires in electrical equipment. The fault transducer under development is based on a gas discharge process that occurs between a high-voltage electrode and a grounded wire conductor in a damaged insulation segment. Oscillographic measurements show that there is a complex signal that appears on the fault transducer when a damaged insulation segment is passed through it. The basic patterns of electrodeposition of dielectric films onto metals were determined. The conditions of this process were optimized. The features of the above-mentioned signal that can be used to monitor the number of faults and lengths of damaged segments, as well as the features of the deposition of dielectric films onto metals were justified. Based on these features, a faulty wire insulation repair process was developed and can be controlled.

Introduction

At present, fault monitoring of enamel wire insulation in Russia is regulated by GOST IEC 60851-5–2011. According to this document, the integrity of insulation is expressed in terms of the number of point faults detected by an electrical testing device per wire of a certain length. Point faults for wires with a rated conductor diameter of up to and including 0.050 mm are identified at a low voltage using a transducer consisting of two felt plates immersed in a solution of aqueous sodium sulfate Na_2SO_4 (30 g/L), between which a controlled wire is passed at a rate of (275 ± 25) mm/s. A test DC voltage (50 ± 3) V is applied between the wire conductor and the solution connected to the electrical circuit. Point faults are recorded by an appropriate relay with a counter. A disadvantage of this monitoring technique is low sensitivity of the transducer and contamination of the insulation with electrolyte during the monitoring. This makes it impossible to test all wires completely, and they are tested selectively on individual segments and from different coils.

For wires with a conductor diameter of 0.050 to 0.25 mm, a transducer electrode made in the form of two rollers is used. A controlled wire is passed through the transducer formed by four rollers, two of which are guides and the other two are electrodes. For wires with a diameter of 0.25 to 1.60 mm, this transducer cannot be used. Instead of it, one high-voltage roller electrode of a larger diameter is used. In this monitoring procedure, the wire is

* Corresponding author: smirnov@main.tusur.ru

repeatedly bent. This leads to high mechanical stresses on the wire insulation on the electrode side, which not only weakens mechanical and electrical strength of the insulation in a controlled wire, but also causes additional faults in the insulation.

The fault monitoring techniques described above are not able to determine the length of damaged segments, another indicator of insulation failure which is equally important as the number of faults.

Another disadvantage of the monitoring technique described in GOST IEC 60851-5–2011 is that it is applied selectively to a segment of the wire from a coil that has been randomly chosen from a batch and, therefore, leaves the bulk of the wire on the chosen coil untested. The same applies to other coils in the batch that have not been chosen for random inspection, thereby reducing its reliability.

Due to the low quality of enamel insulation, electric motors may fail at any time [1–7], potentially leading to an accident. Rejection of faulty wires and their use in products lead to significant economic losses (since such expensive materials as enamel, wire, etc. are wasted) and unreasonable recycling costs.

The studies [8–11] are dedicated to this problem.

The aim of this study is to develop a high-speed fault transducer and, based on it, to create an automated system for monitoring and repairing faulty insulation of winding wires in electrical equipment.

Research Materials and Methods

The tests were performed in two stages. At the first stage, the gas discharge fault transducer was tested. The design of the transducer is justified in detail in [12, 13].

The transducer consists of two cylindrical electrodes that are contiguous to each other along the generatrix and mounted on axes through bearings. The bearing axes are fixed on movable levers (rockers) that allow roller electrodes to make vertical movements in synchrony with wire vibrations. Electrodes are pressed to each other by blade springs. Voltage is applied to electrodes through sliding contacts. For testing purpose, a controlled wire was passed through the above-mentioned electrodes pressed to each other. The signal generated on the transducer when a damaged segment of wire insulation was passed through it, was recorded with one oscilloscope beam from the low-voltage arm of the voltage divider connected to the transducer. The discharge current was measured on the limiting resistance in the discharge circuit of the transducer.

In order to repair faulty wire insulation, it was first necessary to develop a technique for applying an insulating film to a damaged segment. This task was solved at the second stage of the study. The second stage involved searching for and optimizing components that, when combined, would give electrophoretic properties to a PE-939 lacquer designed for enameling PETV grade wires with a temperature index of 140 ÷ 180°C.

It was empirically found that, in case of the film-forming lacquer PE-939, electrophoresis occurs when using dioxane as a solvent and 1% NH_4OH as an electrolyte [14].

Results

A typical oscillogram of the signal from the transducer when passing a damaged segment of the wire insulation is shown in Figure 1.

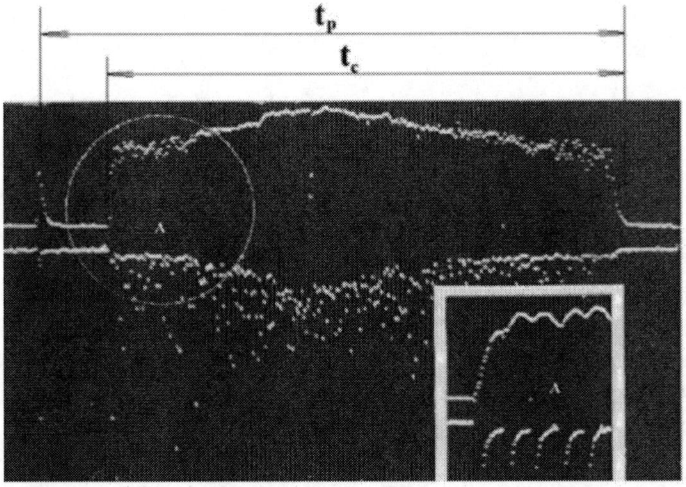

Fig. 1. A typical oscillogram from the fault transducer. t_p is the total burning time of the discharge upon approaching the transducer and until leaving the transducer; t_c is the burning time of the discharge between the transducer and the fault when the fault is just under the transducer; A is the highlighted rising edge of the oscillogram shown on its lower part with a higher temporal resolution.

The oscillogram (Fig. 1) was taken at a negative potential U = -2 kV on the electrode rollers relative to the grounded conductor at wire speed V = 3.07 m/s. Two times are highlighted on the oscillogram: $t_p = 4.1\times10^{-3}$s and $t_c = 3.5\times10^{-3}$s. The pattern of the oscillogram (Fig. 1) can be explained as follows.

The ignition of the corona discharge upon approaching a fault in the wire insulation can be explained as follows. When the fault is moved towards the transducer, the distance between them decreases and, due to this, the electric field strength increases. At a certain critical distance between the transducer and the fault and when the transducer voltage is equal to the initial corona onset voltage, a corona discharge is ignited in the damaged area.

When the fault approaches the transducer, the electric field in the fault area increases, avalanche processes become more intense, and the corona discharge is converted into spark discharge. The capacitance of the circuit begins to discharge and leads to a voltage drop across this gap. The electric field strength in the damaged area falls to a certain value. The current rise rate in the gap decreases and the discharge goes out.

In Figure 1, this process is represented by the first pulse on the current oscillogram. After extinction of the discharge, the capacitance of the discharge circuit begins to be charged from the power source. The voltage across the gap between the fault and the transducer increases. After the transducer voltage is restored to the initial value, spark re-ignition may not occur immediately. It occurs after a certain time Δt, which varies significantly and depends on the wire speed, amplitude, period of transverse wire vibration, and many other factors. (On the oscillogram presented in Figure 1, $\Delta t = t_p - t_c$).

Therefore, when the fault approaches the transducer and leaves it, the spark rate is unstable, and the value Δt varies significantly from fault to fault. When the transducer enters the transducer zone, the discharge stabilizes and, as the distance between the fault and the transducer decreases, subsequent spark ignitions occur at lower voltages than those upon approaching and leaving the fault. On the oscillogram, the residence time of the fault under the roller electrodes is indicated as t_c.

Estimates resulting from the experiments show that the optimal diameter of the transducer rollers should be in the range of 10 to 16 mm.

The second stage of the studies was aimed at optimizing components of the electrophoretic composition. Experiments were carried out using the complete factorial design 2^k.

The levels of variable factors are given in Table 1 where the following notations are introduced: $x_1=(n_1- n_1^0)/2$; $x_2=(n_1- n_2^0)/2$; x_1, x_2 are codes of concentrations of the PE-939 lacquer and liquid ammonia 1% NH_4OH, respectively; n_1, n_2 are real concentrations of the PE-939 lacquer and liquid ammonia 1% NH4OH, respectively; n_1^0, n_2^0 is the center of design determined by the following formulas: $n_1^0= (n_{1max}-n_{1min})/2$; $n_2^0= (n_{2max}-n_{2min})/2$; n_{1max}=390 ml/L, n_{2max}=210 ml/L, n_{1min}= 290 ml/L, n_{2min}=110 ml/L are maximum and minimum values of the selected factors used in the experiments, respectively; Δn_1, Δn_2 are the variability intervals of the factors determined by the formulas: $\Delta n_1=(n_{1max}+n_{1min})/2$; $\Delta n_2=(n_{2max}+n_{2min})/2$.

Table 1. Levels of variable factors.

Components of electrophoretic composition, ml/L	Designation		Levels			Variability interval
	Real	Coded	+ 1	0	- 1	
Content of PE-939	n_1	x_1	390	340	290	50
Content of 1% NH4OH	n_2	x_2	210	160	110	50

After statistical processing of the obtained results and regression analysis, the following adequate regression equation was obtained:

$$U_{forward}= - 5.4375 - 0.26x_1 - 0.24x_2+0.3125x_1x_2-1.973\, x_1^2 - 1.095x_2^2 \quad (1)$$

After the transition to the full scale, by substituting x_1 and x_2 in equation (1) for their full-scale values, we obtain:

$$U_{forward} = - 59.49+0.247\, n_1+ 0.272\, n_2 + 0.000136\, n_1 n_2- 0.00094\, n_1^2 - 0.00044n_2^2 \quad (2)$$

The planning matrix is given in Table 2.

Table 2. Matrix planning and experimental results.

Experiment No.	X_1	X_2	U_1	$U_{1\ reverse}$	U_2	$U_{2\ reverse}$	$\overline{U}$
1	+ 1	+1	2.28	2.16	2.76	2.9	2.53
2	-1	+1	2.24	2.2	1.96	1.62	2.01
3	+ 1	-1	2.88	1.84	1.76	2.04	2.11
4	-1	-1	2.94	2.49	2.54	3.42	2.24
5	+ 1	0	3.22	2.82	2.84	2.28	2.79
6	0	+1	3.94	3.54	3.72	4.06	3.82
7	-1	0	4.4	3.92	3.9	4.36	4.14
8	0	-1	4.96	4.2	5.04	5.28	4.87
9	0	0	2.8	2.16	2.94	2.2	2.85

From equation (2), we can find the optimal ratio of the components n_{1opt} and n_{2opt}

$$\begin{cases} \partial U/\partial n_1 = 0{,}247 + 0{,}000136 n_2 - 2\times 0{,}00094\, n_1 = 0 \\ \partial U/\partial n_2 = 0{,}272 + 0{,}000136 n_1 - 2\times 0{,}00044\, n_2 = 0 \end{cases} \quad (3)$$

By solving the system (3), we obtain n_{1opt} = 156.7 ml/L; $n_{2опт}$ = 333.3 ml/L. The dioxane concentration n_{3opt} = 500 ml/L.

The gas discharge fault transducer and electrodeposition of dielectric films were used as a basis for the wire insulation monitoring and repair system, which is described in detail in the patent [14].

Conclusion

The structure of the signal produced by the gas discharge fault transducer when passing a damaged segment of wire insulation through it was studied and presented in the paper. As shown, this signal has a complex structure. This structure was analyzed. It was shown how to single out a useful zone from this structure to exclude errors when calculating the number of faults and lengths of damaged segments. Concentrations of the components in the electrophoretic composition based on the PE-939 lacquer were justified and optimized. The developed primary fault transducer and the proposed electrodeposition technique for dielectric films can be used not only for designing automated wire repair system, but also for complete monitoring of wire insulations during enameling and production of windings. In addition, the composition and the electrodeposition technique can be used for other purposes, in particular, for wire enameling.

References

1. G.V. Smirnov *Nadezhnost' izoljacii obmotok jelektrotehnicheskih izdelij* [Reliability of coil insulation in electrical products] (Publishing House of the Tomsk State University, Tomsk, 1990) (in Russian)
2. T. Knych, A. Mamala, M. Jablonski, P. Uliasz, Wire Journal International, **45**(1), 55–61 (2012)
3. B. Gupta, Risk in surge testing of turn insulation in windings of rotating machines, in Electrical Insulation Conference and Electrical Manufacturing & Coil Winding Technology Conference, 459– 462 (September 23-25, 2003)
4. J.P. Bellomo, P. Castelan, T. Lebey, IEEE Transactions on Dielectrics and Electrical Insulation, **6**(1), 20–26 (1999)
5. F.R. Bohm, K. Nagel, H. Schindler, A new generation of wire enamel for the production of magnet wires with outstanding corona resistance, in Electrical Insulation Conference and Electrical Manufacturing & Coil Winding Technology Conference, 109–113 (September 23-25, 2003)
6. A.H. Bonnett, G.C. Soukup, IEEE Transactions on Industry Applications, **28**(4), 921–937 (1992)
7. A.V. Andrianov, V.K. Andrianov, E.V. Bykov, Cables and Wires, **5**(342) (2013), (in Russian)
8. M. Chapman, N. Frost, R. Bruetsch, Insulation systems for rotating low-voltage machines , IEEE International Symposium on Electrical Insulation, 257-260 (2008)

9. D. Hwang, K. Lee, Y. Kim, I. Lee, T. Lim, D. Kim, Accessing the insulation characteristics for stator windings of low-voltage induction motors for adjustable-speed drive applications , IEEE Industry Application Conference, 432-438 (2003)
10. D. Fabiani, G.C. Montanari, A. Contin, Aging acceleration of insulating materials for electrical machine windings supplied by PWM in the presence and in the absence of partial discharges, IEEE 7th International Conference on Solid Dielectrics, 283-286 (2001)
11. S.R. Fenger, J. Campbell, IEEE Industry Applications Magazine, **9**(4), 22-31 (2003)
12. G.V. Smirnov, D.G. Smirnov, Russian Journal of Nondestructive Testing, **8** (2016), (in Russian)
13. G.V. Smirnov, D.G. Smirnov, Datchik dlja nepreryvnogo kontrolja izoljacii provodov [Sensor for continuous monitoring of wire insulation], RF Patent No. 2597938 (20.09.2016) (in Russian)
14. G.V. Smirnov, D.G. Smirnov, Sposob kontrolja i remonta izoljacii provodov [Means for monitoring and repair of wire insulation], RF Patent No. 2506602 (10.02.2014) (in Russian)

Structure and preferred orientation of primary phases of Ni-Al-Co-based alloy alloyed by Rhenium after annealing and creep exposure

Elena Nikonenko[1,2,*], *Natalya* Popova[1,3], *Alisa* Nikonenko[4], and *Nina* Koneva[1]

[1]Tomsk State University of Architecture and Building, 634003 Tomsk, Russia
[2]National Research Tomsk Polytechnic University, 634050 Tomsk, Russia
[3]Institute of Strength Physics and Material Science SB RAS, 634021 Tomsk, Russia
[4] National Research Tomsk State University, 634050 Tomsk, Russia

Abstract. The present paper investigates phase composition and fine structure of Ni-Al-Co-based alloy alloyed by Rhenium (Re) (~3 at.%), as well as studies preferred orientation of the given alloy primary phases by means of scanning electron microscopy and X-ray diffraction analysis. The original state was given as the alloy which underwent a complete cycle of heat treatment: stepwise homogenization, cooling, annealing, and cooling at 100°C/min back to room temperature. The alloy was further subject to annealing at 1000°C for 105 h and 968 h. It was also exposed to creep experiments at the same temperature for 105 h (loading 320 MPa), and for 968 h (220 MPa). γ- and γ'-phases were established to be the primary phases in both states of face-centered cubic crystal lattice. γ'-phase (phase with ordered atom arrangement having $L1_2$ superstructure) was the primary phase in all states of the alloy. Investigations showed that Re being able not to dissolve in the primary phases served as a phase-forming element. Phase composition and phase morphology were studied. The influence of high-temperature annealing and creep on superalloy structure and morphology of γ'-phase were defined. The influence of creep on the structure was proven to be different from the annealing effect. First, creep generated larger dissipation of orientations than annealing. Second, anisotropy of γ'-phase cuboids could be observed at meso-level. Cuboids partially joined together. Third, spheroidized cuboids could be often met after creep.

1 Introduction

Development and application of metallic materials to a large extend define advance in modern engineering due to their required operation properties. However the requirement levels applied to these materials constantly grow along with the new objectives engineers are facing while new equipment development, and its operation under extreme conditions in particular – high speeds, high temperatures, etc. Scientists are being in the constant search for metallic materials being able to withstand these extreme conditions. Thus, development

* Corresponding author: vilatomsk@mail.ru

of alloys comprising intermetallic phases is considered as one of the promising research directions. Superalloys based on (γ+γ′)-phases can serve as an example, where γ-phase is a disordered face-centered cubic (fcc) solid solution, while γ′-phase is an ordered phase with $L1_2$ superstructure [1-3]. γ′-phase is the primary phase in these modern superalloys. This is the reason why γ′-phase is largely responsible for the formation of superalloy properties. Currently, superalloys are often produced based on the Ni – Al alloy [1], alloyed by various refractory elements [1, 3, 4]. However, introduction of large amount of different alloying elements as well as annealing and creep of superalloy can change its texture [5]. The present paper investigates the issues mentioned above.

2 Materials and methods

Material of study was a complex metallic alloy. The basic elements of the alloy were Ni from 69 to 59 at.%, Al from 12 to 15 at.% and Co from 10 to 13 at.%. Cr, Ta and Re were used as alloying elements in the amount no more than 3.5 at.% of each, and also W and Mo in smaller amounts.

The structure of the alloy was investigated within the following alloy states: 1) original state (after directional crystallization); 2) after 1000°C annealing for 105 h; 3) after 1000°C annealing for 968 h; 4) after creep tests at 1000°C for 105 h and load 320 MPa; and 5) the state while creep tests 968 h under load 220 MPa and temperature 1000°C.

X-ray diffraction analysis (XRD) on thin foils using EM-125 transmission electron microscope with 125 kV accelerating voltage was used as the main research method. The specimens for investigating structure were cut perpendicularly to the axis of crystal growth [001]. According to microdiffraction images obtained from electron microscope, phase composition of the alloy was defined for all states, also the morphological types of γ′-phase cuboids and appearing stereographic components in the alloy. Orientations distribution considering their "specific mass" was placed on a standard stereographic triangle, the orientations were analyzed.

3 Results and discussion

3.1 Phase composition and γ′-phase state

Electron microscopy investigations have shown a combination of a number of phases in different states of the alloy the qualitative and quantitative composition of which depends on the state (Table 1).

Table 1. Phase composition and volume fractions of the primary phases in the specimens under study.

No. specimen	**Primary phases, % (± 1%)**		**Secondary phases**
	γ′-phase,	**γ-phase,**	
1	80.0	19.7	χ-phase
2	78.0	21.5	Laves phase, σ-phase
3	66.0	33.8	Laves phase , σ-phase
4	79.0	20.7	Laves phase , σ-phase
5	58.0	41.9	Laves phase

Phases observed in the superalloy can be classified into primary and secondary. This classification is based on volume fraction of phases, their contribution to and constant or single presence in the alloy. As can be seen from Table 1, γ'- and γ-phases are the primary. As a rule, they form the primary structure of alloy in almost all of the superalloys [1-4]. They can be found in the alloy under study as γ'-phase quasi-cuboids separated by interlayers of γ-phase. Other phases are secondary [6,7]. Different state of the alloy is marked by different secondary phases [6,7]. Volume fractions of the primary phases present in different states of the alloy are also given in Table 1. As can be seen from Table 1, volume fraction of γ'- phase is kept high in all states of the alloy. It represents cuboids, more often with quite distinct cutting. Their sizes and shape depend on the parameters of heat treatment and creep.

Microphotographs in Figure 1 obtained by XRD show that γ'-phase cuboids can be classified into several types according to their morphology, given as follows:

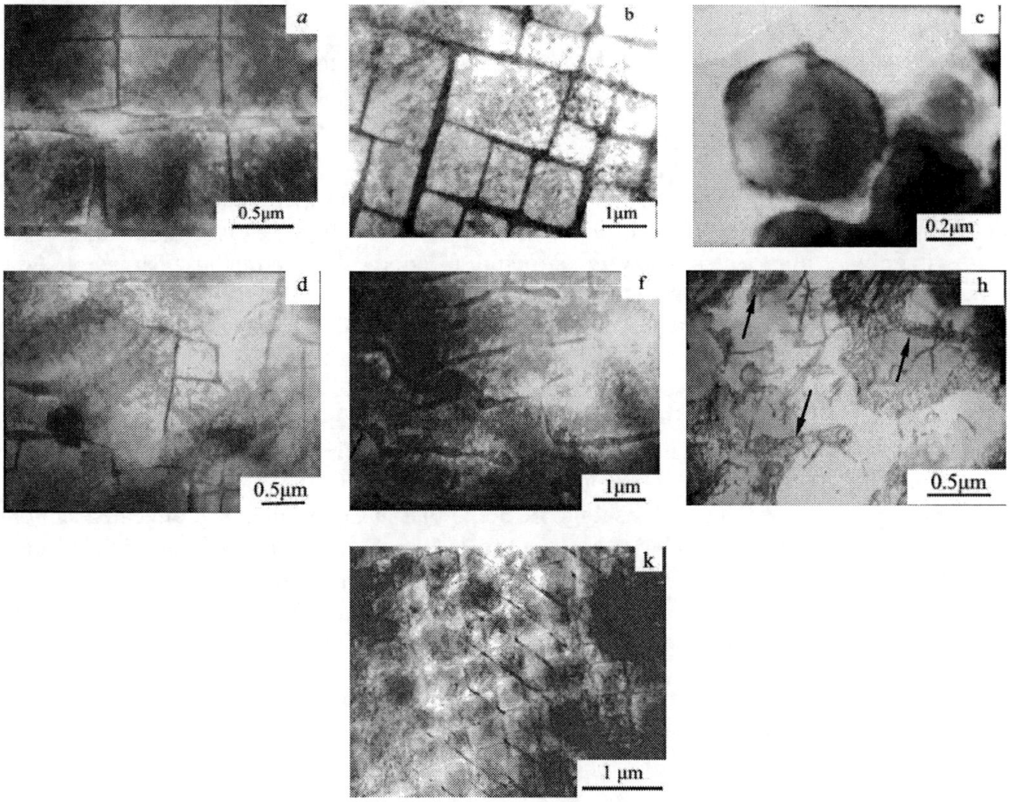

Fig. 1. Morphological types of γ'-phase cuboids: a – homogeneous; b – non-homogeneous; c – spheroidized; d – with damaged structure; f – destructed; h – constrained by dislocation walls; k – with second cuboid orientation.

- Homogeneous cuboids (Fig. 1a) are cuboids which are almost of the same size, i.e. dispersion comprises the value of not more than 0.005 μm.
- Non-homogeneous cuboids (Fig. 1b). The size of cuboids differs significantly from each other. Dispersion of the size in areas of material which contain such cuboids comprises the value which is > 0.2 μm.
- Spheroidized cuboids (Fig. 1c) are cuboids of a round shape. They are indicative of the alloy after creep.

➢ Cuboids with damaged structure. Examples of such cuboids are given in Figure 1d. As it is seen from the given microphotographs structural damage can be defined as presence of cut boundary inside the cuboid or irregular shape of the cuboid, when the cuboid lacks one or more walls. Such cuboids can be found in all alloys studied in the present paper.

➢ Destructed cuboids (Fig. 1f). These areas of structure are marked by only separate walls of cuboids left.

➢ Local location of γ′-phase (Fig. 1h) is when γ′ cuboids are constrained by dislocation walls.

➢ Presence of second cuboid orientation. These are lamellar precipitate (according to packet types of thin long twins) being present in separate locations of structure. In regards to γ′-phase cuboids they are generally located within the direction <110>, or <001> (Fig. 1k).

Damage of γ′-phase cubic structure under creep conditions is apparently connected with the formation of dislocation structure [3, 8].

γ-phase being one of the primary phases is given as relatively thin interlayers placed between the γ′-phase cuboids. It can be found in all states of the alloy.

3.2 Preferred orientation of γ- and γ′-phases

Alloys like (γ+γ′) are generally characterized by complete orientation correspondence between γ- and γ′-phases [3, 9-12]. Some similar phenomenon took place in the alloy under study, mainly in those states where the structure of cuboids was somehow damaged due to directional crystallization by means of which the specimens under study were obtained. In this case well-ordered texture [001] should be developed. In order to investigate that, specimens were cut perpendicularly to the growth axis, then tin foils were made from them and appearing stereographic components were studied. Distribution of orientations was studied considering their specific mass. The key data is presented in Figure. 2.

Compared with specimens having lower Re content which were studied before [3, 8] a significant dispersion of crystal orientation was observed. In the states after annealing the fraction [001] was less than 0.1. It should be noted that in the original and annealed samples cubic orientation of γ- and γ′-phases were not well distinct. Creep test increased the fraction of cubic orientation by 2-5 times. In this case dissipation was significantly eliminated along the angle bisector [001]. It is especially well seen in specimen No. 5. Decrease of diagonal components of texture increased the role of components lying along the triangle side [001] – [011].

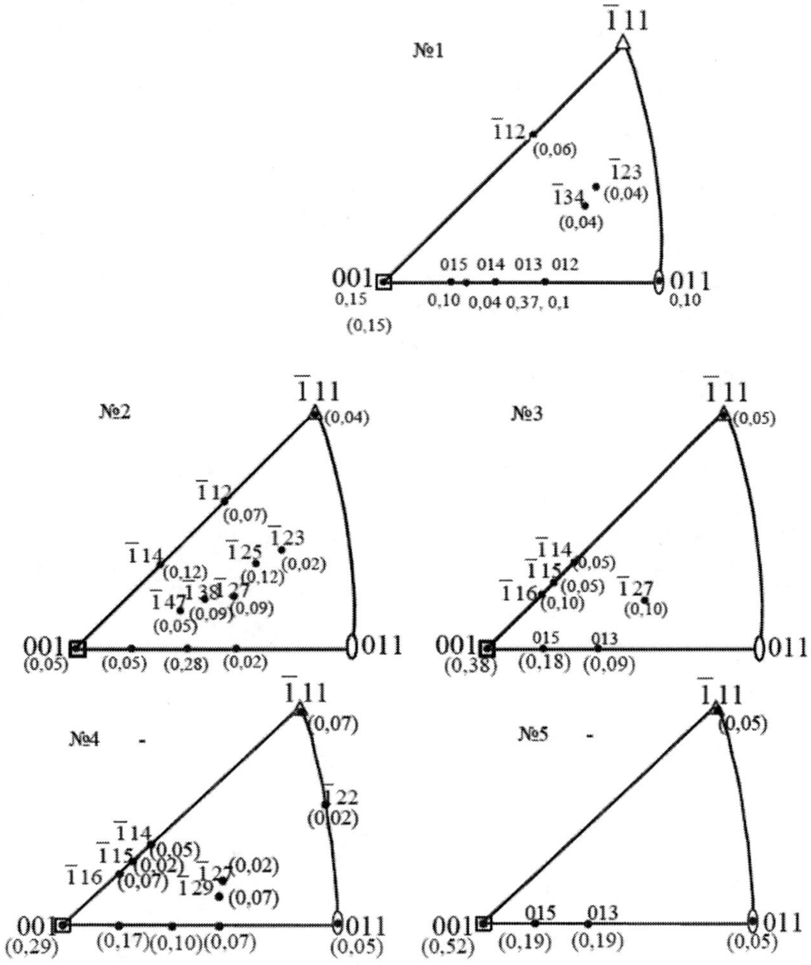

Fig. 2. Distribution of orientations considering their specific mass in the alloy.

4 Conclusion

Resulting from conducted studies it was found that under creep conditions dislocation structure was formed the way that cubic structure was damaged and preferred orientation was changed. The other mechanism of orientation change was diffusive creep mechanism leading to cuboid structure damage at macro level. Cuboid shape turned into round one. The given works highlight the importance of both dislocation and diffusive creep mechanisms.

References

1. N.S. Stoloff, C.T. Sims, W.C. Hagel, *Superalloys II High-Temperature Materials for Aerospace and Industrial Power* (John Wiley & Sons, New York,1987)
2. R.C. Reed, *The Superalloys: Fundamentals and Applications* (University Press, Cambridge, 2006)
3. Yu.R. Kolobov, E.N. Kablov, E.V. Kozlov, *Struktura i svoistva intermetallidnykh*

materialov s nanofaznym uprochneniem [Structure and properties of intermetallic materials with nanophased hardening] (MISiS Publ., Moscow, 2008) (in Russian)
4. R.C. Reed, T. Tao, N. Warnken, Act. Mat., **57**, 5898 (2009)
5. P. Caron, T. Khan, *TMS Fall Meeting* (IN, TMS, Indianapolis, 2001)
6. E.L. Nikonenko, N.A. Popova, T.V. Dement, N.A. Koneva, Rus. Phys. J. **60**, 231 (2017)
7. E.V. Kozlov, A.N. Smirnov, E.L. Nikonenko, N.A. Popova, N.A. Koneva, *Morfologiya faz i fazovye prevrashcheniya pri termicheskoi obrabotke supersplavov na osnove Ni-Al-Cr i Ni-Al-Co. Masshtabnye i kontsentratsionnye effekty* [Phase Morphology and Transformations at Thermal Treatment of Ni–Al–Cr- and Ni–Al–Co-based Superalloys. Scale and Concentration Effects] (Innovatsionnoe mashinostroenie, Moscow, 2016) (in Russian)
8. C.M.F. Rae, R.C. Reed, Acta Mater. **49**, 4113 (2001)
9. E.V. Kozlov, E.L. Nikonenko, N.A. Popova, N.A. Koneva, Bul. Rus. Ac. Sci. Phys. 78, **4**, 267 (2014)
10. J.T. Guo, L.Y. Sheng, Y. Xie, Z.X. Zhang, V.E. Ovcharenko, H.Q. Ye, Intermetallics, 19, **2**, 137 (2011)
11. Y. Yu, J. Zhou, J. Chen, H. Zhou, C. Guo, B. Guo, Intermetallics, 18, **5**, 871 (2010)
12. E.N. Kablov, V.N. Tolorajja, I.M. Demonis, N.G.Orehov, Tehnol. leg. spl., **2**, 60 (2007)

Influence of electrolytic plasma nitriding mode on structural phase state of pearlitic steel

Natalya Popova[1,2,*], *Lyudmila* Erygina[3], *Elena* Nikonenko[1,4], and *Mazhin* Skakov[3]

[1]Tomsk State University of Architecture and Building, 634003 Tomsk, Russia
[2]Institute of Strength Physics and Material Science, Siberian Branch, Russian Academy of Sciences, 634021 Tomsk, Russia
[3]National Nuclear Center of the Republic of Kazakhstan, 071100 Kurchatov, Kazakhstan
[4]National Research Tomsk Polytechnic University, 634050 Tomsk, Russia

Abstract. The paper describes results of studies of phase transitions in structural phase state occurring in the type 0.34C-1Cr-1Ni-1Mo-Fe steel under electrolytic plasma nitriding in nitrogen-containing water solution. The nitriding voltages considered in the given study were 550 and 600 V. The research was conducted by means of X-ray diffraction electron microscopy. The specimens were studied in two states: 1) before modification (original state) and 2) after nitriding in the surface layer of the specimen. The study was conducted on thin foils. It was found that nitriding lead to significant changes in the structure of steel, namely in its phase composition and in the number of existing phases. In the original state the structure of steel was given as lamellar pearlite, ferritic carbide mix and fragmented ferrite. After 550 V nitriding it was lath martensite, plates of α-phase, with colonies of thin parallel plates of γ-phase and coarse grains of α-phase, containing γ-phase grains which were different in size and shape and were various-directional. Increase in nitriding voltage up to 600 V lead to change in the structure given as a lamellar non-fragmented pearlite and fragmented ferrite. The original state was marked by presence of particles of M_3C cementite, after nitriding irrespective of the voltage it had the particles of M_3C alloyed cementite, Fe_3Mo_3N nitride and $Cr_2C_{0.61}N_{0.39}$ carbonitride. The sizes, volume fractions and locations of particles were dependent on nitriding voltage.

1 Introduction

Advancement of technology and development of new branches of industry require application of new high-strength and resistant materials. Special attention is being paid to different ways of surface treatment as the properties of the ready products are to a large extend dependent on the properties of the product surface. Thus it stipulated rapid development of scientific studies as well as engineering solutions aimed at enhancing the traditional and development of new surface treatment techniques in order to provide surface with the required operational properties [1]. Modern methods of materials modification are

* Corresponding author: natalya-popova-44@mail.ru

based on intensification of the applied impact, use of electric discharge in particular. Electrolyte plasma treatment refers to one of the mentioned methods [2-7]. It implies that during the components heating along with quenching the surface layers of metals undergo chemical modification with elements contained in the electrolyte composition. By means of changing the electrolyte composition and electric mode, it is possible to conduct surface treatment under larger velocities, which sufficiently exceeds the velocity of thermal gas phase and electrolytic processes. High heating and cooling rates enable to improve a number of mechanical properties (e.g. hardness and strength), and simplify mechanization and automatization of industrial processes of hardening. In these conditions processing period is considerably declined [8]. And also low-cost compact equipment can be used [8]. Moreover, high heating rates enable to combine both the advantages of ion-beam and high-frequency processing [9-16].

The present work is aimed at studying the influence of electric mode of electrolytic plasma nitriding on the patterns of changing the structure and phase composition of the type 0.34C-1Cr-1Ni-1Mo-Fe pearlitic steel.

2 Materials and methods

The object of the given study was the type 0.34C-1Cr-1Ni-1Mo-Fe steel; its chemical composition is given in Table 1.

Table 1. Chemical composition of the type 0.34C-1Cr-1Ni-1Mo-Fe steel (wt.%).

Mass content of elements, %									
C	**Si**	**Mn**	**Ni**	**Cr**	**Mo**	**S**	**P**	**Cu**	**Fe**
0.3-0.4	0.17-0.37	0.5-0.8	1.3-1.7	1.3-1.7	0.2-0.3	not more 0.035	not more 0.03	not more 0.3	the rest

The study of steel was conducted on thin foils by means of X-ray diffraction (XRD) electron microscopy using EM-125 transmission electron microscope (TEM) at 125 kV accelerating voltage. Operating increase in the microscope column was chosen equal to 8000 up to 50000 volumes. Changes in structural phase composition occuring in 0.34C-1Cr-1Ni-1Mo-Fe steel under electrolytic plasma nitriding in nitrogen-containing water solution were studied at two nitriding modes: 550 V (5 min) and 600 v (5 min). The specimen were examined in two states: 1) before modification (original state) and 2) after nitriding in the surface layer of the specimen.

Foils production technique for TEM observations for each of the exact locations was prior described in [17].

The phase analysis and identification were held according to standard methodologies using the TEM images confirmed by the micro-diffraction images and dark-field images obtained in the respective reflexes. It allowed performing classification of the structure morphology, defining the sizes and volume fractions of secondary phases as well as places of their localization, and parameters of material fine structure.

3 Results and discussion

3.1 Structure and phase composition of steel in the original state

Structural phase composition of the type 0.34C-1Cr-1Ni-1Mo-Fe steel was previously described in details in [18]. It was established that the structure of steel in the original state was first, lamellar pearlite, with its volume fraction comprising 35% of the volume of material matrix. Lamellar pearlite was present in two modifications: non-fragmented pearlite (Fig. 1a) and fragmented pearlite (Fig. 1b). Volume fraction of non-fragmented pearlite was 10%, of fragmented one – 25%. The second structural component of steel was ferritic-carbide mix (Fig. 1c), its volume fraction in the material matrix was 45%. The third structural component of steel was fragmented ferrite (Fig. 1d), its volume fraction was 20%.

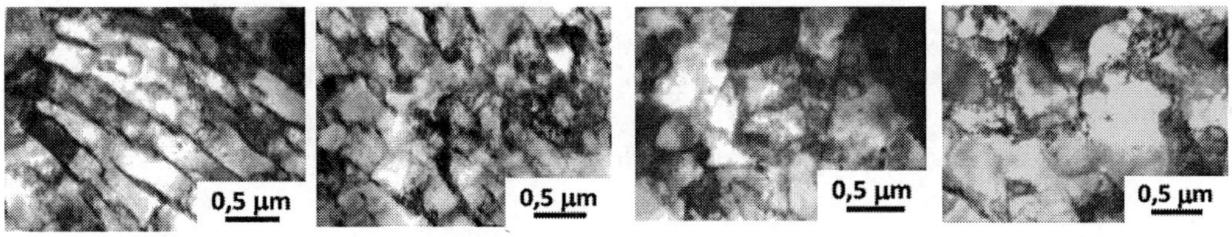

Fig. 1. Electron microscope image of the type 0.34C-1Cr-1Ni-1Mo-Fe steel fine structure in the original state: lamellar pearlite (a), fragmented lamellar pearlite (b), ferritic-carbide mix (c), fragmented ferrite (d).

Carbide component in steel was only present by the particles of $(Fe,Cr)_3C$ or M_3C alloyed cementite, with volume fraction of 5.1%.

3.2 Structure and phase composition of steel nitride at U = 550 V

Steel nitriding 550 V (5 min) lead to a significant change of phase composition, the number of existing phases and their morphology. Thus, both phase composition and fine steel structure in the near-surface layer of nitrided specimen varied considerably from the original state. The obtained structure could be classified into three types: 1) lath martensite; 2) relatively large-size grains (plates) of α-phase, with colonies of thin parallel γ-phase plates inside; 3) mixture of α- and γ-phases grains.

Lath martensite occupied the main fraction of the material (~60%). The typical example is given in Figure 2a. Indexing of microdiffraction images obtained from the area of lath martensite showed that retained austenite (γ-phase) was detected along the boundaries of martensite laths. Retained austenite had the form of elongated interlayers. Dark-field image obtained in coinciding reflexes $[\bar{2}00]\alpha + [\overline{42}\bar{1}]\gamma$ can be seen in Figure 2d. This phase had face-centered cubic (fcc) crystal lattice. It was established that crystal lattices of α-phase and γ-phase were bounded by Kurdyumov-Sachs orientation relationship [19]. The average size of γ-phase interlayers comprised the value of 20×250 µm. Volume fraction of γ-phase in lath martensite was 4.5%.

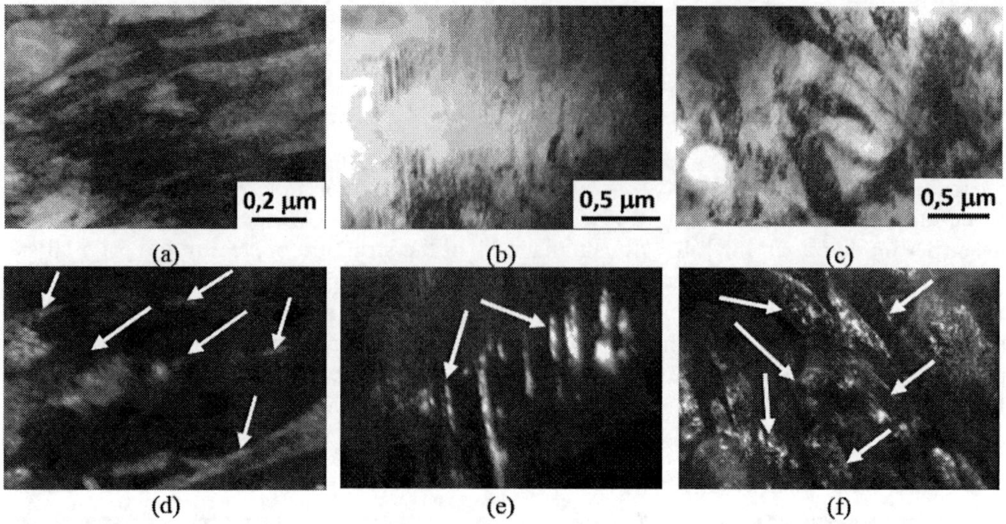

Fig. 2. Electron microscope image of fine structure of the type 0.34C-1Cr-1Ni-1Mo-Fe steel after nitriding at U = 550 V, 5 min: a, d – lath martensite, b, e – grain of α-phase with colony of thin parallel plates of γ-phase, c, f – mix of grains of α- and γ-phases (a-c – bright-field images, d-f – dark-field images). White arrows on dark-field images denote γ-phase precipitation.

Grains (or short wide plates) of α-phase, with colonies of thin parallel plates (like twins) of γ-phase inside were the second component of the steel matrix structure after nitriding at U = 550 V. Volume fraction of this structure comprised 10% of the total material volume. The typical example of such structure is given in Figure 2b where we can see the image of grain with a group of twin-type thin plates inside. Indexing of microdiffraction image obtained from the area of this group of plates showed that these plates were lamellar precipitate of γ-phase. Dark-field image (Fig. 2e) obtained in the reflex $[\bar{1}1\bar{1}]$ of γ-phase serves as a confirmation for that. Crystal lattices of α- and γ-phases, like in lath martensite were bounded by Kurdyumov-Sachs orientation relationship. Volume fraction of γ-phase in this structure was 30%.

The third component of the steel matrix structure after nitriding at U = 550 V was a mix of grains of α- and γ-phases. This mix consisted of large-size grains of α-phases containing elements of γ-phase which were different in size and shape and were various directional, given in the shape of plates varying in width and length, globes or fine grains. The fraction of this structure was 30% in the total material volume. The typical example of this structure is given in Figure 2c; Figure 2f shows dark-field image obtained in the reflex $[\bar{2}20]$ of γ-phase. Between the crystal lattices of α- and γ-phases Kurdyumov-Sachs orientation relationship was also followed. The volume fraction of γ-phase in this structure was 40%.

Indexing of microdiffraction images showed that in all constituent elements of the steel matrix structure except for reflexes belonging to α- and γ-phases, there were reflexes of one more phase – Fe_3Mo_3N nitride, and in lath martensite there was also $Cr_2C_{0,61}N_{0,39}$ carbonitride and M_3C alloyed cementite.

Fe_3Mo_3N nitride had a complex face-centered cubic (fcc) crystal lattice. The particles of Fe_3Mo_3N nitride were inside the interlayers or grains of γ-phase. This was especially well seen on dark-field images, obtained in reflexes referring to crystal lattice of Fe_3Mo_3N phase. Corresponding examples of dark-field images obtained in reflexes $[\bar{1}1\bar{1}]$ of γ-phase + [240] of Fe_3Mo_3N phase and $[\bar{2}20]$ of γ-phase + $[711]$ of Fe_3Mo_3N phase are given in

Figure 3 a,b. Irrespective of the location of Fe_3Mo_3N nitride its particles were of a round shape. Their size was 10-15 μm. Their volume fraction in γ-phase was ~2%.

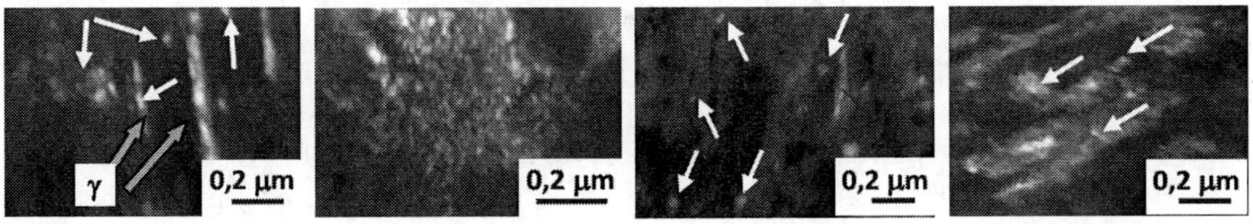

(a) (b) (c) (d)

Fig. 3. Particles of Fe_3Mo_3N nitride inside the interlayers (a) and grains (b) of γ-phase, in α-phase of $Cr_2C_{0,61}N_{0,39}$ carbonitride (c) and M_3C alloyed cementite (d). Dark-field images. Type 0.34C-1Cr-1Ni-1Mo-Fe steel after nitriding at U = 550 V, 5 min.

$Cr_2C_{0,61}N_{0,39}$ carbonitride had orthorhombic crystal lattice. The particles of $Cr_2C_{0,61}N_{0,39}$ carbonitride could be found only in lath martensite, as has been mentioned before. They had a round shape and were randomly located in lath martensite. Dark-field image obtained in the reflex [212] of $Cr_2C_{0,61}N_{0,39}$ phase is given in Figure 3c. The average particles size comprised 50 μm, their volume fraction in lath martensite was ~1%.

The particles of M_3C alloyed cementite were present only in lath martensite. They were located inside the laths. Dark-field image obtained in reflexes [200] of α-phase + [341] of M_3C phase is given in Figure 3d. The average particles size was 30×50 μm, the volume fraction of cementite inside the laths comprised ~2%.

3.3 Structure and phase composition of steel nitride at U = 600 V

Increase in the voltage of steel nitriding up to U = 600 V leads to changes in the steel structure, varying considerably both from the original structure and from the steel nitrided at U = 550 V. In this case, morphology of fine structure was given as lamillar non-fragmented pearlite in the amount of 45% and fragmented ferrite in the amount of 55%. Typical electron microscope images of fine structure obtained from different parts of the material are given in Figure 4.

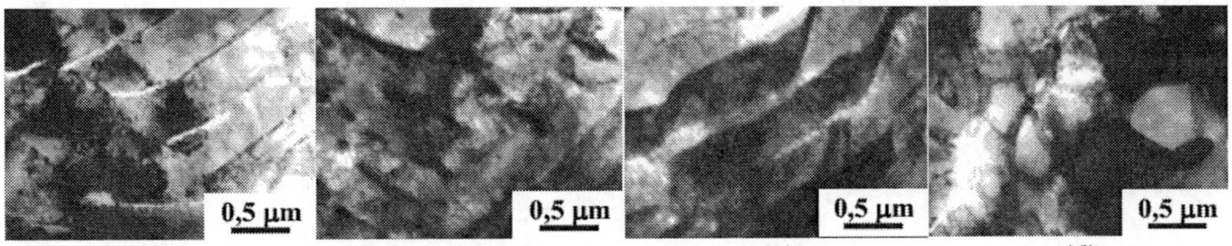

a) b) (c) (d)

Fig. 4. Electron microscope images of fine structure of the type 0.34C-1Cr-1Ni-1Mo-Fe steel after nitriding at U = 600 V, 5 min: a-c – lamellar pearlite (a – ideal, b – with destructed plates of cementite, c – defective); d – fragmented ferrite.

As can be seen from Figures 4a-c, lamellar pearlite was present in three states: 1) nearly ideal lamellar pearlite (Fig. 4a); 2) lamellar pearlite with destructed cementite plates (Fig. 4b); and 3) defective lamellar pearlite (Fig. 4c). Fragmented ferrite (Fig. 4d) like in the original steel state was given as isotropic fragments.

XRD analysis showed that microdiffraction images in all states of lamellar pearlite except for reflexes of α-phase and M_3C alloyed cementite had reflexes referring to Fe_3Mo_3N phase. Moreover, in the first and second states of lamellar pearlite the particles of Fe_3Mo_3N phase were present in α-phase on dislocations, forming 'spotted' contrast and in the third state inside the plates of M_3C. This is well seen on dark-field images obtained in reflexes $[\bar{1}10]$ of α-phase + [022] M_3C + $[00\bar{6}]$ of Fe_3Mo_3N phase (Fig.5a), [200] α-phase + $[\bar{1}30]$ M_3C + $[\bar{1}3\bar{5}]$ of Fe_3Mo_3N phase (Fig. 5b) and $[4\bar{4}0]$ of Fe_3Mo_3N phase (Fig. 5c). The particles possessed round shape. The average size of particles being in dislocations in α-phase was 12 μm, inside the cementite plates – 25 μm. Their volume fraction in α-phase in the first and second states of lamellar pearlite were almost the same and comprised ~1%, in cementite plates (third state of lamellar pearlite) it was 3%.

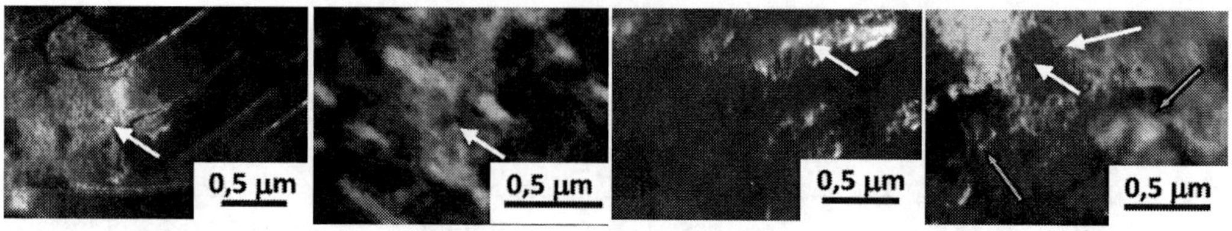

(a) (b) (c) (d)

Fig. 5. Electron microscope image of fine structure of the type 0.34C-1Cr-1Ni-1Mo-Fe steel after nitriding at U = 600 V, 5 min: a-c – lamellar pearlite (a – ideal, b – with destructed cementite plates, c – defective); d – fragmented ferrite. Precipitation of particles of Fe_3Mo_3N nitride (marked by white arrows) in α-phase of lamellar pearlite (a – ideal and b – with destructed cementite plates), in cementite plates in defected lamellar pearlite (c) and inside on the boundaries of fragments in ferrite (d) and $Cr_2C_{0,61}N_{0,39}$ carbonitride particle (marked by grey arrows) on the boundaries and in the joints of fragments in ferrite (d). Dark-field images. Type 0.34C-1Cr-1Ni-1Mo-Fe steel after nitriding at U = 600 V, 5 min.

The particles of Fe_3Mo_3N nitride were also present in fragmented ferrite. They were also of a round shape and were located inside the fragments on dislocations and on the boundaries of fragments (Fig. 5d). The average size of particles both inside and on the boundaries of fragments was the same and comprised 20 μm; the volume fraction in fragmented ferrite was 3%.

The particles of $Cr_2C_{0.61}N_{0.39}$ carbonitride (Fig. 5d) were located on the boundaries and in the joints of fragments, they had irregular shape, and comparatively large size (150 μm) in regards to particles of Fe_3Mo_3N phase. Volume fraction in fragmented ferrite was 2%.

4 Conclusion

From the outcome of the present study it can be concluded that nitriding leads to:

1) Change of the morphological components of the steel matrix;
2) New precipitation of secondary phases: $Cr_2C_{0.61}N_{0.39}$ carbonitride and Fe_3Mo_3N nitride.

It was established that the increase of the nitriding voltage significantly changes the morphology of steel phase structure.

The present study was conducted with the financial support of Russian Foundation for Basic Research within the research project No. 16-48-700198.

References

1. I.V. Suminov, P.N. Belkin, et al., *Mir materialov i tehnologij* [The world of materials and technologies] **1** (Tehnosfera, Moscow, 2011) (in Russian)
2. M.K. Skakov, L.B. Bayatanova, M. Scheffler, Tribologia **5**, 147-153 (2012)
3. J. Martin, A. Melhem, I. Shchedrina, T. Duchanoy, A. Nominé, G. Henrion, T. Czerwiec, T. Belmonte, Surface & Coatings Technology **221**, 70 (2013)
4. V.V. Bedin, L.Kh. Baldaev, D.V. Kolmykov, M.B. Kovynev, Vestnik of Kursk State Agricultural Academy **1**, 77-80 (2013) (in Russian)
5. E. Kozlov, N. Popova, L. Zhurerova, E. Nikonenko, M. Kalashnikov, M. Skakov, AIP Conf. Proc. **1783,** 020112 (2016)
6. K. Rokosz, T. Hryniewicz, World Scientific News **35**, 441 (2016)
7. N. Popova, L. Bayatanova, E. Nikonenko, M. Skakov, Ed. Kozlov, AIP Conf. Proc. **1800**, 030002 (2017)
8. P.N. Belkin, *Elektrohimiko-termicheskaja obrabotka metallov i splavov* [Electrochemical and thermal treatment of metals and alloys] (Mir, Moscow, 2005) (in Russian)
9. R.R. Grin, R.F. Gallyamova, N.Yu. Dudareva, A.A. Sirenko, F.F. Musin, Letters on Materials **3**(15), 1 (2014) (in Russian)
10. S.K. Kiseleva, L.I. Zainullina, M.M. Abramova, A.A. Pinegina, N.Y. Dudareva, I.V. Alexandrov, J. Eng. Sci. Technol. Review **8**, Special Issue 6, 6 (2015)
11. F.F. Komarov, *Ionnaja implantacija v metally* [Ion implantation in metals] (Metallurgija, Moscow, 1990) (in Russian)
12. I.A. Kurzina, E.V. Kozlov, N.A. Popova, M.P. Kalashnikov, E.L. Nikonenko, K.P. Savkin, E.M. Oks, Yu.P. Sharkeev, Bull. Russ. Academy Sci. Physics **76**(11), 1238 (2012)
13. M. Kovaleva, M. Prozorova, M. Arseenko, M. Yapryntsev, Y. Tyurin, O. Kolisnichenko, N. Vasilik, V. Sirota, I. Pavlenko, Ceramics International **41**(10), 15061 (2015)
14. A.G. Grigoryants, I.N. Shiganov, A.I. Misyurov, V.S. Mikhailov, N.P. Kolomeyets, Welding International **29**, 6, 475 (2015)
15. P.N. Belkin, S.Yu. Shadrin, Int. J. Heat and Mass Transfer **55**, 1–3, 179 (2012)
16. S.A. Kusmanov, A.A. Smirnov, Yu.V. Kusmanova, P.N. Belkin, Surface and Coatings Technology **269**, 308 (2015)
17. N.A. Popova, L.G. Zhurerova, E.L. Nikonenko, M.K. Skakov, Vestnik of Tomsk State University. Chemistry **1**, 60 (2016) (in Russian)
18. N.A. Popova, L.A. Erygina, E.L. Nikonenko, M.K. Skakov, N.A. Koneva, E.V. Kozlov, Bull. Russ. Academy Sci. Physics **81**(3), 354 (2017)
19. G.V. Kurdyumov, L.M. Utevskiy, R.I. Entin, *Prevrashhenija v zheleze i stali* [Transformations in iron and steel] (Nauka, Moscow, 1977) (in Russian)

Investigation of steel to dielectric transition using microminiature eddy-current converter

Vladimir Malikov[1,*], *Sergey* Dmitriev[1], *Alexey* Ishkov[2], *Alexander* Katasonov[1], and *Anatoly* Sagalakov[1]

[1]Altai State University, 634049 Barnaul, Russia
[2]Altai State Agricultural University, 634049 Barnaul, Russia

Abstract. The research aims to develop a microminiature converter for electrical steel investigation. The research topic is considered relevant due to the need for evaluation and forecasting of safe operating life of electric steel products. The authors determined the capability to study steel characteristics at different depths based on variations of eddy-current converter amplitude at the steel-dielectric boundary. A microminiature transformer-type converter was designed, which enables to perform local investigations of ferromagnetic materials using eddy-current method based on local studies of the steel electrical conductivity. Having the designed converter as a basis, a hardware-software complex was built to perform experimental studies of steel at the interface boundary. A system was developed for automated converter relocation above the studied object at a specified velocity. Test results are reported for a specimen with continuous and discrete measurements taken at different frequencies. Response dependence of eddy-current converter was found to demonstrate non-linear behavior at steel to dielectric transition. The effect of gap between the eddy-current converter and the test object is investigated.

Introduction

Today, the issue of accurate and reliable evaluation of technical condition of hazardous production facilities has special importance. In modern control methods such evaluation is mainly based on probabilistic approaches and is performed using actual data on operation modes and existing stresses, deformations and temperatures. While structure and surface changes taking place in metals are almost neglected [1, 2]. However, it is known [3] that when stressed metal experiences changes not only in volume, but on its surface. In this regard, it should be reasonable to evaluate changes occurring in metals using methods and devices that allow tracking properties of the surface.

Electric steel is applied for manufacturing various electrical products. Magnetic properties of steel are essential for developing effective control and diagnostic methods for their physical and chemical characteristics which largely define reliability, operating life and remaining useful life of critical joints, components, products in various industries. The works [4-6] demonstrate successful use of hysteretic loop parameters, as well as magnetic

* Corresponding author: osys11@gmail.com

and magnetoacoustic noises for control of physical and chemical properties of ferromagnetic steels.

Eddy-current method is also proved [7, 8] effective for investigation on physico-mechanical properties of steel. This method is a widely used means of non-destructive testing based on analyzing interaction between external electromagnetic field and eddy currents induced by this field.

The works [9-11] suggest the method for brittle fracture evaluation in metal equipment made of 09Г2С low-alloy steel by means of electromagnetic control method with the use of transmission gages and attachable gages.

The eddy-current control method is often used for evaluation of various steel parameters. The investigations [12, 13] demonstrated practicality of using such method for fault location in AISI 304 steels. At the same time, the developed gage positioning system with the pitch of 0.1 mm was used for accurate gage movement. The work [14] highlights that one of the obvious advantages of using eddy-current method for investigation of conductive materials is its applicability for diagnostic of materials with paint and varnish, and other non-conductive coatings.

The purpose of this study is to demonstrate the ability to scan steel to dielectric transition for further investigation of steel surface characteristics. In this regard, it is necessary to define the mode of decrease of induced voltage of eddy-current converter when distancing from the test object, and the cause of such decrease. Another relevant parameter is the mode of induced voltage increase in the converter when approaching the test object. Presentation of induced voltage changes in the form of mathematical dependences enables to elicit the ways of interaction between electromagnetic field of eddy-current converter and residual magnetic field of steel.

The research is aimed at development, investigation, optimization and testing of microminiature eddy-current converter for steel analysis. An important task within the research is to obtain dependences describing the range of response of the eddy-current converter to the change of its parameters and physico-mechanical properties of the studied steels.

Materials and Methods

A unit based on eddy-current converter (ECC) [15-18] and a digital displacement transducer (DDT) were used for measuring stress-strain behavior on the boundary of dielectric and conductive ferromagnetic space.

Eddy-current converter represents a transformer with measuring, transmitting and compensation coils and a coil flux guide located inside a cylindrical platform. Grooves are arranged on the external side of the platform to locate the coils impregnated with compound at the temperature of 200 Co. This prevents destruction when applying ferrite screen meant for localization of electromagnetic field on the test object. On the exterior, ECC is put into alumina bead which protects the core from contacting with the test object.

Parameters of the designed ECC allow effectively localizing magnetic field within the area of 2500 mkm^2 and provide its considerable penetration deep into the test object when operating at low frequencies [16].

Measurement system is based on a sine-wave voltage generator. The voltage is transmitted to the transmitting coils of the ECC and generates electromagnetic field which penetrates into the test object and induces eddy currents in it. Eddy currents induce voltage in the measuring coil, which carries information on the parameters of the test object. The voltage is settled and cleared using selective amplification system with a glitch filter.

After filtration the signal is transmitted to envelope detector and then to the PC through an analog-to-digital converter (ADC). Frequency of the generated signal and cut-off frequency of filtration system change simultaneously. This makes possible to distinguish the useful signal that carries information on the test object parameters. Programmed control enables to vary operating frequency of the measuring system.

Experimental study was performed using two materials located with 1 cm gap between each other. Scanning measurements were initially taken in electric steel of type 1212 (specimen No.1) through a dielectric (paper), and then in steel of type 3414 (specimen No.2). Frequency of signal transmitted to the transmitting coil of the ECC varied within the range of 1000-10000 Hz. The measurable parameter was voltage induced in the measuring coil of the ECC in the process of its displacement from the initial scanning point. The converter displacement was measured using a digital displacement transducer (DDT). The data were digitalized with the help of an ADC and then in real-time mode were transmitted to the software that controls the measuring system. Origin math tool was used for data analysis.

The induced voltage was measured both continuously by moving the EDD with the constant velocity of 1 mm/s, and discretely by moving the EDD with the pitch of 0.1 mm and time of 0.5 s for each pitch. The described measurement method is determined by characteristics of the developed hardware-software complex, such as ADC sampling frequency and DDT operating speed, and is proved effective for steel investigation [13].

Experimental results

For the purpose of detailed study of induced voltage on the ferromagnetic-dielectric boundary, a dependence curve of induced voltage amplitude on the gage position was obtained. This curve demonstrates the behavior of induced voltage amplitude variation while moving the ECC above the steel-dielectric boundary. The variations were observed starting from the point located 5 mm away from the edge of the first specimen up to the point corresponding to the distance of 5 mm from the edge of the second specimen. This way, one can observe ferromagnetic-dielectric interface which is qualitatively related to the distance up to and beyond the edge of the test object. The dependence curve of induced voltage amplitude on the gage position explicitly demonstrates the effects occurring on the interface. The experimental results are presented in Figure 1.

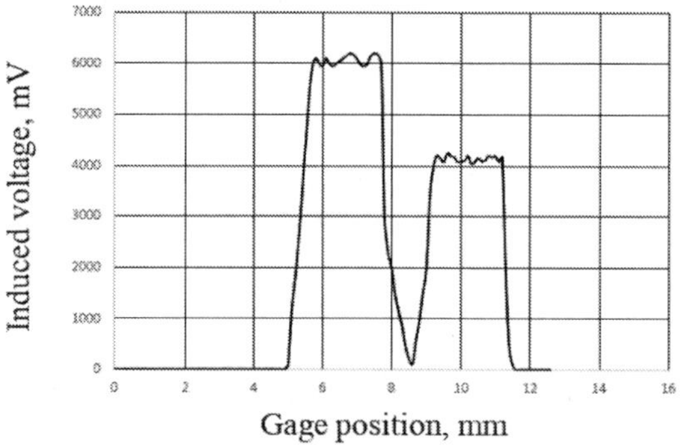

Fig. 1. Dependence curve of induced voltage on the gage position while continuous scanning at 1000 Hz.

Difference of peaks of the curves depends on the permeability of electric steel. The curve demonstrates that in the dielectric area the induced voltage drops from 6000 mV to 2000 mV and then smoothly tends to zero. For more detailed study of induced voltage drop on the ferromagnetic-dielectric boundary, a curve for discrete dependence of induced voltage amplitude on the gage position was obtained (Fig. 2). Discrete scanning of objects was planned within the experiment. Measurements were performed in points located at 0.1 mm from each other.

As a result of the experiment the following conclusion can be drawn: on the interface, ferromagnetic response does not drop to zero value, but decreases according to square law (area 1). The minimum value of induced voltage amplitude is 50 mV. Increase of the signal amplitude when approaching the second steel specimen occurs according to a law similar to exponential one (area 2).

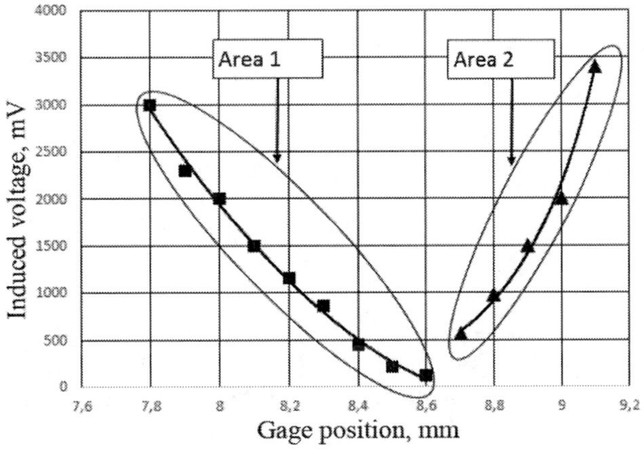

Fig. 2. Dependence curve of induced voltage amplitude on the gage position while discrete scanning at 1000 Hz.

Such dependence can be explained by residual voltage in coils of ECC. Continuous field lines are distinguished in the field of electric steel. Therefore, electromagnetic field that prevents occurrence of self-inductance in the measuring coil and is generated by eddy currents will have little effect on induced voltage. At the same time, the response value will not drop to zero due to ferromagnetic influence on the gage.

Such influence is explained by presence of self-magnetic field of steel which is closed on the ECC even at relatively big distance. In this case, the value of induced voltage is considerably lower than the value of voltage induced in the converter when positioning the gage right above the steel specimen. This voltage keeps reducing with the increase of distance between the gage and the test object.

When approaching the second specimen, induced voltage amplitude increases. In such case, magnetic fields of the first and the second specimens affect the gage simultaneously. Exponential growth of induced voltage amplitude is caused by adding electromotive forces (EMF) of both specimens' fields.

Another experiment was performed to assess the effect of ECC signal frequency on measurement results. The dependence curve of induced voltage amplitude on the gage position was obtained, which demonstrated variation of the amplitude when moving the gage at the frequency of 10000 Hz. The experimental results are presented in Figure 3.

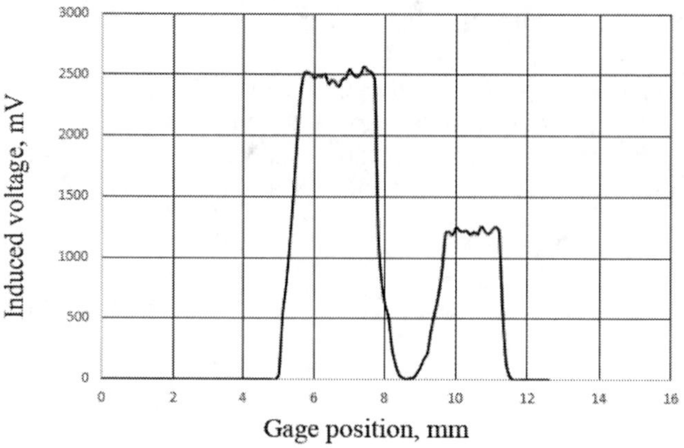

Fig. 3. Dependence curve of ECC induced voltage on the gage position while continuous scanning at 10000 Hz.

The curve shows that the gage position has almost no effect on qualitative dependence of ECC induced voltage. The two peaks of induced voltage correspond to two types of studied steel. Drop of induced voltage is caused by dielectric which separates the two specimens. One can well notice the varied rate of induced voltage drop and the minimum value of induced voltage (5 mV) which corresponds to the moment when the gage comes through the dielectric. Such dependence can be explained by increased magnetic loss in steel with the increase of current frequency in the ECC up to 10000 Hz.

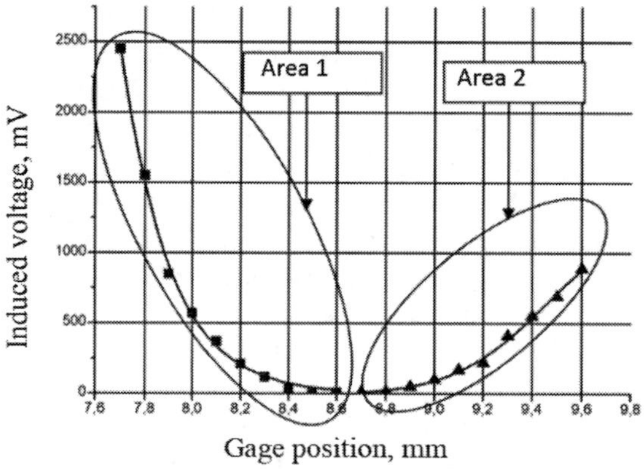

Fig. 4. Dependence curve of induced voltage on the gage position while discrete scanning at 10000 Hz.

Magnetic field of specimens at such frequency has considerably smaller effect on induced voltage compared to that corresponding to 1000 Hz. Figure 4 shows that induced voltage drop in area 1 occurs according to exponential law $= ae^{bx}$. Increase of induced voltage in area 2 also leads to exponential law $y = e^{a+bx+cx^2}$.

During the final experiment the influence of gap between ECC and electric steel on induced voltage was studied. Within the experiment discrete scanning of specimen No.1 was performed. The gap grows with the pitch of 0.1 mm, measurements were taken at the frequency of 1000 Hz. The experimental results are presented in Figure 5.

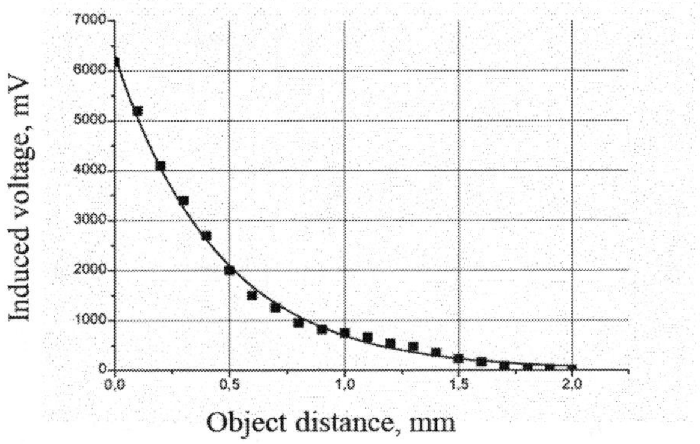

Fig. 5. Dependence curve of induced voltage on gap while discrete scanning at 1000 Hz.

The curve demonstrates that when increasing the distance between the gage and the test object induced voltage decreases according to exponential law. This can be justified by the fact that penetration depth of electromagnetic field in transmitting coil is insufficient for generating counterfield in ferromagnetic which will be strong enough to generate induced voltage in measuring coil of the converter. Besides, the effect of magnetic lines on coils of the ECC becomes weaker as the distance grows.

Conclusion

As a result of the research, a measurement system for electric steel parameters is developed. The investigation revealed practicability of scanning steel to dielectric transition using eddy-current converter. Obtained measurement data allowed describing interaction between magnetic field of ECC and residual magnetic field of steel.

In addition, mathematical functions were obtained to describe variation of induced voltage while scanning similar transitions at various ECC signal frequencies. Changing ECC signal frequency enables to investigate test objects at various depths, but it is required to establish unique dependence between the ECC response and the penetration depth of ECC magnetic field into various types of steel. Moreover, mathematical functions were obtained to describe variation of induced voltage with varied gap, which allows using the developed system for scanning objects with paint and varnish, and other non-conductive coatings.

Based on the conducted research a certain conclusion can be made on the state of the studied object with regard to behavior of ECC induced voltage variation. The investigation also proves that the measurement system is appropriate for thickness measurement and for studying structural properties of steel. Further extension of the method will allow assessing quality of studied material by its comparison with check specimens of corresponding types of steel.

The presented eddy-current converter for structural materials investigation can be used for diagnostic and control of critical joints and structural machine elements used in electrical engineering.

References

1. I.R. Kuzeev, *Jelektromagnitnaja diagnostika oborudovanija neftehimicheskih i neftepererabatyvajushhih proizvodstv* [Electromagnetic diagnostics of equipment at petrochemical and oil refining enterprises] (UGNTU, Ufa, 2001) (in Russian)

2. O.G. Kondrashova, Opredelenie resursa bezopasnoj jekspluatacii neftegazovogo oborudovanija putem ocenki adaptivnyh svojstv metalla po izmeneniju ego magnitnyh harakteristi [Determination of safe operating life of oil and gas equipment by estimating the adaptive properties of metal on changing its magnetic characteristics], PhD thesis (Ufa, 2006) (in Russian)
3. V.S. Ivanova, *Razrushenie metallov* [Fracture of Metals] (Metallurgiya, Moscow, 1979) (in Russian)
4. V.V. Kluyev, *Nerazrushajushhij kontrol'* [Non-Destructive Testing] (Mashinostroenie, Moscow, 2004) (in Russian)
5. V.V. Filinov, V.E. Shaternikov, P.G. Arakelov, Russ. J. Nondestruct. **50**(12), 58 (2014)
6. V.V. Filinov, V.E. Shaternikov, Russ. J. Nondestruct. **50**(10), 47 (2014)
7. N.M. Rodigin, I.E. Korobeynikova, *Kontrol' kachestva izdelij metodom vihrevyh tokov* [Eddy-current testing of products quality] (Mashgiz, Moscow, 1958) (in Russian)
8. V.V. Kluyev, *Pribory dlja nerazrushajushhego kontrolja materialov i izdelij* [Devices for non-destructive testing of materials and products] (Mashinostroenie, Moscow, 1976) (in Russian)
9. R. Ghoni, M. Dollah, A. Sulaiman, F.M. Ibrahim, Advances in Mechanical Engineering **6**, 1 (2015)
10. E.M. Bashirova, *Ocenka predel'nogo sostojanija metalla oborudovanija dlja pererabotki uglevodorodnogo syr'ja s primeneniem jelektromagnitnogo metoda kontrolja* [Using electromagnetic inspection method for evaluation the extreme limit state of the metal of equipment for processing hydrocarbon raw materials] (Ufa, 2005) (in Russian)
11. T.R. Bikbulatov, E.A. Naumkin, V.A. Petrov, Himicheskaja tehnika [Chemical Equipment] **3**, 11 (2010) (in Russian)
12. J.M. Szlagowska-Spychalska, M.M. Spychalski, K.J. Kurzydlowski, NDT and E International **54**, 56 (2013)
13. G. Almeida, J. Gonzalez, L. Rosado, P. Vilaça, T.G. Santos, *Procedia CIRP* **7**, 359 (2013)
14. D. Rifai, A.N. Abdalla, ARPN J. Eng. Appl. Sci. **11**, 1 (2015)
15. S.F. Dmitriev, A.O. Katasonov, V.N. Malikov, A.M. Sagalakov, Russ. Eng. Res. **36**(8), 626 (2016)
16. V.N. Malikov, S.F. Dmitriev, M.A. Davydchenko, A.M. Sagalakov, A.V. Ishkov, AIP Conf. Proc. **1785**, 030014 (2016)
17. V.N. Malikov, S.F. Dmitriev, A.M. Sagalakov, A.O. Katasonov, Russ. J. Nondestruct. **52**(1), 32 (2016)
18. S.F. Dmitriev, V.N. Malikov, A.M. Sagalakov, L.I. Shevtsova, Welding International **31**(8), 608 (2017)

Narrow-aperture electron beam in the forevacuum pressure range as a tool for dimensional processing of silica glass

Aleksey Zenin[1,*], *Aleksandr* Klimov[1], and *Ilya* Bakeev[1]

[1]Tomsk State University of Control Systems and Radioelectronics, 634040 Tomsk, Russia

Abstract. The paper gives overview of the results of using narrow-aperture electron beam generated by a forevacuum plasma electron source as a tool for treatment of electrically non-conductive materials, such as technical glass. Effectiveness of electron beam treatment of non-conductive materials is caused by almost complete neutralization of target charge on the treated surface. Charge neutralization occurs due to ion flow from beam plasma generated while transporting the beam through forevacuum pressure area. The study demonstrates the possibility to control depth and form of milling by changing modes of electron beam treatment. Combined use of electron beam and automated system for beam deflection and sweep makes possible to perform dimensional processing, particularly cutting and milling with complex trajectory. Milling rates were experimentally found depending on treatment time and mode. The suggested method of silica glass treatment represents an alternative for traditional treatment methods.

Introduction

Due to the unique properties, such as high electric strength, low coefficient of thermal expansion, and high chemical resistance, silica glass is widely used in numerous industries [1]. The main silica glass treatment methods consist in either mechanical removal of foreign materials and polish or thermal treatment. Mechanical treatment often causes formation of cracks and microcracks in treated material, which considerably decreases strength parameters of products [2]. Thermal treatment (by torch flame) does not allow producing holes with complex shape and clearly defined dimensions and may add carbon impurities in the material composition, which has negative effect on its optical properties [3].

The technology of laser controlled thermal splitting can be applied for cutting thin glass products with maximum thickness of 3 mm [4]. Whereas electron beam treatment can be effectively used for performing operations with thick-wall silica glass products and products with complex shape, as well as for removing shortcomings of the existing treatment methods [5-8]. However, electron beams which normally function in high

* Corresponding author: Zenin1988@gmail.com

vacuum (up to 10^{-1} Pa) are difficult to use for dielectric treatment due to accumulation of negative surface charge in the treated product [8].

Advance of plasma electron sources able to function in forevacuum pressure areas between 10 and 100 Pa [8] considerably extend opportunities for electron beam treatment of electric non-conductive materials and for their surface modification. Hence, due to presence of dense plasma in the beam transport area such electron sources are capable of performing treatment of both conductive materials and products and dielectrics without any means of surface charge compensation. Effectiveness of electron beam treatment of dielectrics using such sources is showed using the example of welding and vaporization of ultra-high temperature ceramics, sintering of ceramics, and polymer surface modification [9, 10].

Previous studies of the research group on obtaining narrow-aperture electron beams in forevacuum pressure area [11] open up new opportunities of electron beam technology, such as precision milling and dielectric cutting and sizing. This research is focused on studying the interaction between silica glass and electron beam generated by forevacuum plasma source to demonstrate opportunities of using such beam as an alternative sizing tool.

Experimental method

The experiment was performed using a vacuum setup (Fig. 1) with an installed forevacuum electron plasma source.

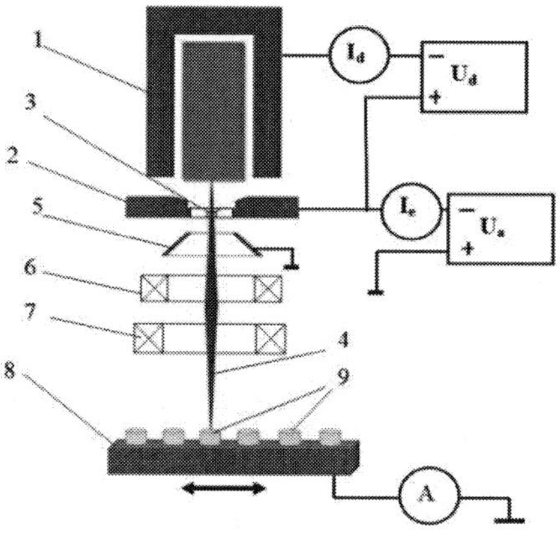

Fig. 1. Experimental unit for measuring shape and depth of milling: 1 – hollow cathode, 2 – anode, 3 – perforated electrode, 4 – electron beam, 5 – accelerating electrode, 6 – focusing system, 7 – deflection system, 8 – carriage, 9 – treated specimens.

Emission plasma in the source was generated by glow discharge with hollow cathode *1*. Electron beam was formed by extracting electrons from plasma boundary through the hole in perforated electrode *3* installed in anode *2*. Accelerated electron beam *4* was focused by magnetic lens *5* and was directed to the treated specimen *9* using magnetic beam deflection system *6*. The electrode system and structural peculiarities of electron source, as well as its parameters and characteristics are described in more detail in the works [8-10]. Silica glass rods having 10 mm in diameter and 30 mm in length were used for investigation of hole form and milling depth. Use of movable carriage allowed performing a series of experiments in a single process cycle [11]. In order to measure the hole form and milling depth, the time of specimen exposure to electron beam was set in the range between 10 and 60 seconds, accelerating voltage was constant and was set in the range between 10 and 19 kV. The beam current was constant during the experiment and was equal to 10 mA. The beam had the diameter of 1 mm and was controlled by the parameters of focusing system. Helium was used as working gas with the pressure of 30 Pa.

Results and discussion

Experimental studies on electron beam treatment of silica glass showed that intense material vaporization occurs at accelerating voltage of 10 kV and a molten pool is formed in the treated area. Due to this effect, deepening of the region of electron beam absorption causes formation of knife-like cavitation, i.e. a capillary the depth of which is several times greater that the orifice diameter. Figure 2 presents various characteristic forms of capillaries formed at different accelerating voltages.

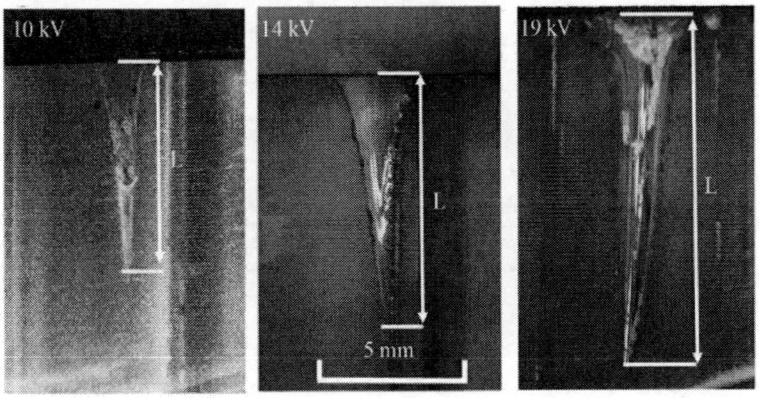

Fig. 2. Cutoffs of obtained cavitations, treatment time 50 s.

The observed intense material evaporation and formation of knife-like cavitation is an effective means of silica glass cutting. Milling depth is precisely controlled by the electron beam parameters and treatment time. This way, the dependence of milling depth on accelerating voltage is almost linear (Fig. 3). However, the dependence of treatment time on penetration depth has a more complex nature (Fig. 4). With treatment time less than 30 seconds the milling depth grows almost independently of accelerating voltage. When increasing treatment time, the depth difference grows for various accelerating voltage values.

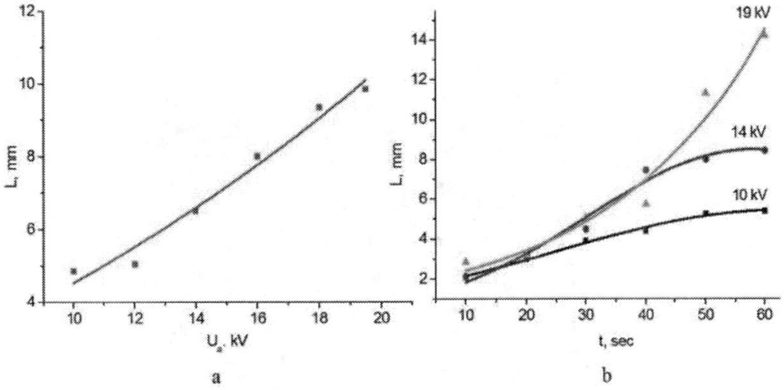

Fig. 3. Dependence of milling depth on electron beam treatment parameters: a) on accelerating voltage with treatment time of 40 seconds, b) on treatment time for various accelerating voltage values.

Figure 4 demonstrates dependences of removed mass on treatment time for various accelerated voltage values. These dependences are obviously similar to the above dependences of penetration depth on treatment time. However, certain differences can be found, such as ablative mass with constant milling depth for accelerating voltage of 10 and 14 kV, which can be related to minor change of electron beam diameter.

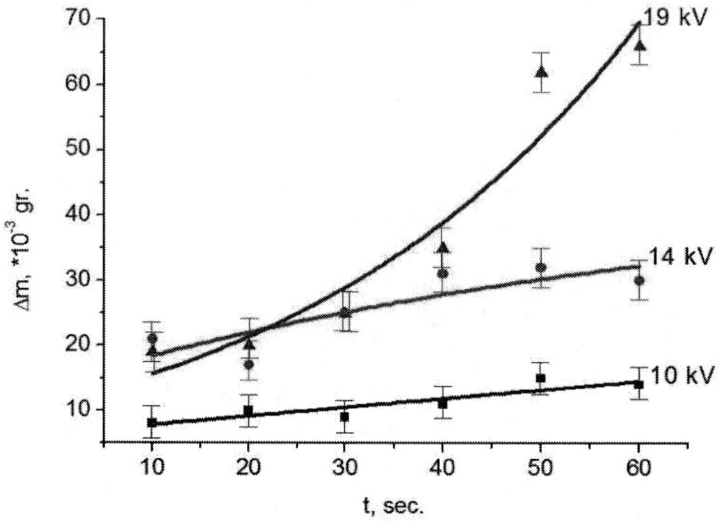

Fig. 4. Dependence of ablative mass on treatment time and electron beam parameters.

The given dependences demonstrate that, as one could expect, increase of accelerating voltage causes proportionate growth of milling depth (Fig. 3). In this way, increase of accelerating voltage from 10 to 19 kV almost doubles the milling depth. Change of duration of electron beam treatment of a silica glass specimen also results in increasing the milling depth.

The use of automated system for electromagnetic beam deflection allows moving the beam with complex trajectory at a given rate, which enables to cut long objects with intricate shape. Below, Figure 5 presents an example of using finely focused electron beam for silica glass sizing. Silica plate 300x50 mm in size and 4 mm thick was used as test material for milling. Accelerated voltage was 18 kV, full milling time was about 4 minutes, the number of passes the electron beam performs over a specimen was 15.

Fig. 5. Example of electron beam milling of a silica plate.

It should be noted that no cracks are observed on the edges of the formed cavitation. The developed treatment method allows avoiding crack formation in treated products, as well as treating objects with intricate shape. Automated beam positioning system allows building complex trajectories while treating objects.

Conclusion

The performed research demonstrates that a knife-like cavitation is formed while treating silica glass specimens with electron beam. The milling depth (provided the transverse beam size is constant) can be controlled by varying accelerating voltage and treatment time. Increase of both accelerating voltage and treatment time results in growth of depth.

Removed mass rate while milling may vary between 10 and 70 mg per minute depending on electron beam treatment parameters. The research demonstrates the possibility of silica glass dimensional cutting and milling with complex trajectory and depth control by means of beam positioning and sweep with the help of automated system for electromagnetic beam deflection. The suggested electron beam treatment method for silica glass represents an alternative for traditional treatment methods. Further extension of the research shall be reduction of electron beam diameter for reducing geometrical parameters of holes, as well as investigation of peculiarities of electron-beam heating of glass with the purpose of welding.

The work has been conducted with the financial support of Russian Foundation for Basic Research, project No.16-38-60032 mol_a_dc.

References

1. N.P. Bansal, R.H. Doremus, *Handbook of glass properties* (Elsevier, New York, 2013)
2. P.V. Volkov, A.N. Bryzgalov, D.E Zhivulin, Scientific achievements in biology, chemistry, physics, Proceedings of VI International Scientific Conference (Novosibirsk, April 4, 2012) (in Russian)
3. A.Y. Legoshin, L.A. Manuylov, *Stekloduvnoe delo* [Glassblowing] (Vysshaya Shkola Publ., Moscow, 1976) (in Russian)
4. V. Kondratenko, P.S. Tchernykh, Phys. Stat. Sol. **7**, 2232 (2007)
5. S. Kornilov, N. Rempe, A. Beniyash, N. Murray, *Journal Of Physics: conference series* **552**(1), 012004 (2014)
6. M.St. Weglowski, S. Błacha, A. Phillips, Vacuum **130,** 72-92 (2016)
7. T.K. Saha, M. Mascarenhas, E. Kandaswamy, The International Symposia on Discharge and Electrical Insulation in Vacuum (ISDEIV) (Mumbai, September 28-October 3, 2014)
8. A. Zenin, A. Klimov, V. Burdovitsin, E. Oks, Technical Physics Letters **39**(10), 454-456 (2013)
9. V.A. Burdovitsin, E.M. Oks, Laser and particle beams **26**, 04, 619-635 (2008)
10. A.A. Zenin, I.Y. Bakeev, Y.A. Burachevskii, A.S. Klimov, E.M. Oks, Technical Physics Letters **42**(7), 712-714 (2016)
11. S. Schiller, U. Heisig, S. Panzer, *Electronenstrahltechnologie* (VEB Verlag Technik, Dresden, 1976)

Aluminum surface modification by electron-ion-plasma methods

Olga Krysina[1,*], *Elizaveta* Petrikova[1], *Vladimir* Shugurov[1], *Pavel* Moskvin[1], and *Yurii* Ivanov[1]

[1]Institute of High Current Electronics, Siberian Branch, Russian Academy of Sciences, 634055 Tomsk, Russia

Abstract. The paper focuses on detection and structural-phase justification of the modes of combined electron-ion plasma treatment of commercially pure A7 grade aluminum carried out in a single vacuum cycle and enabling to enhance mechanical (microhardness) and tribological (wear resistance) properties of the material. Commercially pure A7 grade aluminum underwent combined surface treatment, including deposition of titanium coating by means of vacuum-arc technique and further mixing of the coating/substrate system by intense pulsed electron beam. The varied parameters were energy density of the electron beam (10, 15, 20) J/cm^2 and the number of impact pulses (3-100); the thickness of titanium coating was 0.5 μm. Electron-ion plasma treatment of aluminum was carried out in a single vacuum cycle. Optical and scanning electron microscope investigations, measuring of microhardness and tribological tests allowed defining the modes when hardness and wear resistance of the modified surface layer increases manifold in comparison to the initial properties of commercially pure aluminum.

Introduction

Aluminum and its alloys are widely used in industry due to their relatively high specific strength, satisfactory corrosion resistance and good processing qualities for mechanical treatment methods [1-3]. However, along with those advantages aluminum is also characterized by low hardness and wear resistance [1-4]. Various surface modification techniques are often used to enhance the above mentioned properties of metals and alloys (including aluminum), as well as sintered metal powder materials, such as deposition of hard, superhard and wear resistant coatings, surface treatment with concentrated energy flows, surface saturation with metals and gases atoms, etc. [1-14].

Deposition of various coatings is considered as an advanced way to increase dramatically the strength and tribological properties of metals and alloys. It can be found in recent studies that coatings with complex multi-phase nanocrystalline or amorphous structure possess unique properties, such as superhardness ($\geq$40 GPa) [15, 16], oxidation stability at high temperatures (>1000°C) [17, 18], and low friction coefficient ($\leq$0.1) [19]. Special interest is gained by traditional wear resistant materials coatings based on titanium

[*] Corresponding author: krysina_82@mail.ru

nitride, possessing nanocrystalline structure by introduction of dopants into composition (Si, Al, Cr, Cu etc.). The interest to these systems is conditioned by their high wear resistance and hardness (Ti-Cu-N – up to 45 GPa, Ti-Si-N – up to 80 GPa) while being relatively easy to obtain. The research [20] shows that in coatings of Ti-Si-N system basic phase nanosized crystalline grains (TiN) are incorporated into amorphous silicon nitride matrix (Si_3N_4). In other case when introducing metal coatings that do not form compounds with nitrogen into MeN composition, such as Cu, Ni, Y, Au etc. (the coating of nc-MeN/soft phase system is formed (Me=Ti, Zr, Cr, Ta etc.)), atoms of dopants encircle nitride crystalline grains of the basic element thus restricting their growth in nanometer range [21].

The basic disadvantage of the coating deposition technique is not high enough properties of the coating/substrate system provided by large difference in physical and mechanical properties of the substrate and the coating being formed. The mentioned drawback is compensated by preliminary modification of the surface layer of the substrate by leaving the structure and properties of the material bulk unchanged [22]. Electron-ion plasma treatment is distinctive with high effectiveness level of modifying the structure and properties of the material surface layer. This is especially true for complex processes combining surface impacts of plasma flows, accelerated electron and ion beams [23].

Methods of combined electron-ion plasma treatment implemented so far include consequent modification of the structure and properties of metals and alloys surface layer using the combination of a several number of experimental setups [23, 24]. Drawbacks of the mentioned approach are obvious: high cost of maintenance of a series of separate setups, each of them having duplicating devices (vacuum system, working chamber, power supply and control board), relatively large durability of the cycle vacuuming – de-vacuuming, heating and cooling of the components, violation of purity of the treated surface (pollution of modified surface with atmospheric gases, etc. during samples placement from one setup to the other). The basic criterion in development of complex electron-ion plasma treatment is combination of its processes in a single vacuum cycle. For that purpose researchers of the Institute of High Current Electronics within the framework of the research upon the grant of the Russian Science Foundation (project No.14-29-00091) developed, designed and commissioned the COMPLEX setup of laboratory type. This setup enables to carry out material surface irradiation by intense pulsed electron beam in any sequence in a single vacuum cycle and saturates material with incorporation elements, deposit metal coatings and hard cermet and ceramic coatings. COMPLEX setup is unique in its combination of parameters and possibility of implemented processes and currently has no analogs both in Russia, and abroad. This fact defines undoubted research novelty and practical significance of the obtained results.

The present research focuses on detection and structural-phase justification of the modes of combined electron-ion plasma treatment of commercially pure A7 grade aluminum performed in a single vacuum cycle (COMPLEX setup) and enabling to enhance manifold the material mechanical (microhardness) and tribological (wear resistance) properties.

Materials and methods

Commercially pure A7 grade aluminum (Al – 99.7%; Fe – up to 0.16%; Si – up to 0.16%; Ti – up to 0.02%; Cu – up to 0.01%; Zn – up to 0.01%) was used as a modified material [25]. Combined modification of the structure and properties of commercially pure A7 grade aluminum was carried out on a COMPLEX laboratory setup [26]. The setup is intended for electron-ion plasma finishing modification of metals and alloys in a single vacuum cycle. The basic setup units are gas plasma source based on low pressure non-self-sustained arc discharge with combined thermionic and hollow cathode [27], source of electrons of submillisecond duration based on plasma cathode with grid stabilization of plasma boundary [28], arc evaporation source based on arc discharge with cold cathode [29].

The first stage of electron-ion plasma treatment included deposition of thin titanium coating 0.5 μm thick by vacuum arc method on the substrate made from commercially pure A7 grade aluminum. The second stage included electron-beam mixing of the coating/substrate system. The irradiation parameters were as follows: accelerated electrons energy 18 keV, the energy density of electron beam (10, 15, 20) J/cm^2, pulse duration of 50 μs, repetition rate and the number of impact pulses 0.3 s^{-1}, (3-100) pulses, respectively. Justification of selecting electron-ion plasma treatment for modification of commercially pure aluminum is defined by positive results of earlier studies performed on Al-12% Si alloy [23, 30].

Composition and structure of aluminum samples before and after modification were investigated by optical microscopy (μVizo-MET-221 metallographic microvisor), scanning electron microscopy and X-ray energy-dispersive analysis (Philips SEM-515 scanning electron microscope with EDAX ECON IV microanalyzer). Microhardness was measured using PMT-3 microhardness tester at the loading on the indenter 0.1 N. Tribological studies were carried out using Pin on Disc and Oscillating TRIBOtester (TRIBOtechnic, France) with the following parameters: a ball from steel 100Cr6, with the diameter 6 mm, track radius – 2 mm, load – 1 N, distance – 10 m. The chosen techniques and methodologies are well proved and widely used in material physics to study the structure and properties of metals and alloys. Therefore the validity of the obtained results is proved.

Results

It was established that microhardness of the system "coating (Ti) / (A7) substrate" significantly depends on the mode of intense pulsed electron beam irradiation (Fig. 1*a*). At the energy density of electron beam equal to 10 J/cm^2, microhardness of the surface layer increases with the growth of irradiation pulses number. Maximum value of microhardness was defined at irradiation of the coating/substrate system by electron beam with the following parameters: E_S = 10 J/cm^2, τ=50 μs, n = 50 pulses (Fig. 1*a*). Microhardness was increased by more than 4 times compared to that of the initial samples.

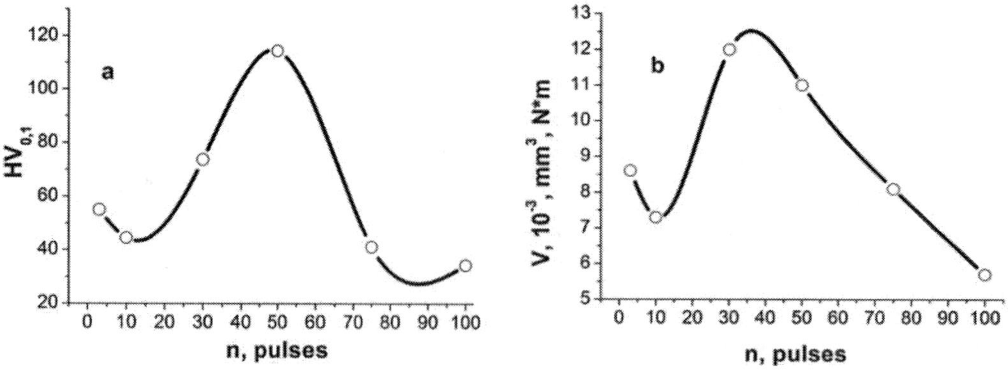

Fig. 1. Dependence of microhardness $HV_{0,1}$ (a) and wear parameter V (b) of modified surface of the commercially pure A7 grade aluminum on the number of irradiation pulses n by intense pulsed electron beam of the system "coating (Ti) / (A7) substrate" at the energy density of 10 J/cm^2.

During irradiation of samples by electron beam with energy density of 15 J/cm^2 microhardness is maximum at three impact pulses (it exceeds aluminum microhardness by more than 3.5 times) and it is reduced with the growth of the number of electron beam pulses. During irradiation of the system "coating (Ti) / (A7) substrate" by intense electron beam with energy density of 20 J/cm^2 microhardness of samples almost does not depend on the number of impact pulses and exceeds aluminum microhardness by 1.7 times.

Wear parameter (the value which is reverse to material wear resistance) is changed the similar way. Figure 1*b* presents dependence of the wear parameter on the number of impact pulses of electron beam by the energy density equal to 10 J/cm^2. As following from the results presented in Figure 1*b* it can be concluded that the maximum wear resistance, exceeding the value of the reference samples of commercially pure A7 grade aluminum by 2.3 times is observed in the samples treated by 100 pulses.

It is obvious that microhardness and wear resistance of material are defined by the state of substructure of modified layer [31-33]. Figure 2 illustrates electron-microscope images of structure formed at irradiating the system "coating (Ti) / (A7) substrate" by electron beam with the energy density of 10 J/cm^2. It is clearly seen that by three irradion pulses the titanium coating on the surface of aluminum sample remains unchanged (Fig. 2*a*). Resulting from thermoelastic stresses the coating is fragmented with microcracks; folded structure is formed on the surface. Increasing the number of irradiation pulses up to 50 results in formation of nanocrystalline structure in the sample surface layer; crystalline grains size being ≈70 nm (Fig. 2*d*). The grain size of the surface layer of these samples varies within the range of 15-20 µm.

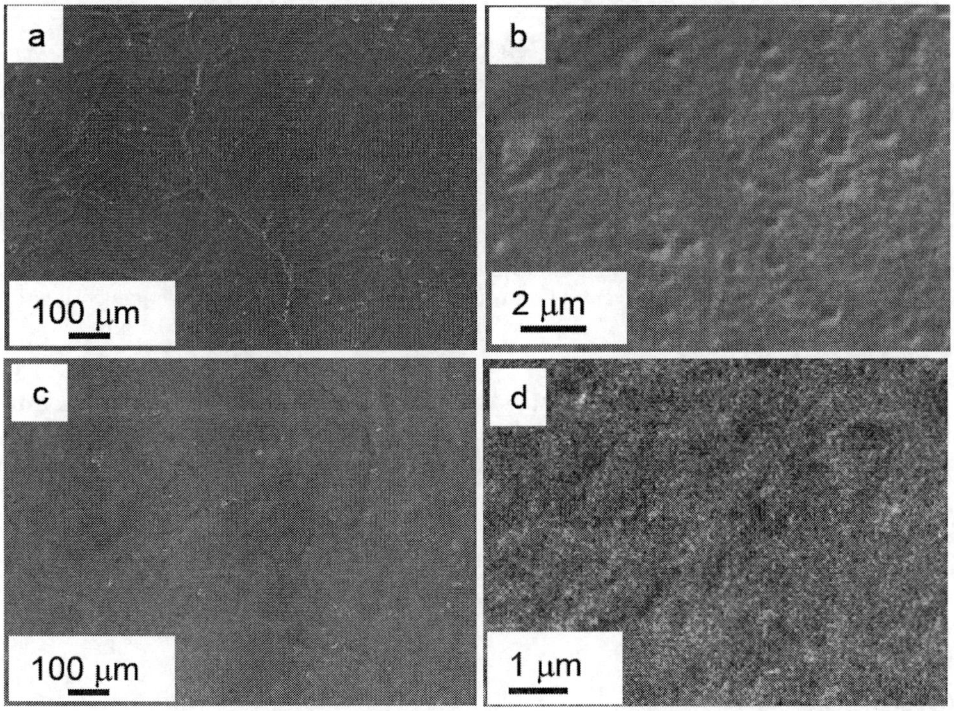

Fig. 2. SEM images: surface structure of the system "coating (Ti) / (A7) substrate" formed resulting from irradiation by intense pulsed electron beam at pulse duration of 50 µs and energy density of 10 J/cm^2; a, b – 3 pulses, c,d – 50 pulses; the thickness of titanium coating of 0.5 µm.

Increasing the number of irradiation pulses up to 100 at the energy density of electron beam of 10 J/cm^2 results in formation of fine-grained structure in the surface layer of commercially pure aluminum; the grain size varies within the range of 2.5-5 μm (Fig. 3*a*). Substructure of cellular crystallization is revealed in grain bulk; it is generally formed in alloys under conditions of ultrahigh speed crystallization [33]. The sizes of crystallization cells vary within the range of 250-500 nm (Fig. 3*b*).

Elemental composition of the surface layer of commercially pure aluminum was defined by electron probe microanalysis. It was established that irrespective of the energy density of electron beam, increasing the number of irradiation pulses leads to reduction of relative concentration of titanium atoms in the aluminum surface layer. It is obviously conditioned by titanium atoms leaving into the bulk of aluminum sample as a consequence of ongoing process of leveling diffusion. Growth of energy density of electron beams stimulates the process of titanium atoms rearrangement in the surface layer of the alloy.

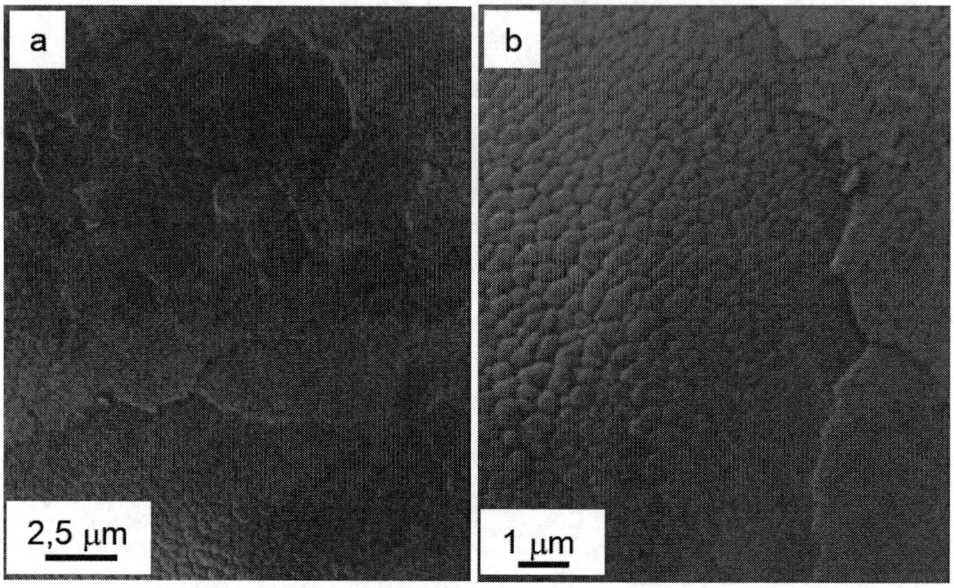

Fig. 3. SEM images: surface structure of the system "coating (Ti) / (A7) substrate" formed resulting from irradiation by intense pulsed electron beam at pulse duration of 50 μs and energy density of 10 J/cm^2; number of irradiation pulses – 100; the thickness of titanium coating of 0.5 μm.

Scanning electron microscopy investigations allow making the following assumptions. Relatively high microhardness of the surface layer of modified aluminum samples is conditioned by the formation of nanocrystalline structure (optimal irradiation mode by electron beam 18 keV, 10 J/cm^2, 50 μs, 50 pulses, 0.3 s^{-1}). Comparatively high wear resistance values taking place during melting of the system "coating (Ti, 0.5 μm) / (A7) substrate" (parameters of electron beam: 18 keV, 10 J/cm^2, 50 μs, 100 pulses, 0.3 s^{-1}) are conditioned by the formation of fine-grained polycrystalline structure with substructure of cellular crystallization.

Conclusion

Modification of the surface layer of commercially pure A7 grade aluminum was for the first time carried out on a COMPLEX laboratory setup in a single vacuum cycle. The process included formation of the system "coating (Ti) / (A7) substrate" and its further irradiation by intense pulsed electron beam. Modification modes and parameters of the formed material structure enabling to enhance microhardness and wear resistance of the modified layer manifold were defined, thus proving successful completion of the set objective of the present study.

The obtained results possess practical and economical significance as they justify new effective direction for forming high operational properties of metals and alloys. The method is based on modifying some surface layer of component or a product by leaving the structure and bulk properties generally unchanged. Further research in this direction will be devoted to search and justification of choosing alloying elements allowing controlling effectively the structure and properties of the component on the whole when applying electron-ion plasma modification to comparatively thin surface layer.

The work was supported by the Russian Science Foundation (grant No. 14-29-00091).

References

1. G.E. Totten, *Handbook of aluminium: Physical metallurgy and processes*, **1** (Marcel Dekker, Inc., 2003)
2. Sh.P. Dwivedi, S. Sharma, R.K. Mishra, Advanced Materials Manufacturing & Characterization, **4**(2), 81 (2014)
3. V.S. Zolotarevsky, N.A. Belov, *Metallovedenie litejnyh aljuminievyh splavov* [Study of aluminum cast alloys] (MISIS Publishing House, Moscow, 2005) (in Russian)
4. N.A. Belov, *Fazovyj sostav promyshlennyh i perspektivnyh aljuminievyh splavov* [Phase composition of industrial and perspective aluminum alloys] (MISIS Publishing House, Moscow, 2010)
5. N.A. Belov, *Fazovyj sostav aljuminievyh splavov* [Phase composition of aluminum alloys] (MISIS Publishing House, Moscow, 2009) (in Russian)
6. G.B. Stroganov, V.A. Rotenberg, G.B. Gershman, *Splavy aljuminija s kremniem* [Aluminum alloys with silicon] (Metallurgija, Moscow, 1977) (in Russian)
7. A.I. Ryabchikov, I.A. Ryabchikov, I.B. Stepanov, Vacuum **78** (2–4), 331 (2005)
8. N.N. Cherenda, V.V. Uglov, A.K. Kuleshov, V.M. Astashynski, A.M. Kuzmitski, Vacuum **129**, 170 (2016)
9. V.V. Uglov, N.N. Cherenda, V.M. Anischik, V.M. Astashinsky, N.T. Kvasov, *Modifikacija materialov kompressionnymi plazmennymi potokami* [Modification of materials by compressed plasma flows] (Belorussian State Univ., Minsk, 2013) (in Russian)
10. A.F. Oliveira, M.C. de Barros, K.R. Cardoso, D.N. Travessa, Materials Science and Engineering A. **379**, 1-2, 321 (2004)
11. K. Abedi, M. Emamy, Materials Science and Engineering: A. **527,** 3733 (2010)
12. *Surface Modification and Mechanisms. Friction, Stress, and Reaction Engineering* / Edited by G.E. Totten and H. Liang (Dekktr Inc, New York – Basel, 2004)
13. I. Schiller, J. Gubicza, Zs. Kovács, N.Q. Chinh, J. Illy, Materials Science Forum **519-521**, 835-840 (2006)
14. B. Vijaya Ramnath, C. Elanchezhian, R.M. Annamalai, S. Aravind, T. Sri Ananda Atreya, V. Vignesh, C. Subramanian, Rev. Adv. Mater. Sci. **38**, 55 (2014)
15. S. Veprek, M.G.J. Veprek-Heijman, P. Karvankova, J. Prochazka, Thin solid coatings **476**, 1 (2005)
16. J. Musil, Surface and coatings technology **125**, 322 (2000)

17. S. PalDey, S.C. Deevi, Materials Science and Engineering **A361**, 1 (2003)
18. J. Musil, J. Vlcek, P. Zeman, Advances in Applied Ceramics **107**, 148 (2008)
19. J. Musil, P. Novák, R.Čerstvý, Z. Soukup, J.Vac.Sci.Technol **A 28(2)**, 244 (2010)
20. S. Veprek, S. Reiprich, Thin Solid Coatings **268**, 64 (1995)
21. J. Musil, P. Baroch, P. Zeman, *Handbook of plasma surface engineering research and its practical applications* (Research Singpost, India, 2008)
22. V.A. Gribkov, F.I. Grigor'ev, B.A. Kalin et al., *Perspektivnye radiacionno-puchkovye tehnologii obrabotki materialov* [Prospective radiation beam material processing technologies] (Kruglyj stol, Moscow, 2001) (in Russian)
23. A.P. Laskovnev, Y.F. Ivanov, E.A. Petrikova et al, *Modifikacija struktury i svojstv jevtekticheskogo silumina jelektronno-ionno-plazmennoj obrabotkoj* [Modification of the structure and properties of eutectic silumin using electron-ion-plasma treatment] (Belorusskaja nauka, Minsk, 2013) (in Russian)
24. V.E. Gromov et al., *Struktura, fazovyj sostav i svojstva titana posle kompleksnyh uprochnjajushhih tehnologij* [Structure, phase composition and properties of titanium after application of integrated strengthening technologies] (SibGIU Publishing, Novokuznetsk, 2015) (in Russian)
25. F.I. Kvasov, I.N. Fridlyander, *Promyshlennye aljuminievye splavy: spravochnik* [Industrial aluminum alloys: Reference book] (Metallurgija, Moscow, 1984) (in Russian)
26. V.N. Devyatkov, Yu.F. Ivanov, O.V. Krysina, N.N. Koval, E.A. Petrikova, V.V. Shugurov, Vacuum **143**, 464 (2017)
27. L.G. Vintizenko, S.V. Grigoriev, N.N. Koval, V.S. Tolkachev, I.V. Lopatin, P.M. Schanin, Russ. Phys. J. **44**(9), 927 (2001)
28. V.N. Devyatkov, N.N. Koval, P.M. Schanin, Russ. Phys. J. **44**(9), 937 (2001)
29. A.A. Andreev, L.P. Sablev, V.M. Shulaev, S.N. Grigoriev, Vakuumno-dugovye ustroistva i pokrytiia [Vacuum-arc devices and coatings] (Kharkov NSC KIPT Press, Kharkov, 2005) (in Russian)
30. Y.F. Ivanov, V.E. Gromov, S.V. Konovalov, K.V. Aksenova, *Ustalost' silumina, modificirovannogo jelektronno-puchkovoj obrabotkoj* [Fatigue life of silumin treated with a high intensity pulsed electron beam] (Poligrafist, Novokuznetsk, 2016) (in Russian)
31. M.A. Shtremel', *Prochnost' splavov. Chast' II. Deformacija. Uchebnik dlja vuzov* [Strength of alloys. Part II. Deformation: Textbook for higher educational institutions] (MISIS Publ., Moscow, 1997) (in Russian)
32. Y.F. Ivanov, V.E. Gromov, N.A. Popova, S.V. Konovalov, N.A. Koneva, *Strukturno-fazovye sostojanija i mehanizmy uprochnenija deformirovannoj stali* [Structural and phase state and strengthening mechanisms for deformed steel] (Poligrafist, Novokuznetsk, 2016) (in Russian)
33. V. Rotshtein, Y. Ivanov, A. Markov, *Materials surface processing by directed energy techniques* (Elsevier, 2006)

Surface treatment by the ion flow from electron beam generated plasma in the forevacuum pressure range

Aleksandr Klimov[1,*], and *Aleksey* Zenin[1]

[1]Tomsk State University of Control Systems and Radioelectronics, 634050 Tomsk, Russia

Abstract. The paper presents research results of peculiarities of gas ion flows usage and their generation from large plasma formation (>50 sq.cm) obtained by electron beam ionization of gas in the forevacuum pressure range. An upgraded source was used for electron beam generation, which allowed obtaining ribbon electron beam with no transmitting magnetic field. Absence of magnetic field in the area of ion flow formation enables to obtain directed ion flows without distorting their trajectories. In this case, independent control of current and ion energy is possible. The influence of electron beam parameters on the parameters of beam plasma and ion flow – current energy and density – was determined. The results of alumina ceramics treatment with a beam plasma ions flow are given.

Introduction

Electron-beam plasma generated while electron beam is transmitted through gaseous atmosphere of a vacuum chamber with the energy of several keV [1] is widely used in various material treatment technologies [2-5]. The parameters of beam plasma can be controlled in a wide range, unlike those of various types of discharge-generated plasma. Variation of electron beam current and energy, as well as composition of gaseous atmosphere, enables variation of plasma density and temperature. Due to low temperature of electron component and relatively high concentration beam plasma is used for etching of thin (monoatomic) surface layer [6-7].

Plasma formation with the electron temperature higher than several tens of eV can be obtained by providing conditions for combined interaction of electron beam and plasma, i.e. beam plasma instability [8]. Because of such instability beam electrons transmit part of their energy to Langmuir waves which while damping transmit their energy to thermal plasma electrons [9]. As a result, chemical activity of plasma increases, and such plasma can be effectively used in plasma-chemical technologies [10].

It should be noted that the optimum pressure range for generating dense beam plasma is between 1 and 100 Pa [11]. High-voltage glow discharge (HVGD) electron sources are more often used for this pressure range [12]. However, while using such sources there is almost no possibility for independent control of electron beam current and energy. Electron

* Corresponding author: klimov@main.tusur.ru

sources with thermionic cathode which generate beams in the pressure range between 0.001 and 0.1 Pa can be used to control beam parameters. But in this case differential pressure must be provided between the electron gun chamber and the chamber where beam plasma is generated [13]. This condition, in the end, makes the technical unit more complex.

The works [14, 15] describe effective usage of plasma cathode electron sources for generating large-area beam plasma. Control of energy spectrum of plasma ion flow in the mentioned works was performed by putting negative bias on the target. Still, it should be noted that this method cannot be used for non-conductive materials treatment, for no potential can be applied in this case. The problem of continuous electron beams generation in the pressure range between 5 and 100 Pa is solved by using the so-called forevacuum plasma electron sources [16]. Such sources use hollow cathode discharge to generate dense discharge plasma. The apparatus also contains accelerating space which prevents from disruption at high pressure values.

During transmission of a ribbon electron beam through gaseous medium of a vacuum chamber in forevacuum pressure range a large-area plasma formation is generated (more than 0.5 sq.m), which is called a plasma sheet [17]. Special structure of accelerating space [18] enables to obtain electron beam without traditionally used transmitting magnetic field. As described in [18, 19], plasma ion flows can compensate for the negative charge transmitted to the target in the process of electron beam treatment. It is obvious that such compensation is possible in case of plasma ion flow treatment due to higher mobility of plasma electron component. At the same time, absence of magnetic field allows electrons to fall on the target without their trajectory distortion. The paper presents investigation results on the influence of electron beam parameters on concentration of the generated beam plasma, as well as on the parameters of plasma ion flows. Plasma in this case is generated when ribbon electron beam is transmitted through gaseous medium at the pressure of 10 Pa. Novelty of the research is based upon absence of magnetic field in the beam transmission region. The paper also gives the results of using beam plasma for ceramic material surface treatment, particularly polycor material.

Experiment

Experimental stusy was performed using the vacuum unit consisting of a rectangular vacuum chamber sized 0.4×0.7×0.4 m with mechanical system of air pumping, a ribbon electron beam source and a power supply and diagnostic system. Pressure in the vacuum chamber was kept on the level of 5-10 Pa with the help of BocEdwards 80 forevacuum pump and was controlled by the inlet valve. The ribbon electron beam was produced by a forevacuum plasma electron source [18]. The source was specially designed for ribbon beam generating while transmission without traditionally used [15, 19, 20] longitudinal magnetic field. The source represented a three-electrode system – hollow cathode *1*, flat anode *2* and extractor *3* (Fig. 1). Structure and operation scheme of the source are described in more detail in [18].

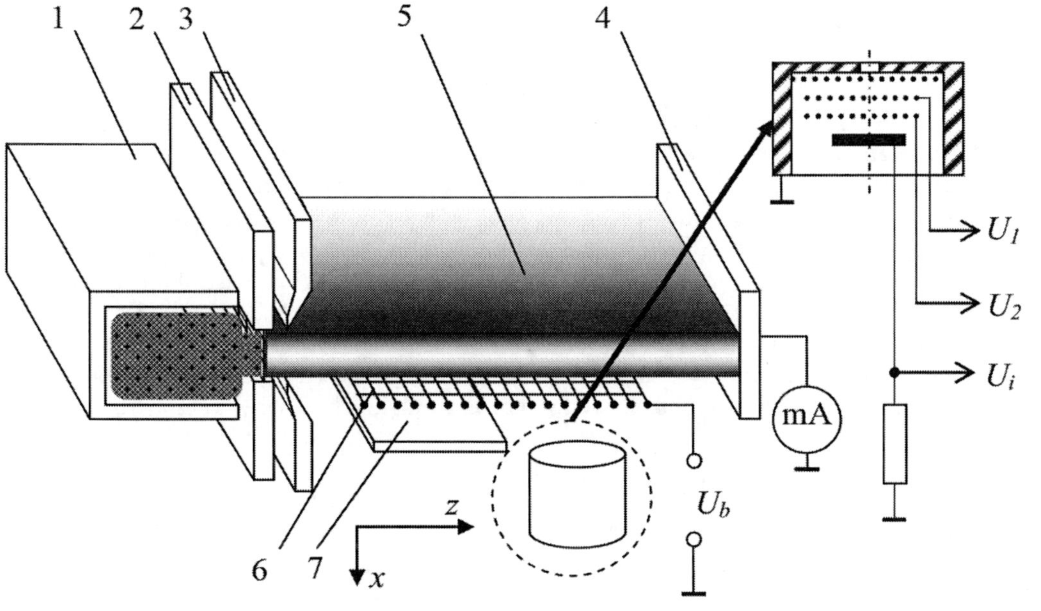

Fig. 1. Scheme of the experiment and the energy analyzer.

After transition in the vacuum chamber the electron beam generated by plasma source is caught by a collector *4*. While making their way to the collector beam electrons were performing ionization of gas molecules and generated beam plasma, the area of which was defined by the beam parameters. A single probe in the form of copper wire 0.3 mm in diameter placed in a ceramic tube was used to measure plasma parameters. 5 mm of the probe receiver part was extruded from the tube. Plasma concentration was evaluated on the basis of probe characteristics using standard method [21]. To extract ions from beam plasma and create clear plasma edge a flat extraction grid *6* was used placed parallel to the beam at the distance of 2 cm from its mean plane. Input of negative bias U_b 10-50 V to the grid enabled to extract ions with controlled energy. Targets *7* represented by ceramic plates made of polycor with thickness of 1 mm and size of 10×10 mm^2 were placed 1 cm behind the grid. Energy spectrum of ions hitting the target was investigated using a multigrid energy analyzer (Fig. 1). It was positioned behind the grid in the way that its axis is perpendicular to the beam propagation plane. Screening grid was placed behind the analyzer inlet with the diameter of 5 mm and depth of 2 mm and was electrically connected with the energy analyzer body which was grounded. Retarding potential U1 in the range between 0 and 100 V was supplied to the second grid. To eliminate the effect of secondary electron emission on the collector placed behind the second grid, negative (in relation to the analyzer body) potential U_2 equal to 10 V was supplied to the second grid. Induced retarding field prevented secondary electrons from scattering from the collector. The grids were made from stainless steel and had the cell size of 0.3×0.3 mm. The distance between the grids was 2 mm. Current signal from the collector was transmitted to a resistor of 10 kOhm and then registered by Tektronix 2024B digital storage oscilloscope. Ions energy spectrum was calculated by derivation of the curve of ion current delay.

Results

The research results showed that beam plasma concentration monotonically decreases while electron beam is transmitted from vacuum source to collector in a vacuum chamber (Fig. 2), as well as while distancing from the mean plane of the beam (Fig. 3). According to the authors, the most probable reason for such concentration behavior is electron beam expansion due to scattering on residual gas molecules. Increase of accelerating voltage allows producing a beam with lower divergence; however, in this case the beam plasma region is moving towards the mean plane of the beam and drops to the periphery. Besides, from the point of view of effective interaction between beam electrons and residual atmosphere atoms and molecules, the use of low-energy electron beam is more advisable given the fact that ionization cross section corresponds to the energy regions of tens and hundreds eV [22].

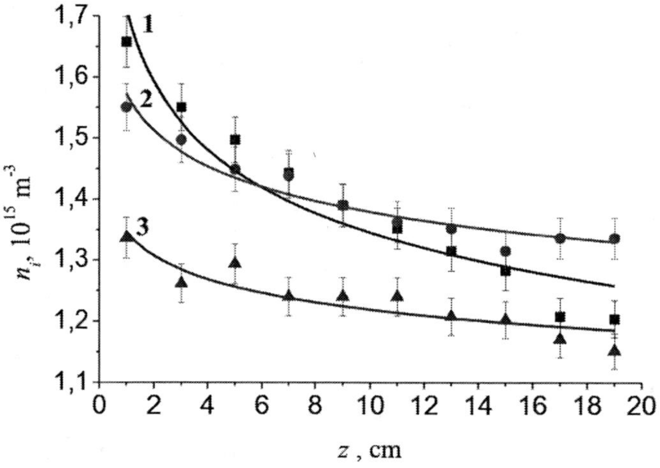

Fig. 2. Dependence of beam plasma concentration n on transmission area z for various distances from mean plane of the beam; distance to the mean plane of the x beam: 1 - 1 cm; 2 - 3 cm; 3 - 5 cm. Accelerating voltage $U_a = 3$ kV and pressure $p = 10$ Pa.

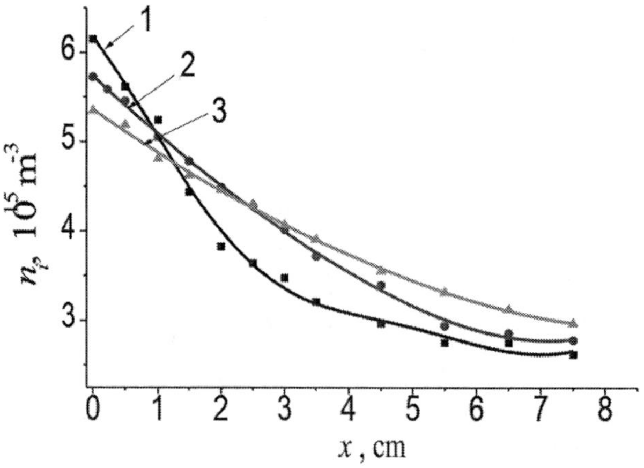

Fig. 3. Beam plasma concentration n depending on the distance to the mean plane of the beam r for various lengths of transmission areas z: 1 - 1 cm; 2 - 10 cm; 3 - 20 cm. Accelerating voltage $U_a = 3$ kV and pressure $p = 10$ Pa.

Decrease of beam plasma concentration in the measurement point with the increase of accelerating voltage is indirectly proved by the dependence of ion current to the collector placed behind the extraction grid *6* (Fig. 1). As can be seen from Figure 4, increase of beam electrons energy leads to decrease of registered ion current; and due to the fact that the collector is placed at a fixed distance from the mean plane of the beam decrease of plasma concentration in proximity to the extraction grid causes decrease of ion current density on collector.

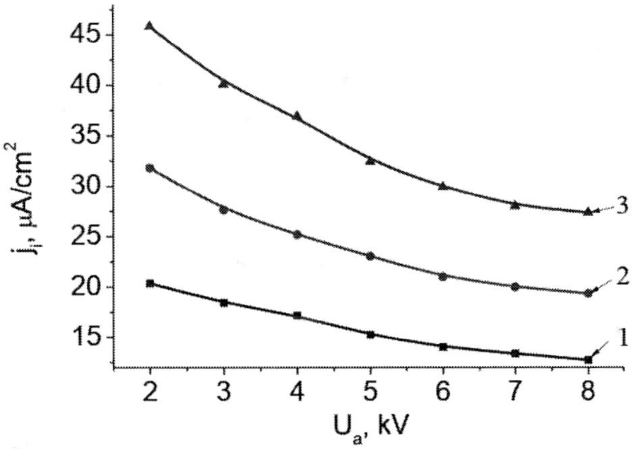

Fig. 4. Dependence of ion current density j_i on accelerating voltage U_a for various beam current values I_b: 1 – 180 mA, 2 – 360 mA, 3 – 550 mA. Pressure equals to 8 Pa.

Dependence of ion current density on the distance to extraction grid presented in Figure 5 enables to evaluate ion loss when propagating in forevacuum pressure region and to determine the extent of ion current control in the process of further surface treatment. As expected, the highest value of ion current density is observed in proximity to extraction grid. Current density decreases while distancing from the grid plane.

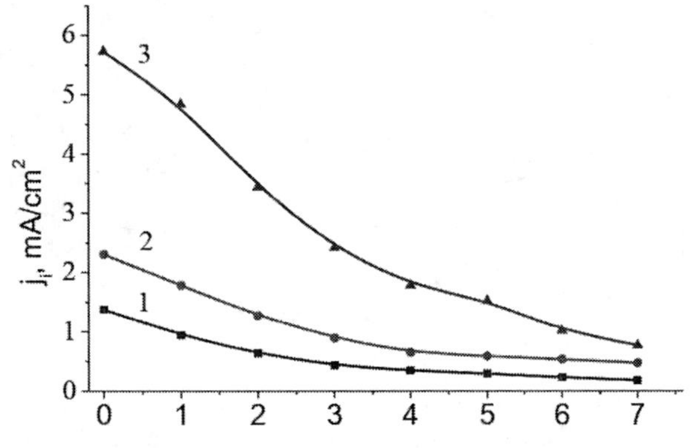

Fig. 5. Dependence of ion current density j_i on the distance to mean plane of the beam h for various beam current values I_b: 1 – 180 mA, 2 – 360 mA, 3 – 550 mA. Pressure equals to 8 Pa.

Another means to control ion current parameters is ion energy control. The experiments showed that ion flow energy is primarily determined by the extraction grid potential and plasma potential. Within the experiments, negative voltage (in relation to grounded chamber walls) in the range of -10 to -50 V was supplied to the extraction grid. Figure 6 shows ion energy distribution depending on the potential supplied to the extraction grid. The peak of energy spectrum is higher than the grid potential by the value of about 10 eV, which can be related to the positive potential of beam plasma. This means that negative potential supplied to the grid causes additional acceleration of ions that approach plasma edge.

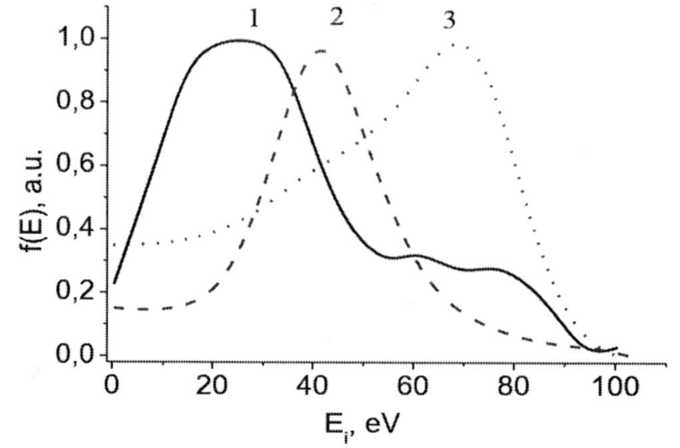

Fig. 6. Normalized functions of ion energies distribution *f(E$_i$)* depending on voltage on the extraction grid U_b: 1 – 10 V, 2 – 30 V, 3 – 60 V.

Plates made from polycor were treated using the ion flow extracted from beam plasma. Ceramic plates were placed at the distance of 1 cm from the extraction grid, ion beam current density was 50 μA/cm^2. The plates were treated for 30 minutes. As a result, the surface became rougher compared to the initial material (Fig. 7).

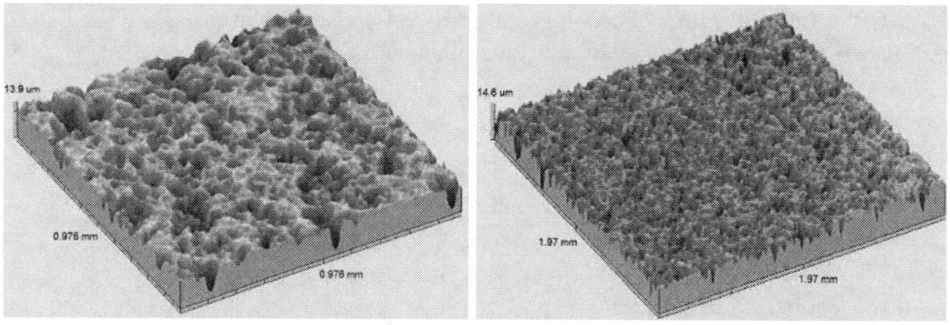

Fig. 7. Ceramic surface before (on the left) and after (on the right) ion beam treatment.

The fact of variation in surface morphology enables to talk about using a forevacuum plasma electron source for beam plasma generation and its application for treatment of long non-conductive products.

Conclusion

The research shows that the use of plasma electron source able to operate in forevacuum pressure regions allows generating large-area beam plasma. Absence of magnetic field in transmission area causes only minor electron beam expansion. Density of beam plasma ion flow is defined by the beam current, accelerating voltage and distance of the beam transmission. While using extraction grid with set potential it is possible to obtain ion flows with controlled energy, which is especially important for treatment of various surfaces. Ion treatment of non-conductive materials surfaces (polycor) leads to variation in their morphology, particularly increases their roughness. This can be used for obtaining durable materials with extended surface.

Further research in this field can be focused on increasing ion flow density. This can be implemented by initiating beam plasma discharge while electron beam transmission through gas.

The work was supported by the Ministry of Education and Science of the Russian Federation, grant No. 3.9605.2017/8.9.

References

1. J.J. Rocca, J.D. Meyer, M.R. Farrell, G.J. Collins, J. Appl. Phys., **56**, 790 (1984)
2. D.R. Borisa, T.B. Petrova, G.M. Petrov, S.G. Walton, J. Vac. Sci. Technol., A **35**, 01A104 (2017)
3. S. Walton, D. Leonhardt, D. Murphy, R. Meger, R. Fernsler, Bull. Am. Phys. Soc., **44**, 8, 58 (1999)
4. D. Leonhardt, C. Muratore, S.G. Walton, IEEE Trans. on Plasma Science, **33**, 2, 783 (2005)
5. T.M. Vasilieva, *Journal of Physics: Conference Series*, **370,** 012012 (2012)
6. A. Agarwal, M.J. Kushner, J. Vac. Sci. Technol., **A 27**, 37 (2009)
7. S.G. Walton, D.R. Boris, S.C. Hernández, E.H. Lock, Tz.B. Petrova, G.M. Petrov, R.F. Fernsler, J. Solid State Sci. Technol., **4**(6), 5033-5040 (2015)
8. I. Alexeff, R.V. Neidigh, et all., Phys. Rev. Letters, **10**(7), 273-276 (1963)
9. W.D. Getty, L.D. Smullin, J. Appl. Phys., **34**, 3421 (1963)
10. E.G. Shustin, N.V. Isaev, M.P. Temiryazeva, Yu.V. Fedorov, Vacuum **83**, 1350-1354 (2009)
11. R.F. Fernsler, W.M. Manheimer, R.A. Meger, J. Mathew, D.P. Murphy, R.E. Pechacek, J.A. Gregorc, Physics of Plasmas, **5**(5), 2137 (1998)
12. I.V. Melnik, Radioelectronics and communication systems, **56**(12), 51 – 61 (2013)
13. G.P. Berezina, V.S. Us, Plasma Physics Reports, **37**(13), 1156–1161 (2011)
14. W.M. Manheimer, R.F. Fernsler, M. Lampe, R.A. Meger, Plasma Sources Sci. Technol., **9**, 370 (2000)
15. S.G. Walton, C. Muratore, D. Leonhardt, R.F. Fernsler, D.D. Blackwell, R.A. Meger, Surf. Coat. Technol., **186**(1-2), 40 (2004)
16. V.A. Burdovitsin, Y.A. Burachevskii, E.M. Oks, M.V. Fedorov, Instrum. Exp. Tech., **46**, 257 (2003)
17. V.A. Burdovitsin, E.M. Oks, M.V. Fedorov, Russ. Phys. J., **47**, 310-314 (2004)
18. A.S. Klimov, M.I. Lomaev, E.M. Oks, A.P. Andreichik, Tech. Phys., **62**, 218 (2017)
19. R.A. Meger et al., Physics of plasmas, **8**(5), 2558 (2001)
20. E.G. Shustin, N.V. Isaev, I.L. Klykov, V.V. Peskov, Vacuum, **85**, 711 (2011)
21. F.F. Chen, Electric Probes, in *Plasma Diagnostic Techniques,* Chapter 4, 113-200 (Academic Press, New York, 1965)
22. Y. Nakamura, M. J. Kurachi, Phys. D, **21**(5), 718 (1988)

Obtaining memristor elements based on non-noble materials

Pavel Troyan[1,*], *Yury* Sakharov[1], and *Ekaterina* Zhidik[1]

[1]Tomsk State University of Control Systems and Radioelectronics, 634003 Tomsk, Russia

Abstract. This study is aimed at creating memristor elements based on TiO_2/TiO_x layers with electrodes that do not contain noble and rare-earth metals, by vacuum deposition method. The characteristics of these elements analyzed by voltammetric methods show that the appearance of an *N*-type region of negative differential resistance on the current–voltage curve can be caused only by metal electrodes whose vacuum work function exceeds that of TiO_2. The appearance of the *N*-type region on the current–voltage curve of a memristor element is possible only after electrically assisted vacuum forming. Mo–TiO_2/TiO_x –Ni/Cu structures, for which the I_{LR}/I_{HR} ratio reaches two orders of magnitude at a voltage of less than 4 V, have the most stable parameters.

Introduction

In the 21st century, there is a trend towards the active development of new hardware implementations of analog neurochips with frequency modulation based on hybridization of CMOS (complementary metal-oxide-semiconductor) technology and memristors — passive elements that are able to change their resistance depending on a charge flowing through it [1]. This implementation paradigm for artificial neural networks makes it possible to copy the brain structure with high accuracy. In this structure, neurons are made of transistors, while synapses are replaced by memristors. Due to this architecture, functioning mechanisms of living neurons can be reproduced much more accurately than DC digital or analog implementations [2].

The following analogy between this type of artificial neural networks and living neurons can be drawn. The synaptic transmission speed depends on activation time of neurons: the shorter the time interval between activations is, the faster the signal is transmitted through a synapse. The memristor array works exactly in the same way: when the current is applied at intervals of 20 ms, the memristor resistance is half that at intervals of 40 ms [3].

The memristor was developed and described by Leon Chua in 1971 [4]. Chua was deeply convinced that there was a fourth device that could ensure a conceptual symmetry with a resistor, inductor and capacitor.

Chua's ideas were brought to life by Hewlett-Packard employees when in 2008 Williams announced the discovery of a memristor based on a nanoscale thin film of titanium dioxide [5]. The development of memristor elements makes it possible to

* Corresponding author: Zhidikyur@mail.ru

implement an "authentic" neural network in practice, while the physical properties of memristors allow them to be reduced up to ten nanometers and scaled up for mass production.

Despite some extensive studies of memristor elements that have been completed to date, e.g. in [5–8], most of them concern the application of TiO_2 films or a TiO_2/TiO_x structure, where TiO_x is non-stoichiometric titanium dioxide. The electrodes in these elements are made of only noble metals — Pt or Au.

This study is aimed at creating and studying characteristics of memristor MDM (metal-dielectric-metal) elements based on TiO_2/TiO_x layers with electrodes that do not contain noble or rare-earth metals.

Experiment

The influence of the upper electrode material on switching properties of memristors was analyzed on the basis of Mo–TiO_2/TiO_x–Ni, Mo–TiO_2/TiO_x–Cu, and Mo–TiO_2/TiO_x–Ni/Cu structures. Core-type structures were used as a substrate [9] (Fig. 1).

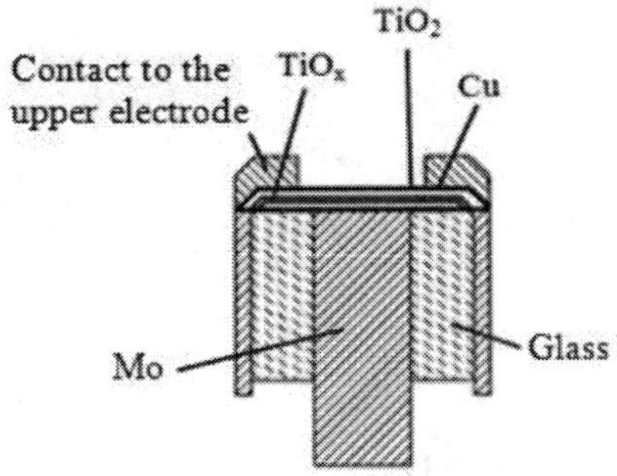

Fig. 1. A schematic representation of the MDM element based on a core-type substrate.

In order to produce the test memristor elements, 50-nm TiO_2 films were deposited by magnetron sputtering of a titanium target in Ar with 35% O_2 and at a gas mixture pressure of $7 \cdot 10^{-2}$ mm Hg. The discharge current was 300 mA, which corresponded to a film deposition rate of 16 nm/min [9]. TiO_x films were deposited by magnetron sputtering of the Ti target in Ar with 10% O_2 under equal identical conditions. The approximate growth rate of the TiO_x film is 24 nm/min.

In order to check the composition of dielectric films, an X-ray diffraction microanalysis was carried out (Fig. 2). The results show that, in terms of oxidation degree, titanium in the produced films is close to TiO_2 and Ti_2O_3 compounds that are necessary for obtaining memristor elements.

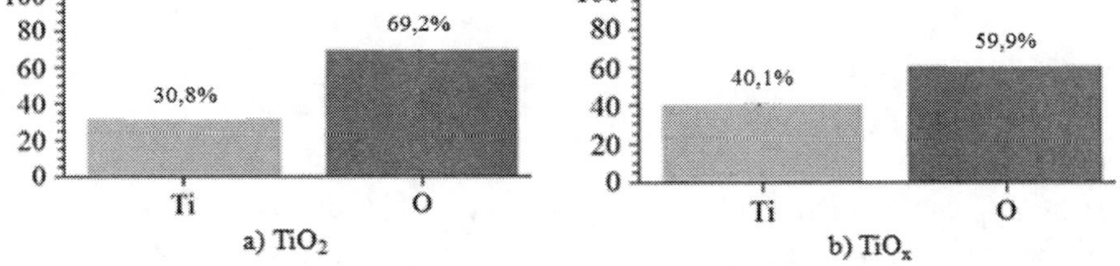

Fig. 2. Composition of the produced dielectric films (atomic percent).

Films of the upper electrode were deposited by thermal evaporation of Ni and Cu in vacuum at a pressure of $2 \cdot 10^{-5}$ mm Hg. A film of the upper electrode is 30 nm thick. For a better contact, a thick metal layer was sputtered onto the upper electrode along the edges of the core so that to avoid covering the working area.

Current–voltage characteristics of the produced MDM elements were measured at a constant voltage in both straight and reverse polarity. The measured current–voltage characteristics show an ohmic linear behavior, without an *N*-type region of negative differential resistance. Apparently, the absence of this region is associated with the insufficient work function of electrons from Cu and Ni, which is 4.34 eV and 4.48 eV, respectively. For materials used as memristor electrodes, this parameter should be no less than the work function of electrons from TiO_2 (4.7 eV) [10]. For example, the work function of electrons from Au and Pt, metals that are traditionally used for memristor contacts, is equal to 5.1 eV and 5.32 eV, respectively. This disadvantage can be compensated for by subjecting the MDM element to the electrically assisted forming procedure described in detail in [11].

To carry out the forming procedure, samples were exposed to vacuum at a pressure of 10^{-2} mm Hg. The forming procedure was conducted at a voltage of 14 V. Mo–TiO_2/TiO_x–Ni/Cu structures were formed almost instantaneously. Mo–TiO_2/TiO_x–Ni structures were formed during 30 minutes. For a shorter forming time, the current–voltage characteristics did not change their pattern. An interesting result was obtained after the deposition of a 20-nm thick copper layer onto Mo–TiO_2/TiO_x–Ni structures. These structures were formed at 14 V during 10 minutes. Then current–voltage characteristics were measured in vacuum, in different states.

Results

The current–voltage characteristics of the produced MDM elements were measured before and after the forming. Before the forming, the current–voltage curve is monotonically increasing, without a region of negative differential resistance (Fig. 3).

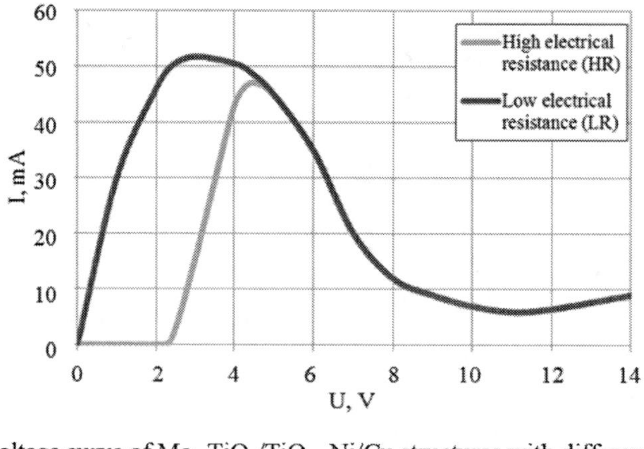

Fig. 3. The current–voltage curve of Mo–TiO_2/TiO_x–Ni/Cu structures with different conductivity values.

The Mo–TiO_2/TiO_x–Ni/Cu structures are formed rapidly and have a shape memory. The I_{LR}/I_{HR} ratio reaches 100 at a voltage of less than 4 V. The current–voltage characteristics of these structures are more stable than those of Mo–TiO_2/TiO_x–Ni, while their switching currents are three orders lower than those in Mo–TiO_2/TiO_x–Cu.

The surface of one of the formed MDM elements is presented in Figure 4.

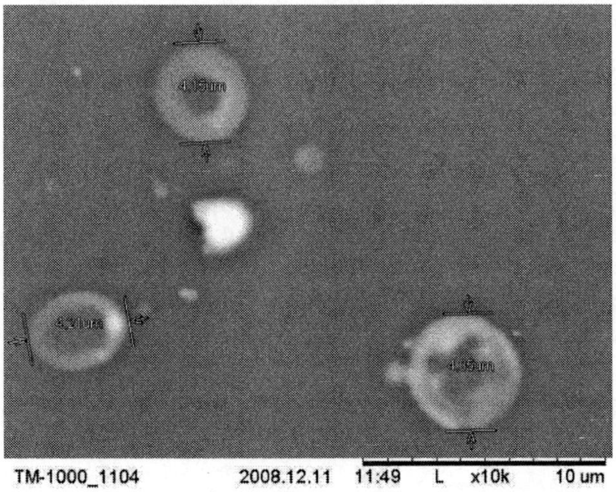

Fig. 4. The surface of the formed MDM structure: Mo–TiO_2/TiO_x–Ni/Cu.

Thus, after the forming process, channels with a diameter of about 4.4 μm are formed on the surface of the MDM structures. The density of these channels is $5 \cdot 10^7$ cm^{-2}. This value is close to the density of the channels obtained in [11].

Conclusion

During the production and research of the memristor elements based on non-noble materials, it was established that these elements could become serviceable only after electrically assisted vacuum forming.

Electrically assisted forming induces some irreversible changes in the test elements that give rise to the *N*-type region of negative differential resistance. At the same time, in the high-resistance state, the currents are reduced by 2–3 orders of magnitude compared to the low-resistance state. It is an acceptable value for a memristor element.

Further research in this area will be aimed at determining a switching mechanism and memory of memristors. As of today, the only proposed switching mechanism is a diffusion of oxygen vacancies from TiO_x to TiO_2 [12]. In our opinion, this switching mechanism is controversial and unproved.

The research was carried out with the financial support of the Ministry of Education and Science of the Russian Federation (grant agreement №14.577.21.0281 of 23.10.2017, unique identifier of the project RFMEF57717X0281).

References

1. Impuls, Uchenye sobrali nejroset' iz plastikovyh memristorov [Scientists have collected a neural network of plastic memristors], available at: https://mipt.ru/newsblog/lenta/madeofplastics (in Russian)
2. 3D News Daily Digital Digest, Memristory: pora li perepisyvat' uchebniki? [Memristors: is it time to rewrite the textbooks?], available at: https://3dnews.ru/906763 (In Russian)
3. Computerra, Memristor: "nedostajushhij jelement" [Memristor: "The missing circuit element"], available at: http://old.computerra.ru/vision/591537/ (in Russian)

4. L. Chua, IEEE Trans. Circuit Theory **18**(5), 507 (1971)
5. D. Strukov, G. Snider, D. Stewart, S. Williams, Nature **453**, 7191, 80 (2008)
6. J.J. Yang, M.D. Pickett, X. Li, D. Stewart, S. Williams, Nature nanotechnology **3**, 429 (2008)
7. J.J. Yang, J.P. Strachan, F. Miao, M.-X. Zhang, M.D. Pickett, W. Yi, A.A. Ohlberg, G. Medeiros-Ribeiro, Appl. Phys. A **102**, 785 (2011)
8. A.V. Emelyanov, V.A. Demin, I.M. Antropov, G.I. Tselikov, Z.V. Lavrukhina, P.K. Kashkarov, Technical Physics **85**(1), 114 (2015) (in Russian)
9. Y.V. Sakharov, P.E. Troyan, Y.S. Zhidik, Reports of TUSUR **38,** 72 (2015) (in Russian)
10. F.Z. Wang, N. Helian, S. Wu, M.-G. Lim, Y. Guo, M.A. Parker, IEEE Electron Device Letters **31**(7), 755 (2010)
11. T.I. Danilina, P.E. Troyan, Russ. Phys. J. **53**, 331 (2010) (in Russian)
12. R.K. Singh, European Scientific Journal **10**(18), 1857 (2014)

Investigation of temperature stability of ITO films characteristics

Pavel Troyan[1,*], *Yury Zhidik*[1], and *Ekaterina* Zhidik[1]

[1]Tomsk State University of Control Systems and Radioelectronics, 634003 Tomsk, Russia

Abstract. The paper represents research of thermal stability of optical and electro-physical parameters of ITO films deposited using various techniques. Variation of optical and electro-physical parameters was recorded using spectroscopy, and Hall's and four-probe measurements. The best thermal stability was demonstrated by ITO films deposited by metal target sputtering In(90%)/Sn(10%) in mixture of gases O_2 (25%) + Ar (75%) with further annealing in air atmosphere. This enables to apply this technique for production of thin film transparent resistive elements capable to heat the translucent structures up to 100°C without deterioration of their parameters.

Introduction

ITO (Indium tin oxide, In_2O_3:Sn) thin films have recently gained much more attention among research teams [1-7], due to their unique combination of high electrical conductivity, high transmission coefficient in the visible portion of the spectrum and reflective power in IR-range.

These films are extensively used in electronics and microelectronics, in transportation industry and architecture. Research in [2] describes how ITO films are used as a layer of current distribution for GaN-based blue light-emitting diodes. The authors demonstrated in their study that application of these ITO films enabled growth of light-emitting diodes efficiency coefficient by 2.5 times. Review paper [3] reports on successful production of ITO-based thin film oxide transistors. Application of ITO films in transportation industry and architecture was discussed in [4]. C.G. Granqvist and A. Hultaker present results of proving out the technology of ITO films deposition on sheet glass for their further use in optimized electrochromic devices for smart windows production. Smart windows find their challenging application as electronic tinted coating of airplane portholes [5].

Apart from the applications mentioned above, ITO films can also be prospectively used as transparent heating elements. Some research and production companies [6, 7] manufacture heated glass used in temperature-proof houses for surveillance cameras, heated sighted devices, observation windows, transparent heating elements of LCD displays enabling to broaden the temperature range of functional devices operation up to -60°C.

Due to large interest in ITO films application as heating elements, its properties are needed to be kept unchanged during long-term high temperature impact. However, the data

*Corresponding author: Zhidikyur@mail.ru

of research papers referring to this issue generally provide the dependences of ongoing variations of films electrical conductivity during their heating in order to define its mechanisms [8]. Variations of electro-physical properties after heating were not considered in previous studies.

The given paper aims to study the stability of optical and electro-physical parameters of ITO films deposited by magnetron sputtering after high temperature treatment.

Experimental

ITO films under study were 130-150 nm thick; they were deposited on preliminary polished sitall substrate by magnetron sputtering technique according to different modes, as specified in Table 1. Films deposition was performed by reactive magnetron sputtering with various working gas composition (modes 1 and 2) according to the technique described in [9]; and also films were obtained by oxide target sputtering in argon atmosphere.

Table 1. Deposition modes of samples under study.

Mode No.	Applied target	Working atmosphere composition	Deposition parameters	In-process annealing of samples
1	In (90%) / Sn (10%)	O_2 (25%) + Ar (75%)	Working pressure $8 \cdot 10^{-3}$ torr; discharge current 0.3 A; temperature of substrate during deposition 25°C	Annealing in air atmosphere at 600°C for 20 min.
2	In (90%) / Sn (10%)	O_2 (29%) + Ar (71%)		Annealing in nitrogen atmosphere at 600°C for 20 min.
3	In_2O_3 (90%) / SnO_2 (10%)	Ar (100%)	Working pressure $6 \cdot 10^{-3}$ torr; discharge current 1.85 A; temperature of substrate during deposition 250°C	Without annealing

Temperature stability of ITO films parameters under high temperature treatment was investigated as described below. Samples with deposited ITO films and the formed metallic islands were placed into muffle furnace and kept for 15 and 30 min at 600°C in air atmosphere. After temporary exposure the surface resistance of films was measured, as well as concentration and charge carrier mobility and transmission coefficient.

Surface resistance measurements were conducted using four-probe method. Parameters of charge carriers in the films were measured using Hall Effect. Both methods of thin films parameters checkout were described in details in [10].

Transmission spectrum of ITO films was recorded by UV-spectrometer USB 2000. The facility is intended for conducting various analytical investigations of optical absorption spectrum, transmission spectrum, reflection and lumenescent spectrum in UV and visible range.

Results

Electro-physical parameters of ITO films were measured using the above mentioned methods right before and after high temperature exposure of samples at 600°C for 15 and 30 min under atmospheric pressure. Measuring results can be found in Tables 2, 3 and 4. Figures 1, 2 and 3 illustrate dependence graphs for transmission coefficient and wavelength during annealing for films obtained using all three modes.

Table 2. Variation of electro-physical parameters of samples obtained using the Mode No.1 depending on the time of exposure in the furnace at 600°C.

Controlled parameter	Duration of high-temperature exposure		
	Before heating	15 minutes exposure	30 minutes exposure
Surface resistance, Ohm/□	5.9	2.5	4.9
Charge carrier concentration, m^{-3}	$5 \cdot 10^{25}$	$4.6 \cdot 10^{25}$	$1.1 \cdot 10^{25}$
Charge carrier mobility, $m^2/(V \cdot s)$	0.16	0.26	0.05

Comparison of the obtained experimental data of the samples under study allowed revealing the following tendencies. During 15 min exposure of samples obtained in accordance with the Mode No.1 in the furnace at 600°C the decrease of surface resistance of the films occurs due to increased charge carrier mobility. It is most probably conditioned by formation of more expressed films crystalline structure. Further exposure results in the decrease in surface resistance of ITO films compared to the films that were not subject to annealing in the furnace. This fact is attributed to the increase in charge carrier concentration originating from activation of the introduced additive and formation of oxygen vacancies.

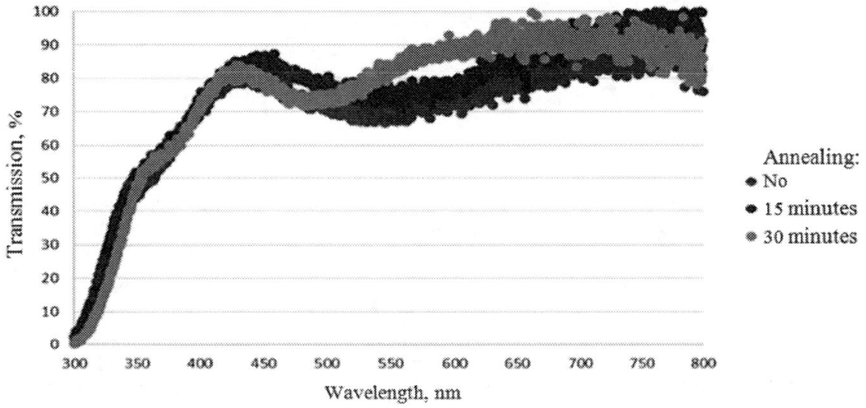

Fig. 1. Dependence graph of transmission spectrum of ITO film on the wavelength obtained using Mode No. 1.

As can be seen from Figure 1, during high temperature treatment of films the average increase of the transmission coefficient by 10% is observed.

Table 3. Variation of electro-physical parameters of samples obtained using the Mode No.2 depending on the time of exposure in the furnace at 600°C.

Controlled parameter	Duration of high-temperature exposure		
	Before heating	15 minutes exposure	30 minutes exposure
Surface resistance, Ohm/□	8.1	21.8	18.9
Charge carrier concentration, m^{-3}	$1.7 \cdot 10^{25}$	$7.2 \cdot 10^{24}$	$1.3 \cdot 10^{25}$
Charge carrier mobility, $m^2/(V \cdot s)$	0.21	0.19	0.12

Comparison of experimental data of samples obtained according to the Mode No.2 showed that the increase of surface resistance is observed during 15 min exposure in the furnace. The increase in surface resistance is attributed to the decrease in charge carrier concentration in ITO film after heating, which probably occurs due to oxygen diffusion from the film into atmosphere during heating. Further exposure leads to continuing decrease in surface resistance. This fact is conditioned by the increased concentration of charge carriers during activation of the introduced stannic additive.

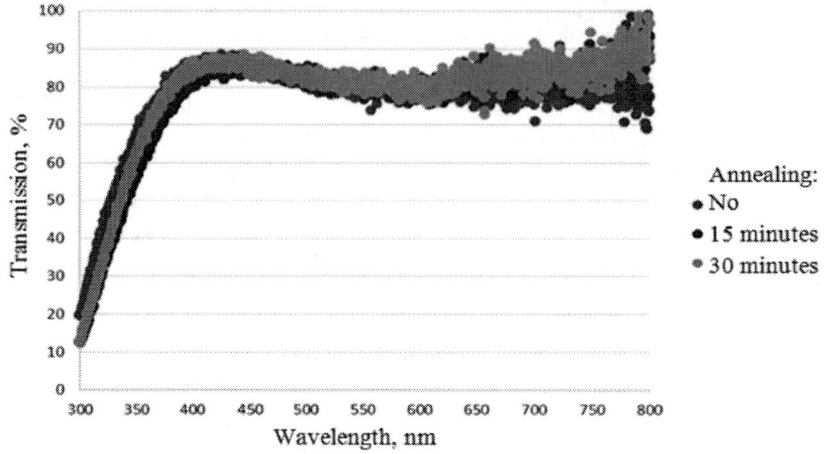

Fig. 2. Dependence graph of transmission spectrum of ITO film on the wavelength obtained using the Mode No. 2.

Figure 2 demonstrates stable dependence of the transmission coefficient on the wavelength irrespective of the exposure duration in the furnace.

Table 4. Variation of electro-physical parameters of samples obtained using the Mode No.3 depending on the time of exposure in the furnace at 600°C.

Controlled parameter	**Duration of high-temperature exposure**		
	Before heating	**15 minutes exposure**	**30 minutes exposure**
Surface resistance, Ohm/□	315.8	25.7	19.9
Charge carrier concentration, m^{-3}	$6 \cdot 10^{24}$	$1.3 \cdot 10^{25}$	$7.67 \cdot 10^{24}$
Charge carrier mobility, $m^2/(V \cdot s)$	0.04	0.11	0.15

Experimental data of samples obtained using the Mode No. 3 shows decrease in surface resistance during high temperature exposure. It is attributed to the increase in concentration and mobility of charge carriers due to formation of more expressed crystalline structure of films and by the activation of dopant stannic atoms.

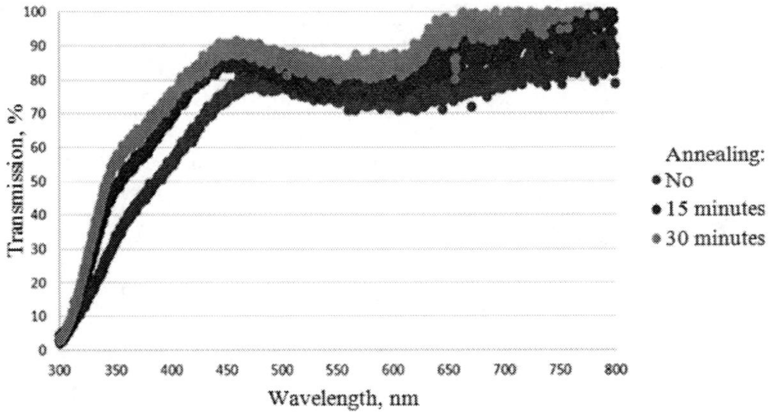

Fig. 3. Dependence graph of transmission spectrum of ITO film on the wavelength obtained using the Mode No. 3.

Figure 3 illustrates the increase of the films transmission coefficient. This is attributed to the increased content of oxygen in the film.

Thus, the Mode No.1 appeared to be the best technique of ITO films deposition on transparent substrates for their further use as a transparent heating element. High temperature heating almost does not alter the surface resistance of these films. At the same time, the transmission coefficient after temperature treatment of films increased from 75 % to 85 %.

Application of this deposition technique for optical transparent electrically conductive ITO films stipulated the design elaboration of thin film transparent heating elements (Fig. 4a). The elements are capable to heat the surface up to 100°C without deterioration of its parameters. The obtained temperature dependence of a heating element based on ITO film (resistance 40 Ohm/□) on the applied power normalized per surface unit is shown in Figure 4b.

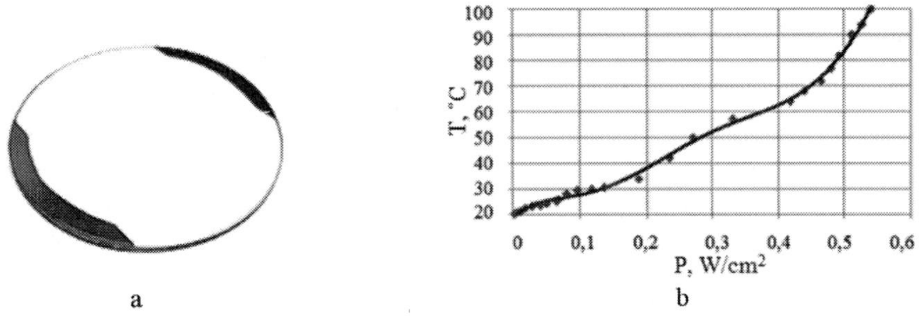

a b

Fig. 4. Heating element based on resistive ITO film (a) and dependence of the temperature of heating on the consumed current power (b).

Conclusion

Therefore, the most rational technique of ITO films deposition for their further use as a transparent heating element is reactive magnetron sputtering technique of metal target In (90%) / Sn (10%) in oxygen-containing atmosphere O_2 (25%) + Ar (75%) with further annealing in air atmosphere. This is the exact mode when the parameters of deposited films under high temperature treatment are not deteriorated.

These results contribute to the more thorough selection of the deposition technique of ITO films further applied in design of reliable thin film heating elements.

The extension of research will be connected with defining the optimal sheet resistance of ITO films providing better effectiveness of heating element.

The research was carried out with the financial support of the Ministry of Education and Science of the Russian Federation (agreement on granting the grant №14.577.21.0281 of 23.10.2017, unique identifier of the project RFMEF57717X0281).

References

1. Review of the market for indium-tin oxides and materials for their production in the worl, available at: http://www.infomine.ru/order-demo/458 (in Russian)
2. I.P. Smirnova, L.K. Markov, A.S. Pavlyuchenko, M.V. Kukushkin, Semiconductors, **46**(3), 384 (2012) (in Russian)
3. E. Fortunato, P. Barquinha, R. Martins, Advanced Materials, **24**(22), 2945 (2012)
4. C.G. Granqvist, A. Hultaker, Thin Solid Films, **411**, 1 (2002)
5. A.I. Cheban, The first European "Plane of Dreams" arrived in Ukraine, available at: http://alexcheban.livejournal.com/102426.html (in Russian)
6. "Polytech" research and production company website, available at: http://www.carmirror.ru/catalogue.php?part=24 (in Russian)
7. "NIKOL" group of companies website, available at: http://www.aonikol.ru (in Russian)
8. V. Brinzari, M. Ivanov, B.K. Cho, M. Kamei, G. Korotcenkov, Sensors Actuators B Chem., **148**, 427 (2010)
9. Y.S. Zhidik, P.E. Troyan, Y.V. Sakharov, Reports of TUSUR, **26**(2) part 2, 169 (2012) (in Russian)
10. S.V. Smirnov, *Metody issledovanija materialov i struktur jelektroniki* [Methods for studying materials and structures of an electron] (TUSUR, Tomsk, 2007) (in Russian)

Features of the structural phase state of a weld produced by electron-beam welding in the submicrocrystalline grade 2 titanium alloy

Kirill Kurgan[1,5*], *Vasily* Klimenov[1, 2], *Anatoly* Klopotov[1, 3], *Sergey* Gnyusov[2], *Yuri* Abzaev[1], *Alexander* Potekaev[3], and *Mikhail* Marzol[4]

[1]Tomsk State University of Architecture and Building, 634003 Tomsk, Russia
[2]National Research Tomsk Polytechnic University, 634050 Tomsk, Russia
[3]National Research Tomsk State University, 634050 Tomsk, Russia
[4]Polyus Scientific and Production Center JSC, 634041 Tomsk, Russia
[5]Gazprom transgaz Surgut, LLC, Technical and Engineering Center, 628404 Surgut, Russia

Abstract. This paper presents the results of structural studies for a welded joint of the Grade 2 titanium alloy in submicrocrystalline and microcrystalline states produced by electron beam welding when joining 2-mm-thick plates. Microhardness distribution patterns of the Grade-2 titanium alloy in micro- and submicrocrystalline states are identified in the weld zone and heat-affected zone. These patterns reflect a difference in structural phase states. It is assumed that one of the key factors affecting both the structural state and microhardness distribution in the weld zone and heat-affected zone during electron-beam welding is high concentration of oxygen atoms embedded into the crystal lattice of α–Ti-based solid solution.

Introduction

Welded joints of titanium alloys are given much attention due to the fact that these alloys are difficult to weld. In this context, laser welding is the most frequently discussed technology [1–4]. Its advantage is the possibility of welding in the atmosphere rather than in vacuum as in the case of electron-beam welding [5, 6]. As compared to electron-beam welding, laser welding is often associated with a higher degree of energy concentration, while the development of the electron-beam technique using plasma cathodes [7] allows highly focused electron beams to be generated. It is important here that the features of structural and phase transitions in welding of titanium alloys are associated with high heating and cooling rates, as well as with polymorphous transitions $\alpha \rightarrow \beta$ which are typical of titanium alloys.

The purpose of this study is to investigate the structural features, phase composition and microhardness in the weld zone produced by electron-beam welding of the Grade 2 titanium alloy in the submicrocrystalline state as compared to alloys in the ordinary state.

* Corresponding author: kirill_k2.777@mail.ru

Materials and Methods

For the purpose of this study, the Grade 2 titanium alloy was chosen. Samples in the submicrocrystalline state were made of this alloy. The submicrocrystalline state was obtained by rolling bars with a cross section of 6×6 mm produced by one of severe plastic deformation methods (multi-axial forging) [8]. 2×12 mm plates with a length of 30 mm were used as samples for welding. The samples were pre-cleaned and clamped in a special tool in such a way to avoid contact of the weld zone with the tool, for welding in a vacuum chamber of an ELS-0,5-6 industrial-scale plant. Electron beam welding of the samples was performed by a plasma cathode electron gun [7] with an accelerating voltage (V) of 28 kV, beam current (In) of (I_n) 35 mA, focus coil current up to 551 mA and a beam travel speed of 25 mm/s.

The phase composition of the initial samples and welds was determined using a DRON-4-07 X-ray diffractometer with $Cu_{K\alpha}$ radiation. The quantitative phase analysis was performed by Rietveld method based on processed diffractograms of the Grade 2 alloy [9].

Samples for metallographic studies were cut from welded plates perpendicular to the weld, then ground, polished and etched with a solution HF(1 part) + HNO_3(2 parts) + H_2O(7 parts).

The weld metal structure was examined using an Olympus GX-51 optical microscope (OM) equipped with a SIAMS700 image analyzer.

The structure of the samples was analyzed by scanning electron microscopy. A TescanVega 3 SB scanning electron microscope was used for the analysis.

The microhardness was determined on cross sections of the samples at different points of the weld using an Instron Wilson Hardness Tukon 2500-6 tester with a load of 25–50 g. Microhardness was measured along five parallel "paths" in the cross-section plane. These "paths" were perpendicular to the weld. The distance between the paths is 200 μm.

Results and Discussion

Microstructures of the analyzed Grade 2 alloy welds in different structural states are presented below on the basis of optical metallography (Fig. 1) and scanning electron microscopy (Fig. 2) data.

According to the standard method, the weld structure is described by identifying three zones in it: I – weld zone (WZ); II – heat-affected zone (HAZ); III – parent metal zone (PM). There are interfaces between the zones that may be clear or blurred. The presence of such boundaries is typical of welding by concentrated energy flows [4].

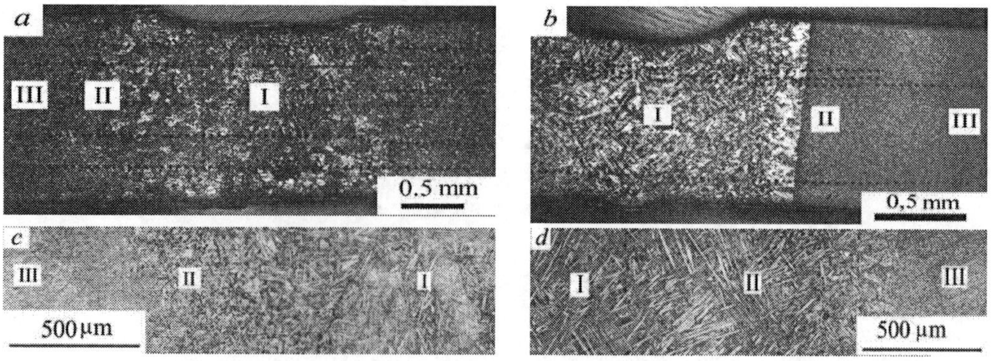

Fig. 1. The microstructure in the cross section of the Grade 2 alloy weld: *a, b* is the general view of welds (I – weld zone; II – heat-affected zone; III – parent metal zone).

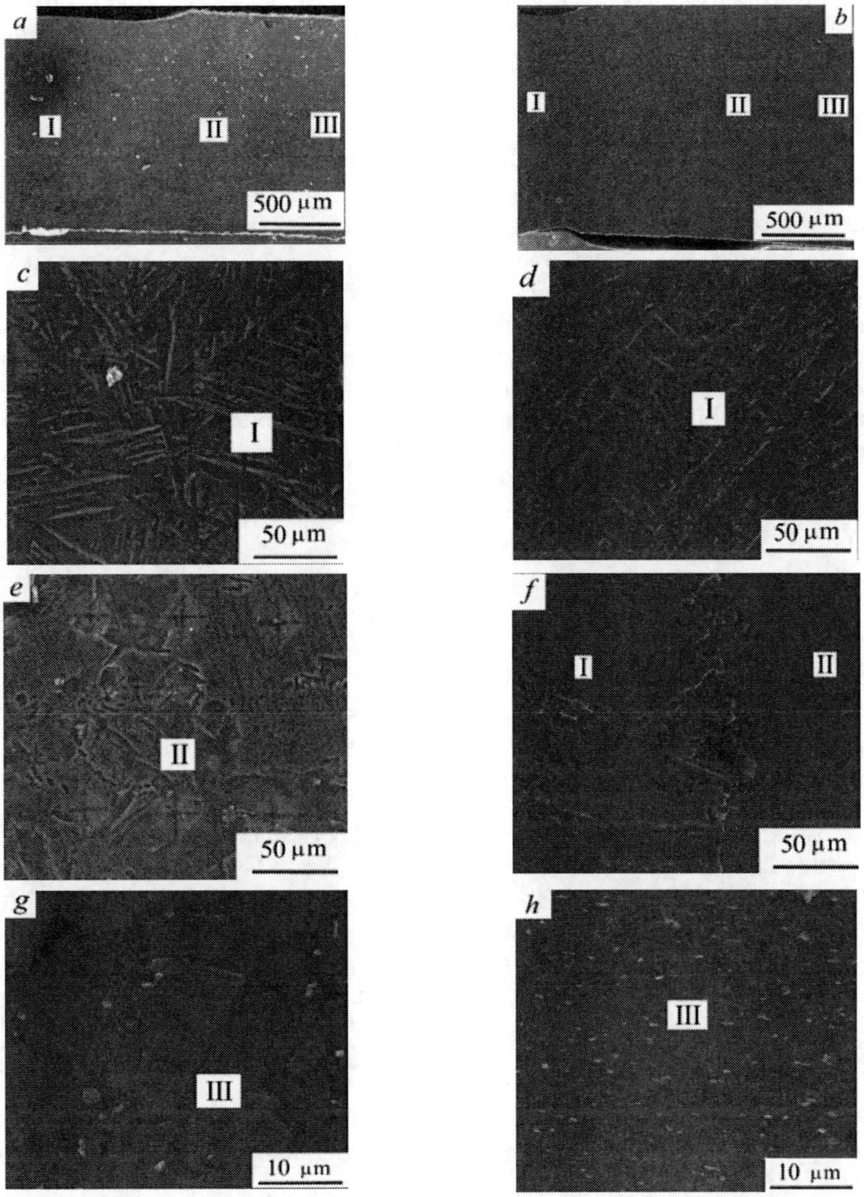

Fig. 2. Scanning electron microscopy of different zones of the Grade 2 alloy weld: general view (*a, b*); weld zones, heat-affected and base metal zones of microcrystalline (*c, e, g*) and submicrocrystalline (*d, f, h*) titanium (I – weld zone, II – heat-affected zone, III – base metal zone).

When describing the weld structures for initial microcrystalline and submicrocrystalline states of the Grade 2 titanium alloy, three zones can be identified as well (Fig. 1, 2). By analyzing the structure of these microstructural images, we determined the following features. The boundary region between the HAZ and the PM zone is more blurred than that between the HAZ and the WZ. The formation of the weld is accompanied with rapid cooling both in the weld zone and in the heat-affected zone. This leads to states in weld zones that are specific in terms of structure and composition. Martensitic transformations that occur in titanium alloys are an effective strengthening mechanism that ensures sufficiently high strength characteristics of the weld. The presence of the β-phase is one of the features of the crystalline structure formed in the melt zone. This phase is formed as a result of quenching from a temperature above the temperature of transition to the β-phase [10].

In our case, the structural analysis of the titanium alloy weld in initial microcrystalline and submicrocrystalline states revealed a number of features (Fig. 1, 2). When a welded joint is formed under melt cooling conditions from a high-temperature β–region, the weld structure has two morphological features (Fig. 1, Fig. 2 *c, d*): large polyhedral grains of the transformed β–phase with an average grain size of 60 ÷ 90 μm and a plate-like intragranular structure of the α–phase. A similar structure was observed in [1] when welding a titanium alloy in a weld after laser welding. Structural changes induced by laser welding of the Grade 2 titanium alloy led to a slight increase in the average microhardness value in the weld zone [1]. In our case, the HAZ microstructure induced by electron beam welding (Fig. 1 and Fig. 2 *e, f*) is representative for the structure formation under conditions of gradient temperature distribution. While a mix of martensitic α-phase and PM α-phase grains is observed on the WZ boundary (Fig. 1), the HAZ structure on the PM boundary (Fig. 1) is mainly represented by α-phase grains, as in the PM zone, but these grains are larger in size than that in the PM zone. In the PM zone, the titanium alloy is in the single-phase state with the α-phase. This finding correlates with the data in [1]. According to X-ray diffraction data, the α-phase is found to have a hexagonal close-packed lattice with parameters $a = 0.29761$ nm and c=0.47367 nm. Using the optical metallography technique, we established that the average grain size of the α-phase is 11.2±1.5 μm.

The microstructural analysis of the submicrocrystalline titanium weld joint (Fig. 1) revealed a number of features both in the PM structure and the marthensite phase structure in the weld zone. In addition, the β-phase was detected in all zones of the welded joint of the Grade 2 submicrocrystalline titanium alloy. Despite relatively rapid cooling, the WZ structure is similar to the structure formed by welding of microstructural titanium, i.e. there is a growth of grains as compared to PM grains (Fig. 1 and Fig. 2 *g, h*). The specific feature of the PM zone in submicrocrystalline titanium is that the structure has been formed as a result of step-by-step grinding of initial titanium alloy grains by multi-axial forging with appropriate heat treatment [11, 12]. The average grain size varies from 0.5 to 0.8 μm. In the weld zone, the grain size increases to 60 ÷ 90 μm. It should be noted that resistance spot welding of the Grade 2 submicrocrystalline titanium creates the quenching conditions in the WZ under which a submicrocrystalline cooling structure is formed. Growth of grains is noted in a small PM zone [13].

The structural differences of the Grade 2 titanium alloy in the microcrystalline and submicrocrystalline states are reflected in the microhardness distribution in the weld zone (Fig. 3 and 4).

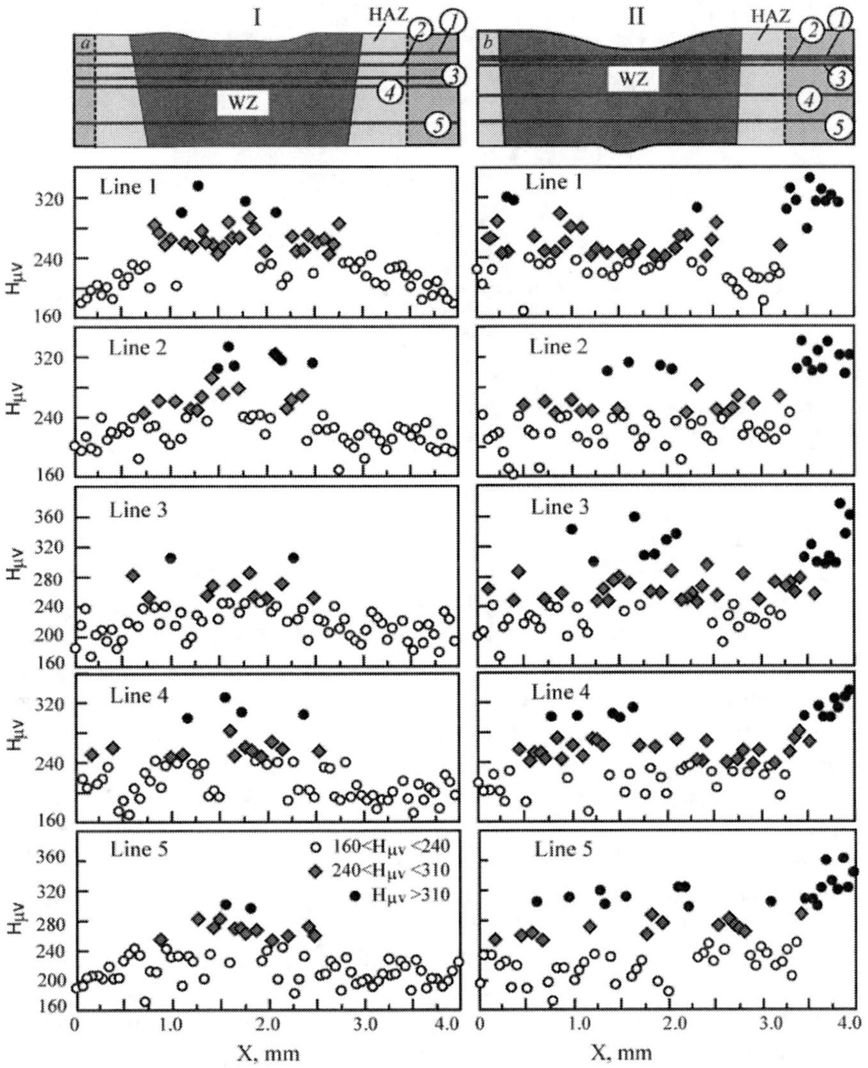

Fig. 3. Schematic representation of heat-affected zones (*a, b*) and microhardness distribution depending on the distance along the lines 1, 2, 3, 4, and 5 in the weld zone in the microcrystalline (I) and submicrocrystalline (II) states of the Grade 2 alloy .

The microhardness was determined on the cross sections of the samples in the micro– and submicrocrystalline states of the Grade 2 alloy using different measurement procedures. In the first procedure, microhardness was measured from different weld sections along five "paths" perpendicular to the welded joint. Figure 3 shows the layout of the heat-affected zone and the microhardness distribution depending on the distance on the cross sections along the five lines.

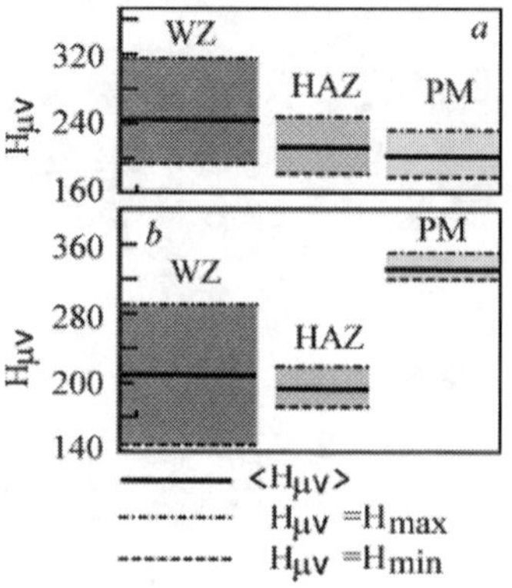

Fig. 4. Diagrams based on statistically processed microhardness data along the five lines in Figure 3 with identified zones in the weld zone in the micro- (*a*) and submicrocrystalline (*b*) states of the Grade 2 alloy.

A significant spread of microhardness values is observed in the weld zone and the HAZ of the tested alloy for different structural states. It can be used as an indirect indicator of structural heterogeneity of the alloy. Based on the obtained data, three typical groups can be distinguished in terms of microhardness. In Figure 3, they are highlighted by different markers. The first microhardness group includes low values with the maximum value $H_{\mu V}$ ~230. These microhardness values are typical of the Grade 2 allow [14]. The second group includes intermediate values of $H_{\mu V}$ between high and low ones. The third group contains high values of $H_{\mu V}$ ($H_{\mu V}$>310). These values are typical of the Grade 2 alloy in the submicrocrystalline state [12]. Such a spread of microhardness values is due to the fact that an intender forced into the surface of the sample falls into different phases or into interphase boundaries every time.

By statistically processing $H_{\mu V}$ values obtained as a result of microhardness measurements along all paths of the test samples, we distinguished specific microhardness zones in terms of magnitude and location in the weld zone (Fig. 4). Microhardness distribution in the micro- and submicrocrystalline states of the Grade 2 alloy is characterized by close average values in the HAZ. The significant differences are observed in the microhardness distribution in the weld zone of the Grade 2 alloy in the microcrystalline state, where the average microhardness $H_{\mu V}$ is higher than that in the submicrocrystalline state. The spread of microhardness values (H_{max}–H_{min}) in the submicrocrystalline state is much higher. This reflects a higher structural heterogeneity of the alloy in the submicrocrystalline state in the weld zone as compared to the microcrystalline state. Another situation is observed in the HAZ area and in the area far from the weld zone. Here, the spread of microhardness values (H_{max}–H_{min}) in the microcrystalline state is slightly higher than that in the submicrocrystalline state (Fig. 4).

It is important to note that the structural heterogeneity of the Grade 2 alloy in these zones in the microcrystalline state is higher than that in the submicrocrystalline state. The diagrams in Figure 4 reflect the fact that a spectrum of structural phase states in the submicrocrystalline state of the Grade 2 alloy in the HAZ zone is different from the states in the microcrystalline titanium alloy. All these findings are consistent with the metallographic data presented above.

This is also reflected in the microhardness distribution in the test samples determined using another measurement procedure. According to this procedure, microhardness in the weld zone was measured on three "paths" inside and along the welded joint. Microhardness on the fourth "path" parallel to the first three ones was measured far from the weld zone in the base metal (Fig. 5). The above microhardness distributions depending on distances along the welded joint clearly demonstrate that the microhardness values in the weld zone are not widely scattered and differ significantly in numerical values (Fig. 5). These microhardness values far from the weld joint are typical of the Grade 2 titanium alloy in the corresponding structural states [1, 10, 12, 14].

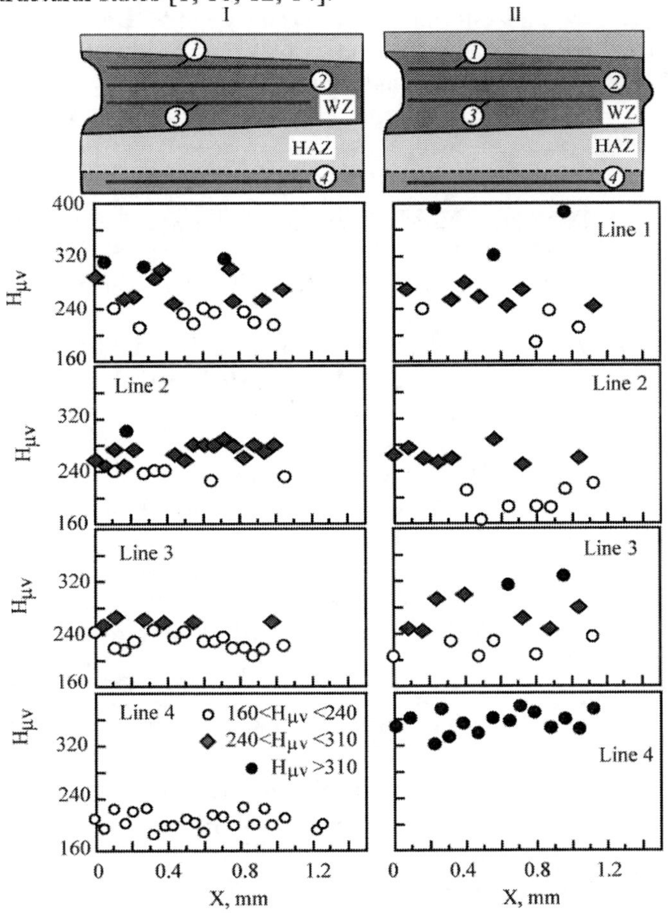

Fig. 5. The microhardness distribution depending on the distance along lines 1, 2, 3 and 4 in the weld zone in the microcrystalline (I) and submicrocrystalline (II) states of the Grade 2 alloy. Different symbols denote microhardness values from different ranges.

The above microhardness distributions depending on distances in the base metal zone far from the welded joint clearly show that the microhardness of the microcrystalline alloy at the center of the weld joint (lines 2 and 3, Fig.5) has a smaller spread of values as compared to the alloy in the submicrocrystalline state, while average microhardness values in both states are close to each other. This reflects the fact that the structural heterogeneity is larger in the central part of the weld joint when the titanium alloy is in the submicrocrystalline state as compared to the microcrystalline state.

Conclusion

The welded joints produced by electron beam welding of Grade 2 titanium alloys in the microcrystalline and submicrocrystalline states were examined using structural methods and microhardness measurements. The findings lead to an important conclusion. In the submicrocrystalline state, the spectrum of structural phase states of the Grade 2 alloy in the weld zone and the HAZ zone is different from the states in the microcrystalline titanium alloy. The structural features of all weld zones that have been identified taking phase transitions into account determine the hardness distribution pattern and the related strength properties of compounds. This information is necessary for developing a technique aimed at optimizing physical and mechanical properties of welds produced in submicrocrystalline Grade 2 titanium alloys by electron-beam welding.

References

1. S. Gnyusov, V. Klimenov, Yu. Alkhimov, A. Budnicky, A. Orishich, A. Cherepanov, Yu. Afonin, Welding International, **1**, 17 (2012)
2. H. Liu, K. Nakata, N. Yamamoto, J. Liao, Journal of Materials Science, **47**, 1460 (2012)
3. T.S. Balasubramanian, V. Balasubramanian, M.A. Muthumanikkam, Journal of Materials Engineering and Performance, **20(9)**, 1620 (2011)
4. E. Akman, A. Damir, T. Canel, T. Sinmazçelik, J. Materials Processing technology, **209**, 3705 (2009)
5. M. Weglowski, S. Błacha, A. Phillips, Vacuum, **130**, 72 (2016)
6. A. Smirnova, V. Panin, Yu. Pochivalov, A. Gorbunov, A. Malikov, A. Orishich, AIP Conference Proceedings, **1785**, 040071 (2016)
7. S. Wang, J. Liu, D. Chen, Materials Science and Engineering, **A 584**, 47 (2013)
8. J.E. Posr, D.L. Bisn, American Mineralogist, **78**, 932 (1993)
9. V. Ljasockaja, *Termicheskaja obrabotka svarnyh soedinenij titanovyh splavov* [Thermal treatment of welded joints of titanium alloys] (Ekomet, Moscow, 2003) (in Russian)
10. G.A. Salishchev, S.Yu. Mirinov, S.V. Zherebtsov, Rev. Adv. Mater. Sci., **11**, 152 (2006)
11. V. Klimenov, A. Potekaev, A. Tabachenko, M. Marzol, A. Klopotov, Reshetnevskie chteniya: proceedings of the XVIII International scientific conference (Krasnoyarsk, Siberian State Aerospace University), **1**, 296 (2014)
12. K. Kurgan, V. Klimenov, A. Klopotov, Vestnik of TSUAB, **1**, 140 (2016) (in Russian)
13. I. Kurzina, I. Bozhko, M. Kalashnikov, A. Eroshenko, Yu. Sharkeev, Materialovedenie [Inorganic Materials: Applied Research], **5**, 48 (2010) (in Russian)
14. I. Korolev, S. Gorbunov, A. Bogdanov, G. Nadutenko, Metallovedenie titana [Metallurgy of titanium], Proceedings of the 5th Conference on Metallurgy, Metallurgical Science and Application of Titanium and its Alloys, 117-123 (Nauka, Moscow, 1964) (in Russian)

MATEC Web of Conferences 143, 04001 (2018) https://doi.org/10.1051/matecconf/201814304001
YSSIP-2017

Housing sector development in socio-economic space of agglomeration

Yulia Podoprigora[1,*], *Irina* Volchkova[1], *Maria* Danilova[1], and *Evgeniya* Ufimtseva[1]

[1]Tomsk State University of Architecture and Building, 634003 Tomsk, Russia

Abstract. The paper is aimed at studying the influence of balanced housing sector development on the cohesion level of the socio-economic space (SES) in agglomerations. Statistical and comparative analytical techniques are used as the basic methods in this research. The paper presents indicators that measure the level of integration and balance in the housing sector. It proposes a set of mechanisms aimed at promoting a more balanced development of the housing sector in order to achieve SES cohesion in the Tomsk agglomeration. The relevance of the study is associated with intensification of agglomeration processes and the need for new methods and techniques to assess the balanced state of housing sector development in agglomeration SES.

Introduction

Ensuring the balanced socio-economic development of the country, including the housing sector, is one of the priorities in the current national policy of the Russian Federation. Nevertheless, inconsistent housing policies of communities have a number of destructive consequences, including increasing differentiation in the level of socio-economic development across communities and less coherent socio-economic space (SES) of an agglomeration. In this context, the relevance of this study is conditioned by theoretical and practical importance of studying the balanced development processes of the housing sector taking into account the need to increase the cohesion level of agglomeration SES. The conceptual scientific problem of the study is the lack of an updated theoretical and methodological framework necessary for studying and measuring the influence of housing sector development on the cohesion level of agglomeration SES.

Theoretical and applied studies of agglomeration SES were conducted by V. S. Antonyuk, E. Boze, V. L. Glazychev, E. E. Goryachenko, N. V. Demchuk, A. M. Lola, N. L. Mosienko, Z. Z. Mullagaleeva, T. Nelson, I. V. Starodubrovskaya, H. Uchida, A. Franchini, V. A. Shabashev, A. V. Schmidt, etc. [1–7]. Notably, the international experience shows that the level of agglomeration development is often estimated together with the agglomeration effect. A set of criteria used by foreign researchers is usually close to the criteria taken for the Soviet methods developed in the 1970s and 1980s by the Central Scientific Research and Design Institute of Urban Planning and the Institute of Geography of the USSR Academy of Sciences (population density, population size of the core city,

[1] Corresponding author: y.v.p@rambler.ru

temporal availability of the core city) [7]. In the last decade, experts of the Institute for Urban Economics (Moscow) made a valuable contribution to improving methods for assessing the development level of Russian agglomerations. They developed two types of assessment indicators: integration and balanced ones [8].

This study is aimed at assessing the influence of balanced housing sector development on the cohesion level of agglomeration SES. The scientific and practical novelty of the study is associated with the creation of mechanisms aimed at promoting a more balanced development of the housing sector in order to achieve SES cohesion in the Tomsk agglomeration.

Materials and Methods

The systems theory, the theory of spatial development, concepts proposed by Russian and foreign scientists in the field of agglomeration development serve as a methodological framework of the study. Statistical and comparative analytical techniques are used as the basic methods and approaches in this research. In particular, the statistical analysis is required to assess internal inequality in the agglomeration on the basis of numerical indicators. The comparative analysis is used to compare and analyze dynamic factors of individual indicators that represent the housing sector development in the Tomsk agglomeration.

In our opinion, among the available methodological tools for assessing the balanced state of the housing sector, a technique proposed by the Institute of Urban Economics (Moscow) is the most interesting one. According to the developers of this technique, "when accessing the development level of agglomerations, it is important to focus on indicators that reflect the variability of certain parameters across the agglomeration by using the coefficient of variation" [8].

The Institute's specialists developed the following indicators to assess the balanced state of the housing sector:

– the coefficient of variation in housing availability (in square meters) across the agglomeration;

– the coefficient of variation in housing availability (in housing units) across the agglomeration;

– the coefficient of variation in a proportion of persons who have improved their housing conditions among those who have applied for improving their housing conditions;

– relationship between the ratio of the price per square meter of housing to per capita income in the core city of the agglomeration and the weighted average ratio of price to per capita income in other municipalities of the agglomeration [8].

It is worth noting that the practical use of the indicators listed above is limited and depends on available statistical data.

However, in our opinion, in addition to the coefficients of variation for estimating the inequality within the agglomeration, it is important to use the Theil entropy index, which allows us to decompose the interregional inequality index into two components to reflect intergroup and intragroup inequality [9]. In the case of absolute interregional parity, the Theil index takes a minimum value (zero). As the degree of interregional inequality grows, the Theil index increases. The Theil index can be determined by the following formula:

$$IT = \sum_{j=1}^{n} \frac{X_j}{X} \cdot \ln\left(\frac{X_j / P_j}{X / P}\right) = \sum_{j=1}^{n} \frac{X_j}{X} \cdot \ln\left(\frac{x_j}{x}\right), \quad (1)$$

where X_j is the absolute value of the indicator in the j-th territory (municipality); X is the total value of the indicator for all territories (region or districts); P_j is the population size of the j-th territory; P is the total population size; x_j is the value of the indicator in the j-th territory per capita; x is the average value of the indicator for all territories per capita; n is the number of territories; $\ln(x_j/x)$ is the natural logarithm of quotient.

Experiment

We will estimate the influence of the balanced housing sector development on the SES cohesion level in the Tomsk agglomeration, one of the fastest growing agglomerations in Siberia. In recent years, there is a trend towards increasing new housing construction rates in most communities of the agglomeration (Fig. 1). All municipalities of the Tomsk agglomeration have been steadily expanding their housing stock over the last few years. New housing construction rates are increasing due to various sources of financing, making it possible to reduce the housing deficit significantly and increase housing availability for the agglomeration residents.

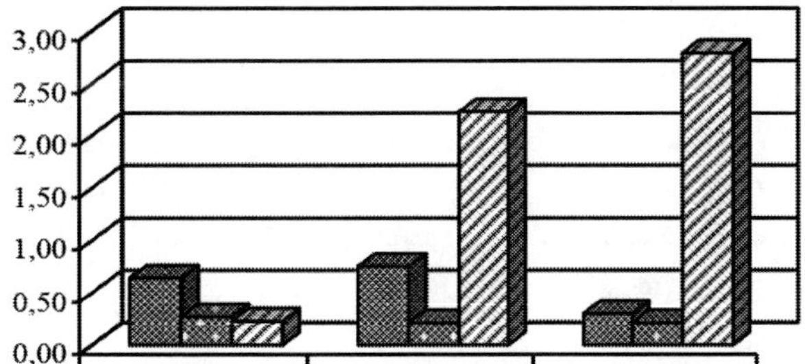

	2010	2015	2016
Tomsk	0,64	0,75	0,30
Seversk	0,27	0,21	0,21
Tomsk District	0,23	2,23	2,79

Fig. 1. Commissioning of the total housing area, m^2 per capita.

Agglomeration interactions in the housing sector are becoming more intense as evidenced by growing individual housing construction rates in the Tomsk District. The real estate and land market of this district is particularly attractive for the agglomeration residents. However, the development of interactions in the housing sector of the Tomsk agglomeration is highly constrained by limited availability of the housing stock in Seversk closed administrative and territorial entity [10].

Next, the balance of the housing sector development in the Tomsk agglomeration is assessed. For this purpose, we calculate the coefficient of variation and the Theil index for the indicator of residential buildings commissioned in the agglomeration per capita (Table 1). The choice of this indicator is associated with its analytical properties and the availability of statistical data.

Table 1. Analysis of dynamics of residential buildings commissioned in the Tomsk agglomeration per capita.

Index	2011	2012	2013	2014	2015	2016
Coefficient of variation, %	23.8282	22.9928	**45.5976**	**45.9345**	16.6626	22.8965
Theil index	0.0691	0.0379	0.0163	0.1022	0.6064	0.5876

The calculated coefficient of variation shows that the indicator of residential buildings commissioned in the Tomsk agglomeration in 2013–2014 per capita is distributed unevenly (since the coefficient of variation exceeds 33%). The calculated Theil index indicates a significant increase in the differentiation within the agglomeration in 2015 as compared to 2014, which negatively affects the coherence of the social and economic space in the Tomsk agglomeration.

Results

The results of the analysis demonstrate an increased differentiation within the Tomsk agglomeration in terms of the residential houses commissioned per capita in 2015. This has negative effect on the cohesion of the agglomeration SES and requires development of mechanisms aimed at promoting a more balanced development of the housing sector (Fig. 2).

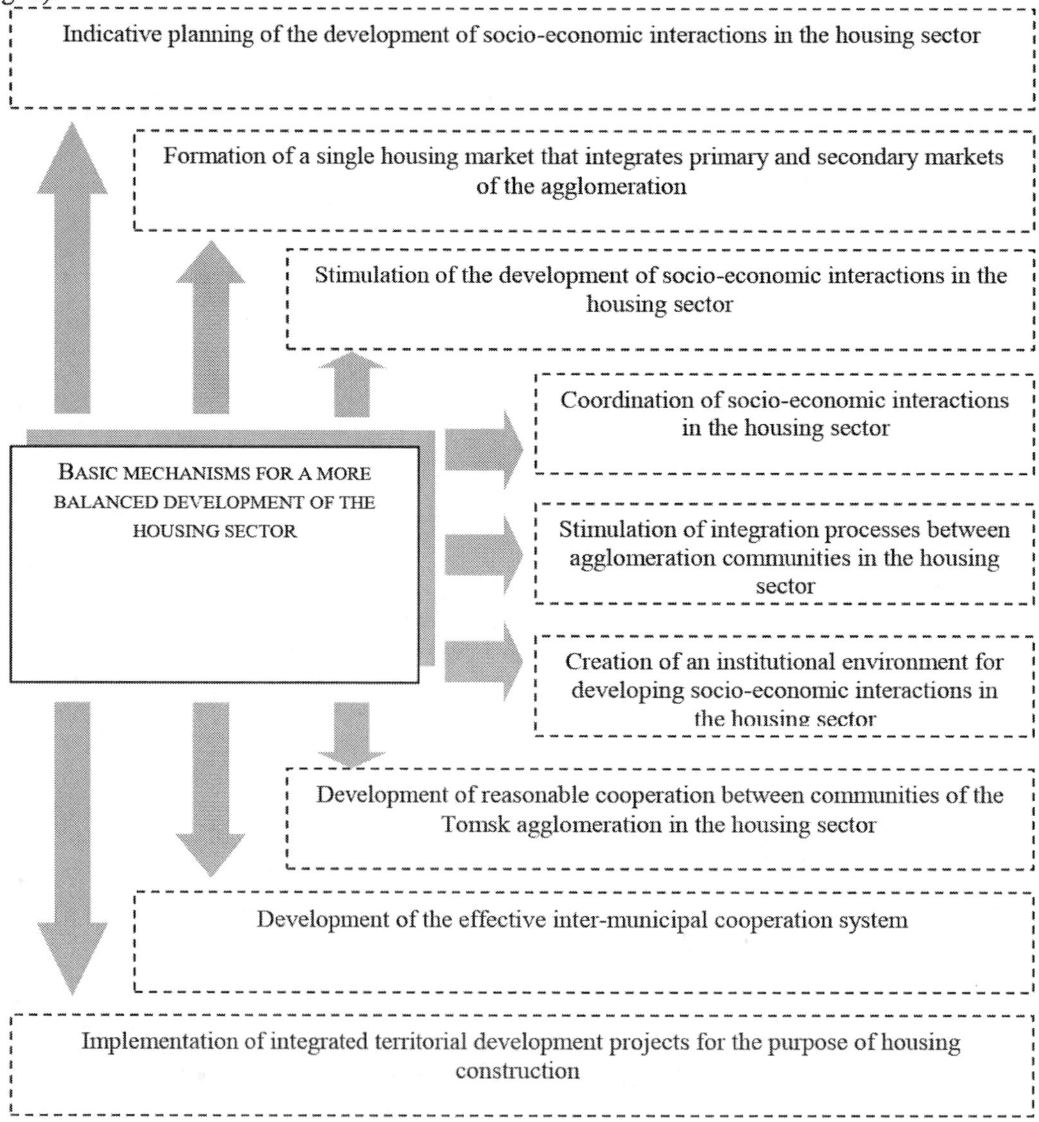

Fig. 2. Mechanisms for a more balanced development of the housing sector.

Conclusion

The analysis shows that assessing the balance in the housing sector development in the agglomeration SES is an important task. We assume that by converging levels of socio-economic development across communities and enhancing SES cohesion, it is possible to make the economic sector of the agglomeration more sustainable, reduce social inequality, increase the competitiveness of the agglomeration, as well as to create single agglomeration markets, strengthen integration, and eventually improve the welfare of residents.

In our opinion, the mechanisms aimed at balancing the housing sector in the Tomsk agglomeration should be based not only on methods regulating socio-economic interactions, but also on methods that focus on creating an environment for such interactions. It is also important to emphasize that the balance can be achieved both through self-regulation and direct regulation by local authorities of municipalities belonging to the agglomeration. Taken together, the efforts to achieve more balanced agglomeration processes in the housing sector enhance the quality of these processes and increase the general level of agglomeration development.

The presented mechanisms, in turn, imply using a certain set of tools for their implementation. For example, stable relations between SES agents based on equal opportunities of the agents in their socio-economic interactions will contribute to the balanced development of the housing sector in the Tomsk agglomeration. Equal opportunities should be provided through such basic tools as coordination of goals, opportunities, interests, responsibilities, as well as the coordination of partner relations among SES agents.

The results of the study can serve as a basis for improving the methodological framework that is necessary for studying and measuring the influence of the single housing market development on the cohesion level of the socio-economic space in agglomerations.

References

1. V. Glazychev, I. Starodubrovskaya, et al, *Chelyabinskaya aglomeratsiya: potentsial razvitiya* [Chelyabinsk agglomeration: the development potential] (The Institute of Economics and Law, Chelyabinsk, 2008) (in Russian)
2. E.E. Goryatchenko, N.L. Mosienko, N.V. Demchuk, Region: Economics and sociology, **3**, 94-112 (2011)
3. A.M. Lola, *Urban and Agglomeration Control in Russia: state and what to do* (Moscow, 2013)
4. E. Bose, Russian expert review, **4-5**(22), 13-16 (2007)
5. Z.Z. Mullagaleeva, V.A. Shabashev, *Teoretiko-metodologicheskie osnovy jekonomicheskoj politiki gorodov, obrazujushhih aglomeraciju* [Theoretical and methodological foundations of economic policy of the cities, forming the Metropolitan area] (Kuzbassvuzizdat, Kemerovo, 2009) (in Russian)
6. A.V. Schmidt, V.S. Antonyuk, A. Franchini, Economy of Region, **12**(3), 776-789 (2016)
7. H. Uchida, T. Nelson, Working Paper, World Institute for Development Economics Research, **29**, (2010)
8. Institute for Urban Economics, Koncepcija ocenki urovnja razvitija gorodskih aglomeracij [Evaluation concept of urban agglomerations development], available at: http://www.urbaneconomics.ru/sites/default/files/koncepciya_ocenki_urovnya_razvitiya_gorodskih_aglomeraciy_13.01.17.pdf (2017) (in Russian)
9. N.V. Voroshilov, Problems of Territory's Development, **3**(65), 31-41 (2013) (in Russian)

10. I.V. Volchkova, M.N. Danilova,Y.V. Podoprigora, A.A. Seliverstov, E.V. Ufimtseva, N.R. Shadeyko, *Razvitie aglomeracionnyh processov v aspekte svjazannosti social'no-jekonomicheskogo prostranstva (po materialam Tomskoj aglomeracii)* [The development of the agglomeration process in the aspect of connectedness socio-economic space (on materials of the Tomsk agglomeration)] (Integral cover, Tomsk, 2016) (in Russian)

Surface models of granular materials in storage tanks

Nikolaj Zamyatin[1,*], *Vladimir* Makovkin[1], and *Gennadij* Smirnov[1]

[1]Tomsk State University of Control Systems and Radioelectronics, 634045 Tomsk, Russia

Abstract. The paper focuses on the problem of measuring the volume of granular materials in storage tanks where, due to uneven surfaces of these materials, the error in measurement by conventional methods reaches 15–20 percent. To increase the accuracy of volume measurement, a technique for modeling surfaces formed by granular materials in tanks is proposed.

Introduction

Enterprises of the construction industry, chemical plants, agricultural companies and enterprises of other industries use a variety of tanks (tank batteries) to store raw materials, finished products, various granular substances and materials, and should operate these tanks as efficiently as possible [1–4].

If a material is granular, viscous or has a consistency other than liquid, it forms an uneven surface of convex or concave shapes (depending on whether materials are supplied to or pumped from a tank). A curved surface of materials is formed by cohesive forces that make particles coalesce with each other and with walls of the tank (Fig. 1). The most common method of volume measurement in tanks is to measure the surface level at only one point and then average the data, but this method should generally be used for liquid materials [5, 6]. Such a volume measurement method leads to low accuracy of data for tank batteries and, therefore, has a direct influence on economic losses of enterprises. This results in downtime due to miscalculated dynamics of consumption or production of raw materials; low quality of mixtures due to errors in calculating ratios of ingredients in different formulations; additional costs arising from related accounting errors; miscalculated production volumes and consumption of raw materials, and other problems.

* Corresponding author: zamnv47@gmail.com

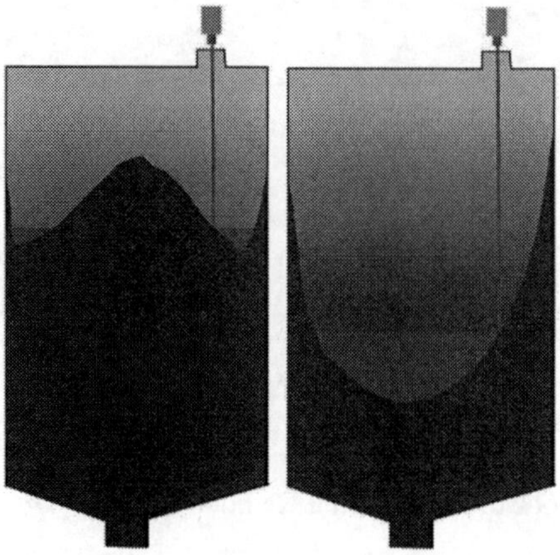

Fig. 1. Shape of the surface when supplying the material to/pumping it from the tank.

In order to increase the accuracy in practice, researchers model the surface of materials using known coordinates and, for this purpose, rely on mathematical surface relief models, in particular by approximating heights between adjacent points of the surface model [7]. More general statistical models include the "floating square" and "floating circle" described in [8].

Volume calculation is based on different techniques and depends on required accuracy [9]. This study uses a laser method to measure the surface level and three modeling techniques: by smooth surface, by cone, and by uneven surface, with neural network approximation and volume calculation through numerical integration. Similar studies are performed in Russia and abroad for other level measurement tools [5, 6, 9].

Materials and Methods

According to the proposed technique, volume measurement for granular materials in a storage tank is divided into the following stages:

1) determining a discrete set of values of surface levels;
2) approximating the surface to determine intermediate unknown values of coordinates and its nonlinear function;
3) calculating the volume of material under the formed surface.

The volume measurement method by smooth surface lies in measuring the level at a certain location and calculating the volume of material. This is the simplest method that is usually applied to control the volume of liquid materials in case of manual measurement. It is the fastest method among the considered ones and does not require complex calculation. However, since this method ignores rheological and flow properties of mixtures, it has the lowest accuracy for granular materials as compared to all other methods proposed above.

The volume measurement method by cone lies in measuring the level of material at a specific location and modeling the cone or conical depression (for cases when the material is supplied to and pumped from the tank, respectively). The angle between generatrices and base is equal to the angle of repose of the material. The angle of repose (sometimes the angle of internal friction or angle of slope) is the angle formed by a free surface of loose mass or other granular material with a horizontal plane.

By using the angle of repose and calculating the cone volume, it is possible to significantly increase the accuracy as compared to volume measurement by smooth surface. However, this method has the following disadvantages:

– it requires more data and preliminary calculations for different types of tanks;

– the angle of repose varies depending on moisture, type and grade of granular materials and is determined experimentally;
– the angle between the base and loose mass is equal to the angle of repose only when supplying to or pumping materials from the tank. However, when switching from the first process to the second one and vice versa, the angle will change too.

The volume under an uneven surface in a vertical cylindrical tank is determined subject to the geometric meaning of the double integral. If the $f(x,y)$ function takes only positive values in the D domain, the double integral of this function is numerically equal to the volume V of the vertical cylindrical body constructed on the basis of D and bounded above by the corresponding part of the surface $z=f(x,y)$ [10].

$$V = \iint_D z dx dy \tag{1}$$

Values of integrals can be calculated in two common ways: either using quadrature or cubature formulas, or using the Monte Carlo method.

For mathematical modeling of surfaces, it is suggested to use neural networks that have high speed and accuracy of models. In [7, 11], various neural networks were compared, in particular for approximation of input values. In [11], the output values of neural networks were compared according to the coefficients of determination presented in Table 1.

Table 1. Quality of approximation based on the coefficients of determination.

	GRNN network	RBF network	Linear network
Coefficient (R^2)	0.9473	0.9918	0.6422

When smoothing the function on the basis of models other than neural networks, such as Fourier series and a multiplicative model, the coefficients of determination are equal to 0.8747 and 0.7262, respectively.

From the data shown above, it is clear that RBF and GRNN neural networks are best suited for approximating surfaces. The RBF network gives the coefficient of determination $R^2 = 0.9918$. This means that the actual output of the neural network and the desired output are almost identical. Using analytical or numerical methods, it is practically impossible to achieve such a high value of the coefficient of determination. In terms of processing speed and network learning, RBF networks are best suited for approximating the surface model, since they do not have the second intermediate layer that would be responsible for estimating the weighted mean, as in case of GRNN. Therefore, RBF networks can be used to model the surface of materials.

In 1956, Kolmogorov proved that any continuous function can be implemented as a superposition of functions of smaller dimension [12]. The implicit function of three-dimensional surface can be represented in the form (2):

$$f(x,y) = \sum_{i=1}^{n} \omega_i \varphi(r), \tag{2}$$

where ω_i are the constants; $\varphi(r)$ is the radial basis function.

As a function of $\varphi(r)$, we can choose any radial basis function (3) that is used to solve problems of multidimensional interpolation and approximation.

$$\varphi(x, y, c_i^x, c_i^y) = 1 - \sqrt{(x - c_i^x)^2 + (y - c_i^y)^2} \tag{3}$$

where c^x_i, c^y_i are some constants, the so called RBF kernels.

The implicit surface function can be implemented using a neural network consisting of two layers: a hidden nonlinear layer with n neurons and an output linear layer with one neuron that transmits a weighted sum of neuron outputs of the first layer to the output.

The RBF neural network learning algorithm is as follows:

- to select the number of neurons in the hidden layer so that it is equal to the number of input points;
- to specify a RBF kernel for each neuron (c^x_i, c^y_i) so that $c^x_i = x_i$, $c^y_i = y_i$;
- to determine weights ω_i.

The output value of the network for the ith input data is determined by the formula (4)

$$D_i = w_1 f(x_i, y_i, c_1^x, c_1^y) + w_2 f(x_i, y_i, c_2^x, c_2^y) + ... + w_n f(x_i, y_i, c_n^x, c_n^y) \quad (4)$$

Thus, the surface relief is represented in the form of a nonlinear function — a superposition of the learned RBF neural network. Then the volume of a certain material can be determined by solving a double or triple integral of an already known function.

The error in calculating the additional volume depends only on surface irregularities, since the volume of material below the minimum surface level of the mixture is determined by the shape and dimensions of the tank. The volume of material can be measured by the following methods.

Results

The possibility of using the proposed method was tested experimentally based on a model of vertical cylindrical tank in which M400 cement was used as a material to be measured. To measure the level, we used a cylindrical coordinate system with the **0z** axis passing through the center of the model. In real conditions, non-uniformly distributed data can be transmitted to the input of a RBF neural network. However, for the sake of simplicity, the points were equally spaced from each other, while level values were measured at equal intervals φ and r from each other.

A SICKDT50 (error ± 1mm) laser rangefinder was used as a level meter.

Since the applied radial basis function does not work with cylindrical coordinates directly because of the Euclidean distance between the center and the points, it is necessary to change variables through a Jacobian.

A cubic RBF function is used as an activation neuron function. The activation function was chosen experimentally while comparing with the real surface by the appearance of the model:

$$a = n^3 \quad (5)$$

When teaching the RBF neural network with zero error, the number of neurons in the hidden layer is always equal to the number of elements in the learning set (values transmitted to the network input). With such an error, it is impossible to obtain an acceptable (in terms of speed) solution for large training sets. Therefore, additional manipulations are required to minimize the error of the model. Moreover, the zero error does not guarantee that the intermediate values for which the neural network has not been learned will give an accurate result. On the contrary, the surface model will be more likely to have a complex shape that is different from the real one.

As shown by experiments, twelve neurons were sufficient to learn a neural network with an error of less than one percent, while significantly reducing the calculation time. The performance of the neural network with a cubic activation function is shown in Figure 2.

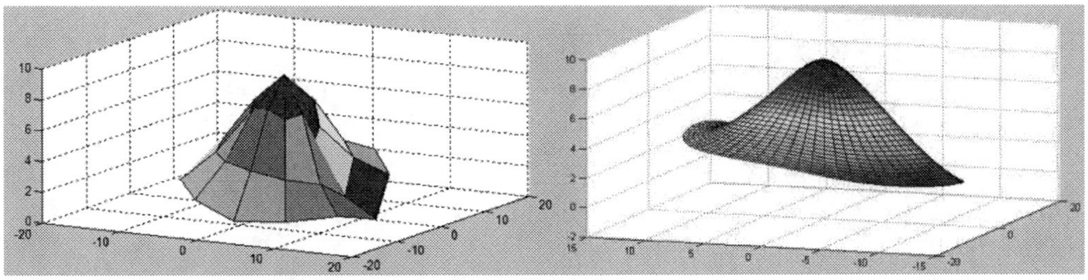

Fig. 2. A graphical representation of the obtained data can be seen on the left. The performance of the cubic activation function in the neural network is presented on the right.

The learning time of the neural network is tenths of a second. This is sufficient when using this method for tank batteries.

In this paper, the authors investigate the surface in a static state (effects of supplying the mixture to or pumping it from the tank in dynamics were neglected). They focus on measuring the volume of the "heap" formed by the material (the volume of the mixture under the heap remains unchanged), while estimating level values from its lower boundary. The volume of the mixture under the "heap", $V_{осн}$, is calculated by the formula for the vertical cylinder volume.

The measurement of the volume by smooth surface was based on the average level of material with a smooth surface.

For the conical surface, the average angle of repose was equal to thirty-nine degrees due to unevenness and lumpiness, which is close to the tabulated values, and the angle of repose is chosen to be equal to forty degrees. When measuring the volume by the nonlinear function, the double integral was calculated by the Gauss-Kronrod method. The estimation of the errors for different volume measurement methods is presented in Table 2.

Table 2. Errors of methods.

Smooth surface	Cone	Nonlinear function
25.8%	21.7%	8%

Conclusion

According to the experimental results, the method of volume measurement by smooth surface is found to have the greatest error. The error produced by the measurement method by cone is slightly less. By analyzing characteristics of materials in a more detail and developing the appropriate formulas, it is possible to increase the accuracy of measurement by cone. However, its main weaknesses will still exist — a narrow focus on the type of material and labour-intensive research.

Among three available methods, the method with approximation of a nonlinear function is the best one, since it allows the error to be controlled by the number of partitions of function responses or the number of experiments. Moreover, this method takes the surface model into account.

It is reasonable to use a RBF neural network for obtaining the surface model of granular substances in order to improve the accuracy of measuring volumes in tank batteries and storehouses of production enterprises.

The most effective techniques are those that are aimed at modeling surface functions and measuring their volume by quadrature or Monte Carlo methods, since these techniques, according to the obtained results, have the smallest error. In this case, the Gauss-Kronrod method is preferable, since its accuracy remains several times greater even if the difference in the number of operations is thousands of times.

The proposed method of volume measurement that takes a nonlinear surface function into account can also be applied to any other material. The measurement error will remain the same regardless of moisture content, flow and other rheological properties of materials.

References

1. T.S. Namazbaev, S.H. Esenbaev, V.A. Kolesnikov, *Sredstva izmerenija i kontrolja tehnologicheskih processov na predprijatijah gorno-metallurgicheskogo kompleksa* [Measurement and control tools of technological processes at the enterprises of the mining and metallurgical complex] (2012) (in Russian)
2. E.A. Yakimovich, N.V. Zamyatin, Information technologies in territorial management, industry, education, 165 (2002) (in Russian)
3. G.V. Smirnov, D.G. Smirnov, Proceedings of TUSUR University, **19(2)**, 99 (2016) (in Russian)
4. V.I. Makovkin, N.V. Zamyatin, Proceedings of TUSUR University, **19(4)**, 95 (2016) (in Russian)
5. Y. Huang, B. Chen, E. Pienkowski, G. Chen, H. Xiao. Sensor and Accurator A: Physical (2011)
6. S. Khaliq, S.W. James, R.P. Tatam, Opt. Lett., **26**, 1224 (2001)
7. A.V. Kulazhskiy, Vestnik RGUPS, 95 (2009) (in Russian)
8. N.E. Konovalov, N.N. Sonne, I.G. Melnik, A.K. Yakimovich, Proceedings of GIPRODORNII, **18** (1976) (in Russian)
9. A.G. Rybakov, E.V. Lukyanova, I.S. Burmistrova, Yu.F. Muratova, M.V. Fyodorova, Journal "Power and Electrotechnical Systems", 430 (2015) (in Russian)
10. M.Ya. Vygodskiy, *Spravochnik po vysshej matematike* [Handbook of Higher Mathematics], 622 (AST, Moscow, 2006) (in Russian)
11. T.V. Filatova, Tomsk State University Journal, **284** (2004) (in Russian)
12. A.N. Kolmogorov, Doklady AN SSSR [Proceedings of the USSR Academy of Sciencies], **108**, 179 (1956) (in Russian)

Fuzzy batch controller for granular materials

Nikolaj Zamyatin[1,*], *Gennadij* Smirnov[1], *Yuri* Fedorchuk[2], and *Olga* Rusina[2]

[1]Tomsk State University of Control Systems and Radioelectronics, 634050 Tomsk, Russia
[2]National Research Tomsk Polytechnic University, 634050 Tomsk, Russia

Abstract. The paper focuses on batch control of granular materials in production of building materials from fluorine anhydrite. Batching equipment is intended for smooth operation and timely feeding of supply hoppers at a required level. Level sensors and a controller of an asynchronous screw drive motor are used to control filling of the hopper with industrial anhydrite binders. The controller generates a required frequency and ensures required productivity of a feed conveyor. Mamdani-type fuzzy inference is proposed for controlling the speed of the screw that feeds mixture components. As related to production of building materials based on fluoride anhydrite, this method is used for the first time. A fuzzy controller is proven to be effective in controlling the filling level of the supply hopper. In addition, the authors determined optimal parameters of the batching process to ensure smooth operation and production of fluorine anhydrite materials of specified properties that can compete with gypsum-based products.

Introduction

To ensure the smooth production of building products based on fluorine anhydrite, it is necessary to monitor continuous filling of supply hoppers with a raw material — industrial anhydrite binder. This is essential for timely batching at a specified, carefully adjusted ratio of initial components and, therefore, production of anhydrite-based building products with required strength and waterproof characteristics. Industrial anhydrite binder is produced from fluorine anhydrite, a waste product of chemical industry, which can replace natural gypsum in the traditional building materials industry [1–7].

Batch control of granular materials with a fuzzy controller is proposed for production based on fluorine anhydrite. This solution can be used under conditions of incomplete information where smoother switching with more accurate timing is required. Therefore, the aim of our study is to analyze the possibility of using a controller based on principles of fuzzy logic to maintain a required level of industrial anhydrite binder in a supply hopper.

Fuzzy controllers based on fuzzy inference systems [8, 9] are able to produce control actions, just like a human operator who makes decisions and controls stages of the production process. Researchers proposed different applications of the fuzzy controller in production processes, for example, maintaining constant temperature, water level in a tank, etc. [10–12]. For the first time, the fuzzy controller is proposed for the production of

* Corresponding author: zamnv47@gmail.com

building materials based on fluorine anhydrite. It can be used as an element of the automatic level control system that controls filling of supply hoppers with industrial anhydrite binder.

Research Methods

Different studies and system modeling in the field of artificial intelligence are often founded on principles of fuzzy logic and the fuzzy-set theory. Fuzzy control processes are modeled mostly in the popular MATLAB environment using a special Fuzzy Logic Toolbox package [13–14].

Fuzzy inference processes are modeled on the basis of an if-then rule base and membership functions of linguistic terms.

Modeling can be performed under the following conditions:

- there is at least one rule for the relevant linguistic terms of the output variable;
- any term of the input variable contains at least one rule, in which it is a premise.

This paper considers fuzzy inference based on the Mamdani mechanism, the most common inference approach in fuzzy systems.

A supply hopper is used as a fuzzy controlled element.

Construction of the Fuzzy Controller

A frequency converter of the signal transmitted to the screw drive motor is used as a controlling element in the supply hopper. A controller of the asynchronous screw drive motor is designed to monitor filling of the hopper with industrial anhydrite binder. The controller generates a required frequency, thereby ensuring required productivity of the feed conveyor.

Relying on the accumulated experience of studying production processes based on fluorine anhydrite and our own studies [15–18], we have established that levels of the supply hopper depend on its productivity for fluorine anhydrite (G), while the normal level of the supply hopper, $Level_{norm}$, is G x 8 hours + 0.5 hours of margin; the maximum level, $Level_{max}$, is G x 8 + 1.0 x G = 9 x G; the minimum level, $Level_{min}$, is 1.0 x G.

The knowledge base of the fuzzy controller includes the following rules:

1. If the level of industrial anhydrite binder in the supply hopper is below the minimum, the controller of the asynchronous screw drive motor generates 60 Hz (the maximum productivity of the feed conveyor is 100%);
2. If the level of industrial anhydrite binder in the supply hopper is above the minimum but below the normal level, the controller generates 54 Hz (the productivity of the feed conveyor is 90%);
3. If the level of industrial anhydrite binder in the supply hopper is normal, the controller generates 30 Hz (the productivity of the feed conveyor is 50 %);
4. If the level of industrial anhydrite binder in the supply hopper is above the normal level but below the maximum, the controller generates 6 Hz (the productivity of the feed conveyor is 10 %);
5. If the level of industrial anhydrite binder in the supply hopper achieves the critical level but is still below the maximum, the controller generates 0 Hz (the productivity of the feed conveyor is 0 %);

We take 8.5xG as the normal level; 9xG as the maximum level, and 1.0xG as the minimum level, where G is the productivity of the feed conveyor. Consumption of industrial anhydrite is assumed to be constant.

We choose frequency and productivity as input variables of the fuzzy inference system, and the fuzzy linguistic variable “level” as an output variable.

The range of input variables (Range and Display Range) is from 1 Hz to 60 Hz for the variable “frequency” and from 1% to 100% for the variable “productivity”.

Figure 1 shows a membership function editor after making changes for the input frequency variable. A membership function editor for the input variable “productivity” has the similar form.

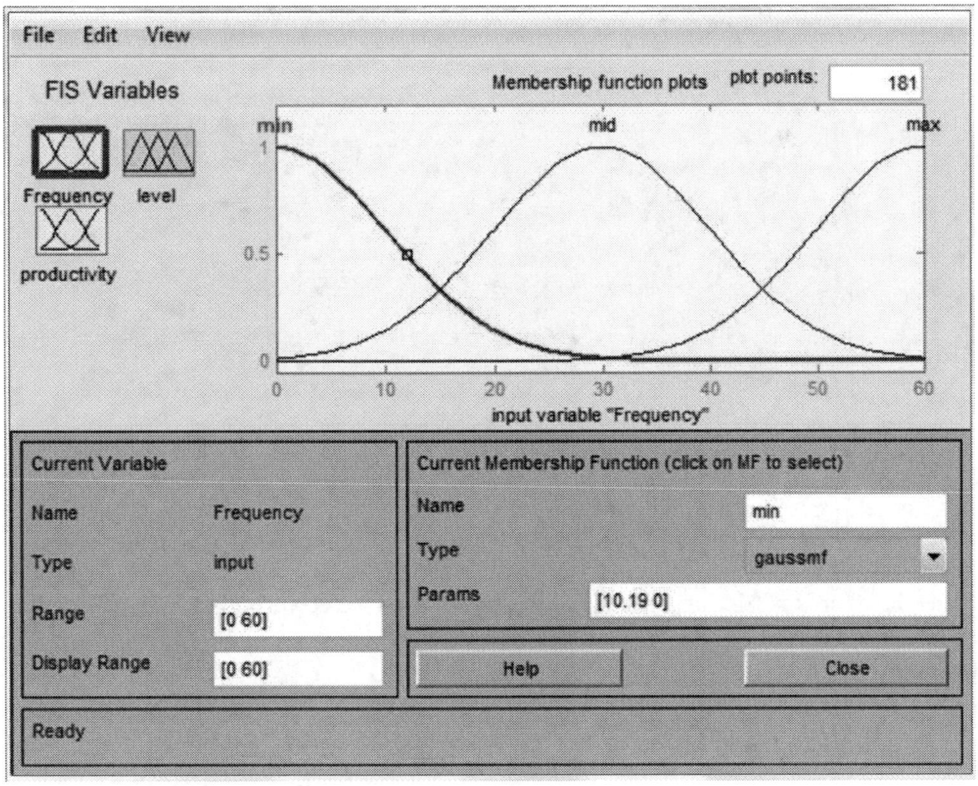

Fig. 1. Membership functions of the input signal “frequency”.

We also change the names of terms for the output variable “level”. Then we choose a membership function of triangular type. We take values [0 0.02644 1] for the term “productivity”; [1 3.75 8.3] for the term “min”; [8.3 8.4 8.5] for the term “norm”, [8.3 8.5 9.33] for the term “more min”; and [8.6 8.82 8.9] for the term “more norm”. The membership function editor after making changes for the output variable “level” is shown in Figure 2.

The fuzzy inference rules for the fuzzy system under development are generated in the rule editor of the Fuzzy Logic Toolbox package.

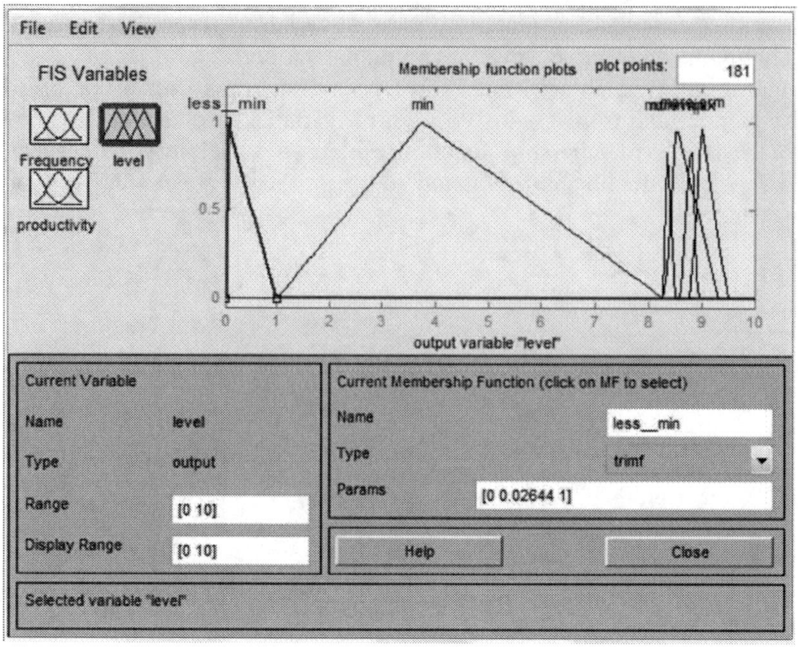

Fig. 2. Membership functions of the output signal "level".

Figure 3 presents a window of the rule viewer, which shows that average values from the allowable range (values [30 50] in the Input field) are proposed as default values for the input variables. This means that the controller produces 30% when the productivity of the feed conveyor is 50%. These values of the input variables correspond to 8.52xG level of the supply hopper, which is the normal filling level of the hopper. Therefore, the model is valid (Fig. 3).

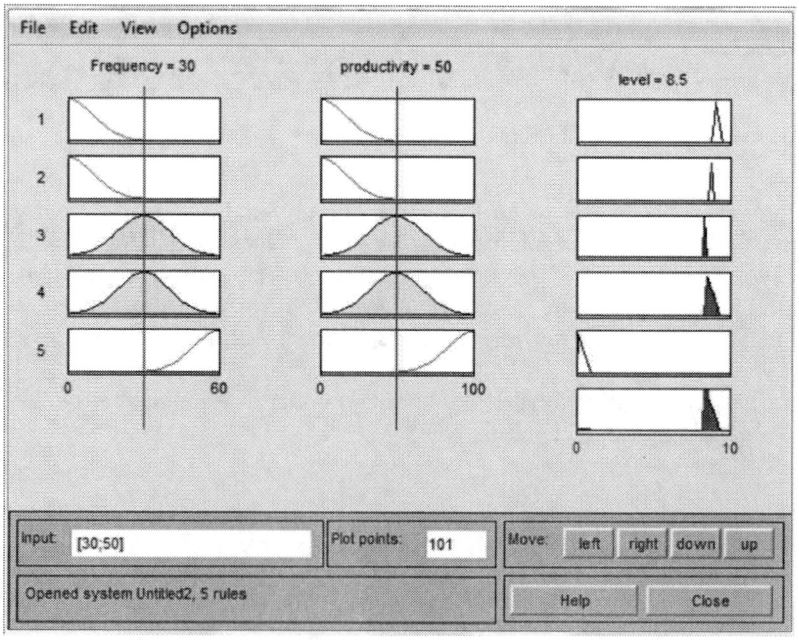

Fig. 3. Interface of the software.

Results

The logical inference algorithm usually consists of the following stages: fuzzification, fuzzy inference, composition and defuzzification (Figure 4).

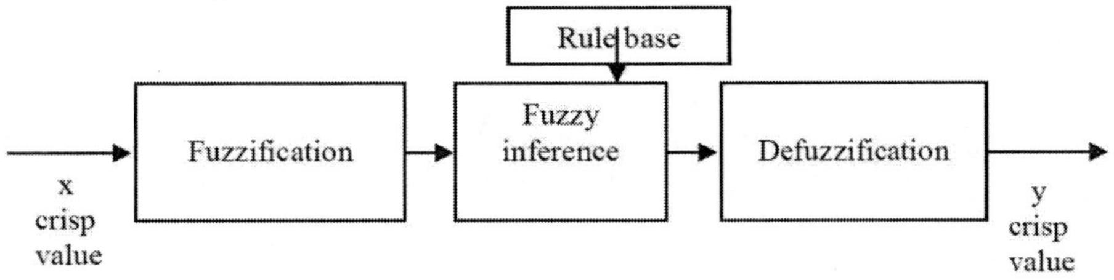

Fig. 4. Fuzzy logic inference system.

Fuzzification is conversion of x values into fuzzy format. Defuzzification is a procedure reverse to fuzzification.

Each fuzzy inference algorithm has its own set of logical operations and rules, as well as its own type of defuzzification method. Fuzzy inference models of Sugeno, Tsukamoto, Larsen and Mamdani are known.

Functioning of fuzzy logic systems is based on a single principle: readings from measuring devices are fuzzificated, processed, defuzzificated and transmitted to actuators in the form of usual signals.

This paper considers a fuzzy inference based on the Mamdani mechanism. The feature of this mechanism is that the minimax composition of fuzzy sets is used in modeling.

For final analysis of the developed fuzzy model, we constructed a fuzzy inference surface using a fuzzy inference surface viewer (Fig. 5). This viewer allows us to analyze the adequacy of the constructed fuzzy model and to assess how a change in one of the input linguistic variables affects the output variable. As can be seen from the figure, the response surface is adequate and sensitive to input effects.

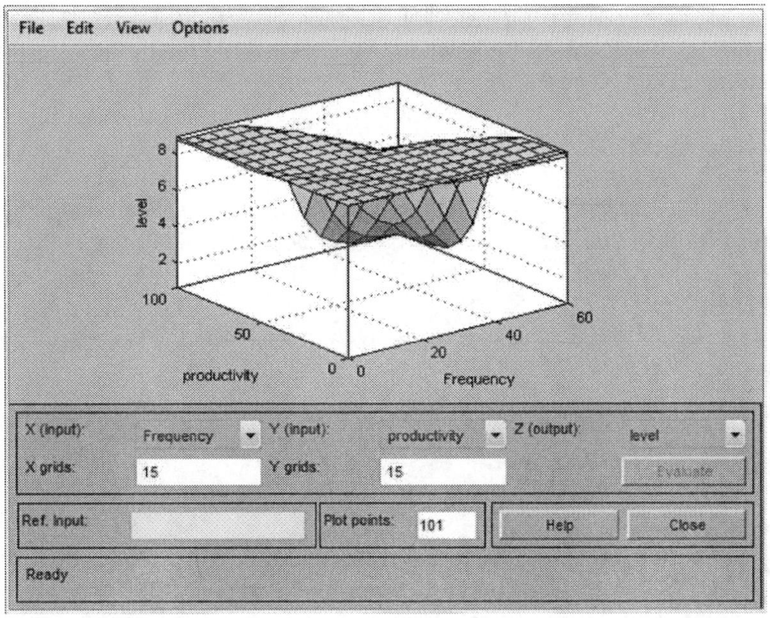

Fig. 5. Interface of the fuzzy inference surface viewer for the developed fuzzy model.

Thus, it is possible to determine optimal parameters of production processes in the building materials industry through modeling of the fuzzy controller, without additional laboratory tests and experiments.

Conclusion

Thus, it was found that the proposed modeling of the fuzzy controller could be successfully applied by producers of building materials from fluorine anhydrite waste. Based on chosen input linguistic parameters, frequency and productivity, we created rules to generate a fuzzy linguistic variable "level", and determined their values and relationships. Using these results, we tried out control actions to control the required filling level of the supply hopper at an existing production facility where fluorine anhydrite was used.

This approach minimizes the so-called "human factor" when making decisions about batching of industrial anhydrite binder, reduces the time of production operations and, therefore, contributes to bringing an enterprise to a new, modern level where it can compete with enterprises that use natural gypsum.

It is reasonable to further increase the accuracy of filling and batching control in relation to industrial anhydrite binder by improving the fuzzy inference rule base. The calculations performed in this study suggest that this kind of modeling is applicable to other stages of production based on fluorine anhydrite, such as batching of initial components of the building fluorine anhydrite mixture — ash and slag waste, water and improvers.

References

1. U. Kuhlmann, DE Patent 3023666 (1981)
2. H. Scsmidt, G. Fietsch, R. Grohmann, H-J. Grun, P. Pfutzner, FR Patent 2495131 (1982)
3. T. Gaedt, H. Grassl, M. Winklbauer, O. Mazanec, US Patent 2017022108 (2017)
4. D. Hums, K. Lippe, DE Patent 19619263 (1997)
5. K. Harr, M. Tax, G. Bolte, I. Unsin, DE Patent 4313148 (1994)
6. E. Stave, E. A. BurKard, R. Finkelstein, AU Patent 2124799 (1999)
7. P.A. Carlucci, FR Patent 2727675 (1996)
8. M.V. Krivenkov, S. R. Zalyaleev, A. N. Pakhomov, Optimizacija raboty jelektrotehnicheskih sistem [Optimization of electrical systems], 315 (Krasnoyarsk, 2006)
9. Y. Peng, J. Luo, J. Zhuang, C. Wu, Proceedings of the IEEE International Conference on Automation and Logistics (Qingdao, China September 2008), 896-901
10. D. Wu, F. Karray, I. Song, Proceedings of the IEEE Conference on Control Applications, 3134-39 (2005)
11. G. Zhiqiang, T.A. Trautzsch, J.G. Dawson, IEEE Transactions on Industry Applications, **38(2)**, 414 - 424 (2002)
12. Z. Zhi, H. Li-Sheng, Annals of Nuclear Energy, **45**, 94-105 (2012)
13. T. Hung, M. Sugeno, R. Tong, R.R. Yager, *Theoretical aspects of fuzzy control* (John Wiley and Sons Inc., New York, 1995)
14. J. Yen, R. Langari, F*uzzy logic: intelligence, control, and information* (Prentice Hall, New Jersey, 1998)
15. Yu.M. Fedorchuk, Scientific basis and methods for environmental impact reduction in hydrogen fluoride production areas, doctoral thesis (Tomsk, 2004)
16. T.S. Tsygankova, Development of methods for reducing environmental impact of hydrogen fluoride production sites, PhD thesis (Tomsk, 2009)

17. O.V. Volkova, Finishing dry building mixes based on fluoride acid binders, PhD thesis (Tomsk, 2015)
18. L.A. Anikanova, Fluoride acid composites for low-rise buildings, PhD thesis (Tomsk, 1997)

Analysis of drinking water quality from the standpoint of health risks

Valentina Nikiforova[1,*], and *Elena* Vidishcheva[1]

[1]Bratsk State University, 665709 Bratsk, Russia

Abstract. The paper presents a hygienic quality assessment of water from sources in the Irkutsk Region (Russia) used for public drinking water supply. The aim of the study is to assess health risks to the population of Bratsk and the Bratsk District, which is associated with the exposure to chemicals coming with drinking water from the public water supply system. Risk of occurrence and development of non-carcinogenic effects is assessed for separate age groups using a special method. Based on the risk assessment methodology, non-carcinogenic risks for children and adults are described in quantitative terms. The total health risk of the test groups associated with the oral intake of chemicals with drinking water is estimated as negligible and, therefore, does not need to be reduced through additional measures. Lead and fluoride contents in drinking water are the major contributors to the risk level (59.5–68.4%). There is a risk of developing blood disorders and cardiovascular diseases in children of Bratsk (HI = 1.19), as well as bone tissue and teeth disorders in children of Bikey village, Bratsk District (HI = 1.10).

Introduction

Water is one of essential elements that are vital to every living creature on the Earth. With the development of new technologies, the need for water is growing day by day. The volume of world's water resources is estimated to be 1.3 billion km^3, but strategically important fresh drinking water makes up only 2 to 2.6% [1].

South America and Asia have the largest reserves of available fresh water on the planet. Twenty-nine countries grouped together in terms of economy have a fifth of the available water resources. The post-Soviet states account for more than 20% of world's water resources. The rest of water resources are found in the Middle East and North Africa. It should be noted that India, China, the USA, Pakistan, Japan, Thailand, Indonesia, Bangladesh, Mexico and Russia demonstrate the highest consumption levels [2–4]. There is acute shortage of water in China, India and the USA. Currently, there are more than 260 international river basins shared by 145 nations. Such rivers as the Nile, Danube, Tigris and Euphrates, Ganges and Brahmaputra provide sufficient water to all consumers [5].

The CIS states have the largest water resources in the world and rank second (after Brazil) in terms of average annual river runoff. Moreover, the CIS states have significant potential groundwater reserves. However, these resources are distributed across CIS

* Corresponding author: nikiforovabr@mail.ru

countries extremely unevenly due to various geographic, climatic, geological and hydrogeological conditions of regions.

The main sources of drinking water supply are river runoff, surface water and groundwater, the quality of which is estimated by researchers as unsatisfactory [5–7]. The major rivers of Russia (the Volga, Don, Kuban, Ob, Yenisei, Lena and Pechora), are rated as "heavily polluted", while water in the Neva, Tom, Oka and Ural rivers is unsuitable for drinking and domestic purposes. Due to pollution, water taken from such rivers as the Volga, Dnieper, Don, Kuban, Northern Dvina, Pechora, Ob, Yenisei, Lena, Kama, Amur and many lakes, needs to be specially treated.

Regions with the worst quality of drinking water include the republics of Dagestan, Ingushetia, Kalmykia, Karachay-Cherkessia, Karelia and Tatarstan, Altai Krai and Primorsky Krai, Arkhangelsk, Ivanovo, Kirov, Kostroma, Ryazan, Smolensk and Ulyanovsk regions. In the context of ever deteriorating water quality in sources, the available water treatment techniques cannot ensure the treatment of drinking water to the required level of purity.

In some regions of the world, including Russia, people have to use the water that does not meet the applicable standards. For example, in some water supply sources of the Orenburg Region, cadmium levels may reach 0.007 mg/L, lead – 0.04 mg/L, arsenic – 0.023 mg/L [7]. Uganda uses groundwater with an iron content of 0.3 to 4.9 mg/L, manganese – up to 2.0 mg/L, arsenic – 0.005–0.009 mg/L, nickel – up to 0.12 mg/L, lead – up to 0.011 mg/L, cadmium – up to 0.013 mg/L [8]. In a mining region of Bangladesh, high concentrations of arsenic (from 0.01 to 0.049 mg/L) were found in groundwater [9].

Thus, the problem of water quality is the focus of attention for the world community, legislative and executive bodies in all civilized countries and regions, which is associated with growing water consumption, qualitative changes in water sources that are often exposed to uncontrolled anthropogenic impact, and inadequate water treatment methods [10].

One of the most important tasks in the field of sanitary and epidemiological public welfare in the Russian Federation is to provide the population with high-quality drinking water that is epidemiologically safe and harmless in terms of chemical composition [11, 12]. Quality of drinking water supplied from public sources is one of the most important environmental factors that affect public health in the country, including the region under study.

To date, the development of epidemiological risk to public health have been studied with regard to drinking water treatment methods, the sanitary and technical condition of distribution networks, and the possibility of restoring microorganisms during transportation of water [13]. Transformation of chemical substances caused by various physical and chemical methods of water treatment, and the possibility of changing the chemical composition of water and its organoleptic properties in the distribution network have been studied as well [14]. However, the water factor has its own specific effects on human health, as evidenced by a number of studies in which drinking water quality and public health risks have been analyzed from a hygienic perspective [15–17].

In order to preserve the health of the nation, while creating favorable conditions for business entities, it is necessary to use new approaches to the management of sanitary and epidemiological supervisory bodies. One of these approaches is a risk-based model of surveillance in the field of sanitary and epidemiological public welfare as related to drinking water of proper quality [18]. In a specific sanitary situation, the relevant and effective tool is to assess the public health risk arising from the water factor by calculating doses to which certain population groups are exposed [19].

The aim of this study is to assess the public health risk associated with the exposure to chemicals coming with drinking water from the public drinking water supply system (based on cities and settlements of the Irkutsk Region). The scientific novelty is associated with comparative compositional analysis of drinking water taken from quality control points. Low carcinogenic risk of drinking water for children of Bratsk and low non-carcinogenic risk for children of the Bikey village (Bratsk District) were identified.

Materials and Methods

When selecting a test object, we assessed all controlled underground sources of the Irkutsk region located in rural (28 districts) and urban municipalities (9 cities). Bratsk and other settlements of the Bratsk District were selected as test objects.

To accomplish this task, we undertook a retrospective comprehensive assessment and analysis of laboratory data included in a database of the regional social and hygienic monitoring system in the Bratsk Branch of the Irkutsk Region Center of Hygiene and Epidemiology (Federal Budgetary Healthcare Institution) for the period from 2000 to 2015.

The quality of drinking water from underground sources was assessed for compliance with SanPiN 2.1.4.1074-01 Drinking Water. Hygienic Requirements for Water Quality of Public Drinking Water Supply Systems. Quality Control. Hygienic Requirements for Safety of Hot Water Supply Systems [20]. The quantitative analysis of water pollution was carried out in accordance with GN 2.2.5.1315-03 Maximum Allowable Concentrations (MACs) of Chemicals in Water of Water Facilities Used for Household and Domestic Purposes [21].

The risk was assessed using the classical procedure [22] on the basis of data from systematic monitoring of water quality performed by Teplovodokanal (municipal water supply enterprise) of Bratsk municipality under the production monitoring program, and on the basis of laboratory data provided by the Bratsk Branch of the Irkutsk Region Center of Hygiene and Epidemiology as part of social and hygienic monitoring efforts.

The research consists of the following stages: health hazard identification, determination of dose-response relationships, exposure assessment, and risk characterization. Subject to the aim of the study, we considered an exposure scenario that implies the oral intake of harmful substances with water. Oral non-carcinogenic effects were assessed in terms of hazard quotients and indices. Hazard indicators of the chemicals identified through monitoring were estimated on the basis of reference levels for chronic exposure to chemicals. At the same time, we identified critical organs/systems and effects that corresponded to the established reference doses.

The risk assessment was carried out at four water quality control points of the public water supply system, which receives water from water intakes providing the population of Bratsk with water for domestic and drinking purposes. The risk indicators estimated on the basis of drinking water quality data from the control points were extended to people who used this water for drinking purposes. When estimating the non-carcinogenic risk, we used exposures of 12 chemicals, while calculating average daily doses of toxic agents separately for adults and children.

Results and Discussion

238 samples (data from the regional social and hygienic monitoring system) taken from public drinking water supply systems in 41 municipalities of the Irkutsk Region were tested. In terms of chemical composition (iron, manganese, nitrates), organoleptic indicators (color, turbidity) and total hardness, 15.5% of the tested samples did not meet the standard. Unsatisfactory samples were recorded in Bokhansky (100% of samples), Alarsky (50%), Ekhirit-Bulagatsky (42.9%), Chunsky (66.7%), Kachugsky (66.7%) districts,

Bodaybo town and district (42.9%), Irkutsky (35%), Usolsky, Ust-Ilimsky and Bratsk districts.

In terms of microbiological safety, 3.3% of drinking water samples water quality did not meet hygienic requirements (including samples from Bokhansky (Buret village), Tulunsky (Badar village), Irkutsky (Revyakin village, Mamony village, Sosnovy Bor village, Nikolsk village) districts, as well as in Irkutsk city and Shelekhov town.

Monitoring of drinking water quality in non-public water supply in the Irkutsk Region was carried out in 33 municipalities. The tested samples do not comply with hygienic standards in terms of sanitary and chemical indicators (Fig. 1), including chemical composition (iron, manganese, nitrates) and total hardness, as well as in terms of organoleptic indicators (color, turbidity).

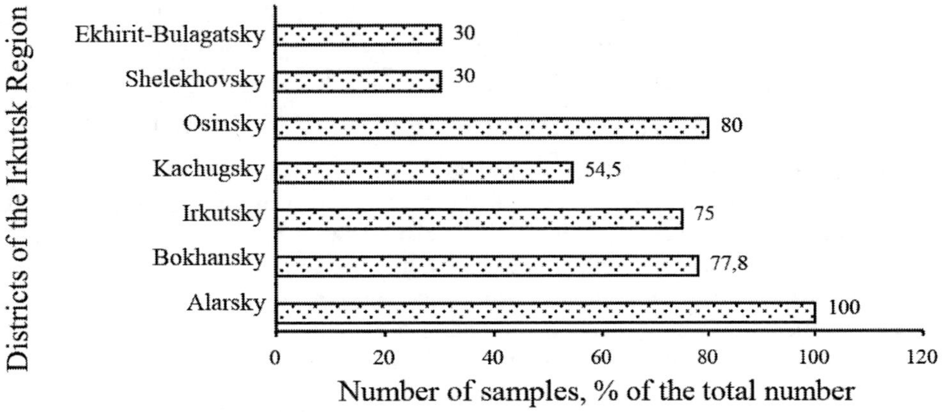

Fig. 1. The number of samples taken in the Irkutsk Region that do not comply with standard sanitary and chemical indicators.

In terms of sanitary and chemical indicators, from 30 to 100% of the samples taken in the districts of the Irkutsk Region under study do not meet the standard. In terms of microbiological safety, 3.7% of water samples are not compliant.

Thus, according to data of the regional social and hygienic monitoring system, there is a high proportion of water test samples from some areas of the Irkutsk Region that do not meet hygienic standards in terms of sanitary and hygienic indicators.

The health risk assessment gives more reliable information about the exposure of the population to the toxicological hazard of drinking water. The chemical composition of drinking water from the Bratsk public drinking water supply system is presented in Table 1.

The chemical analysis of drinking water at all control points shows that the average concentrations do not exceed the limits of the standards, but the maximum levels are often above the MACs.

Table 1. Chemical Composition of Drinking Water from the Bratsk Public Drinking Water Supply System.

Water quality control points	Chemical composition							
	Bratsk (city)		Padun (village)		Bikey (village)		Zayarsk (village)	
Substance	2005–2009	2010–2014	2005–2009	2010–2014	2005–2009	2010–2014	2005–2009	2010–2014
NO_3^-, mg/dm³	2.41	2.52	0.54	0.87	0.71	2.08	6.96	6.73
Fe^{3+}, mg/dm³	0.04	0.02	0.03	0.02	0.05	0.04	0.02	0.05
F^-, mg/dm³	0.48	0.48	0.43	0.41	0.54	0.53	0.52	0.54
Mn^{2+}, mg/dm³	0.003	0	0.005	0.009	0.001	0.006	0.003	0.005
As^{2+}, mg/dm³	0	0.002	0	0.002	0	0.001	0.0001	0.0001
B^-, mg/dm³	0.13	0.15	0.05	0.09	0.24	0.27	0.015	0.016
Mo^{2+}, mg/dm³	0.001	0.006	0.001	0.006	0.002	0.002	0.004	0.003
Cu^{2+}, mg/dm³	0.01	0.01	0.01	0.02	0.01	0.01	0.02	0.01
Zn^{2+}, mg/dm³	0.004	0.013	0.022	0.013	0.021	0.018	0.013	0.012
Pb^{2+}, mg/dm³	0.012	0.009	0.002	0.002	0.003	0.007	0.006	0.006
phenols, mg/dm³	0.001	0.0009	0.0005	0.0016	0.0021	0.001	0.0020	0.035

Hazard quotients (HQ) of the substances contained in drinking water from the Bratsk public drinking water supply system for the adult population was found not to exceed the permissible level (1). For children, high level of hazard quotients was observed for lead (HQ = 1.095) at a water quality control point (Bratsk) (Table 2).

With the additive unidirectional effect of the substances that enter the human body with drinking water, the hazard index for the adult population is estimated as negligible (HI = 0.43–0.48) at water quality control points (Fig. 2).

Table 2. Hazard Quotient of Chemical Substances Contained in Drinking Water from the Bratsk Public Drinking Water Supply System.

Water quality control points	Hazard quotient of chemical substances								Average values	
	Bratsk (city)		Zayarsk (village)		Zayarsk (village)		Zayarsk (village)			
Substance	Children	Adults	Children	Adults	Children	Adults	Children	Adults	Children	Adults
1	2	3	4	5	6	7	8	9	10	11
NO_3^-, mg/dm³	0.0922	0.0395	0.0208	0.0089	0.0393	0.0168	0.2351	0.1008	0.09685	0.04151
Fe^{3+}, mg/dm³	0.0067	0.0029	0.0046	0.002	0.0107	0.0046	0.0067	0.0029	0.00718	0.00308
F^-, mg/dm³	0.554	0.2374	0.5601	0.2401	0.411	0.1761	0.4764	0.2042	0.50038	0.21445
Mn^{2+}, mg/dm³	0.0012	0.0005	0.0032	0.0014	0.0008	0.0004	0.0003	0.0001	0.00137	0.00059
As^{2+}, mg/dm³	0.0487	0.0209	0.1705	0.0731	0.1826	0.0783	0.0974	0.0417	0.12481	0.05349
B^-, mg/dm³	0.0852	0.0365	0.0493	0.0211	0.2314	0.0992	0.0805	0.0345	0.11160	0.04783
Mo^{2+}, mg/dm³	0.042	0.018	0.0393	0.0168	0.0438	0.0188	0.0402	0.0172	0.04132	0.01771
Cu^{2+}, mg/dm³	0.0336	0.0144	0.0264	0.0113	0.0336	0.0144	0.0224	0.0096	0.02904	0.01245
Zn^{2+}, mg/dm³	0.0032	0.0014	0.0041	0.0017	0.0037	0.0016	0.008	0.0034	0.00475	0.00204
Pb^{2+}, mg/dm³	1.0959	0.4697	0.0462	0.0198	0.1035	0.0444	0.1644	0.0705	0.35249	0.15107
phenols, mg/dm³	0.0001	0	0.0002	0	0.0001	0	0.0002	0	0.00017	0.0

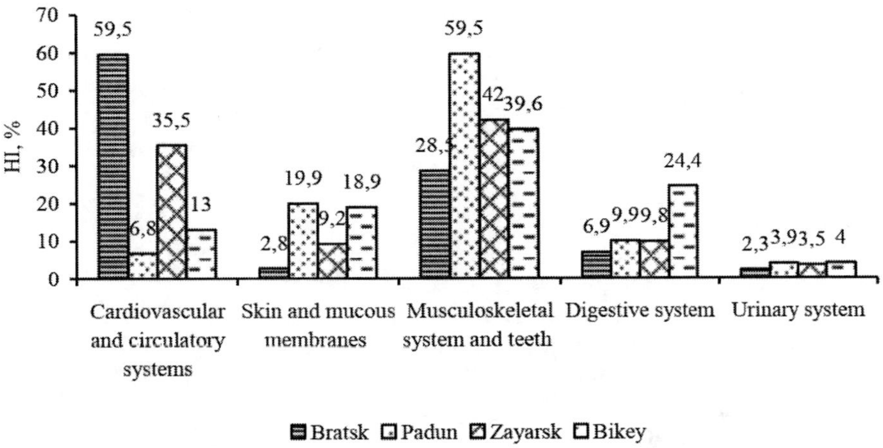

Fig. 2. Hazard indices for critical systems of the human body exposed to drinking water from the Bratsk drinking water supply system.

Taking into account that toxic agents are tropic in relation to organs and systems of the human body, we found that the digestive system, circulatory and hematopoietic system, skin, mucous membranes, and the musculoskeletal system of children are under the greatest toxic load. There is an increased risk of bone tissue and teeth disorders in this group of Bikey village (HI = 1.19), where fluorides are the major contributor. High hazard indices (HI = 1.19) are caused by nitrates contained in drinking water in concentrations exceeding the MAC in 27.9% of samples, and by fluorides (HI = 1.19) contained in drinking water in concentrations exceeding the MPC in 24.6% of samples in Zayarsk village.

As compared to adults, children have more organs and systems affected by polluted drinking water. For example, in relation to drinking water from pumping stations of the Bratsk public drinking water supply system, hazard indices for the cardiovascular and circulatory systems were found to exceed the permissible level (HI = 1.19). Among the tested chemical substances contained in drinking water, nitrates and fluorine (in terms of non-carcinogenic (general toxic) effects) and lead (in terms of carcinogenic effects) have the highest levels of hazard to public health in the Irkutsk Region, including Bratsk.

Thus, the public health risk of drinking water directly depends on a water treatment technique, its compliance with a water source category and the sanitary condition of the city's distribution networks. In order to provide the population of urban districts and municipal areas of Bratsk with high-quality drinking water, it is necessary to implement a variety of measures, as well as to develop and implement programs aimed at improving water supply to settlements.

Conclusion

The public health risk assessment shows that the toxicological hazard of drinking water from the public drinking water supply system in certain areas of the Irkutsk Region is acceptable for adults, but exceeds the permissible level for children and is associated with exposure to lead.

Long-term quality monitoring of raw water in water sources and drinking water from public water supply systems, as well as calculated quantitative and qualitative characteristics of the public health risk associated with water consumption, are important information sources for introducing the risk-based model of surveillance in the field of sanitary and epidemiological public welfare. Depending on hazard level of activities of business entities, characteristics of the public health risk are an additional criterion for selecting priority projects at the planning stage. Subject to the estimated risk levels and in the context of the exposure to the water factor that can lead to adverse effects on critical organs and systems of the human body, it is necessary to develop a set of measures for managing the public health risk.

References

1. L.K. Kazanceva, T.O. Tagaeva, ECO **9**, 30-37 (2006)
2. R. Shehzadi, H.M. Rafique, I. Abbas, M.A. Sohl, Sh.M. Ramay, A. Mahmood, Desalination and Water Treatment **55 (8)**, 2253-2264 (2015)
3. L.E.A. Bradford, L.A. Bharadwaj, U. Okpalauwaekwe, Ch.L, Waldner, International Journal of Circumpolar Health **75(1)** (2016)
4. L. Benneyworth, J. Gilligan, J.C. Ayers, S. Goodbred, G. George, A. Carrico, International Journal of Environmental Health Research **26 (5-6)**, 508-524 (2016)
5. V.A. Vasilenko, ECO **2**, 128-142 (2006)
6. N.V. Efimova, N.I. Matorova, N.N. Jushkov, V.A. Nikiforova, T.G. Perceva, Mediko-jekologicheskie riski sovremennogo goroda [Medical and ecological risks of the modern city], 200 (BrSU, Bratsk, 2008)
7. L.G. Konshina, M.V. Sergeeva, L.L. Lipanova, A.V. Solonin, Hygiene and sanitation **2**, 22-24 (2004)
8. R.G. Taylor, K.W.F. Howard, Groundwater quality in rural Uganda: hydrochemical consideration for the development of aquifers within basement complex of Africa, In: Nash H., McCall G.J.H., eds. *Groundwater Quality*, 31-44 (Chapman and Hall, London, 1994)
9. N. Sohel, L.A. Persson, M. Rahman, P.K. Streatfield, M. Yunus, E.C. Ekström, M. Vahter, Epidemiology **20(6),** 824–830 (2009)
10. G.G. Onishchenko, Health Care of the Russian Federation **4**, 3-9 (2007)
11. N.V. Zaytceva, I.V. Mai, S.Yu. Balashov, Izvestia of Samara Scientific Center of the Russian Academy of Sciences **1–6 (1)**, 1144-1148 (2009)
12. S. Rapant, K. Kremova, Environ. Geochem. Health **29,** 131-141 (2007)
13. A.V. Zagaynova, Yu.A. Rahmanin, Yu.G. Talaeva et al., Hygiene and sanitation **3**, 28-31 (2010)
14. A.A. Sekunda, L.P. Ignat'eva, O.A. Projdakova, L.A. Nikolaeva, Materials of the Plenum of the Scientific Council on Human Ecology and Environmental Hygiene of the Russian Academy of Medical Sciences: *Methodical issues of studying the assessment and regulation of physical factors in environmental health*, 71-72 (2008)
15. L.G. Konshina, V.L. Lezhnin, Hygiene and sanitation **3**, 5-10, (2014)
16. D.V. Goryaev, I.V. Tikhonova, N.N. Torotenkova, Health Risk Analysis **3**, 35-43, (2016)
17. A.E. Oparin, Sibirskij medicinskij zhurnal [Siberian Medical Journal] **5,** 99-102, (2012)
18. G.G. Onishchenko, N.V. Zaytceva, *Analiz riska zdorov'ju v strategii gosudarstvennogo social'no-jekonomicheskogo razvitija* [Analysis of the health risk in the strategy of state social and economic development], 738(Perm national research polytechnic university, Perm, 2014)
19. G.I. Rumyantcev, Health Risk Analysis **2**, 4-11 (2015)
20. SANPIN 2.1.4.1074-01. Potable Water - Hygienic Requirements for Water Quality in Central Potable Water Supply Systems - Quality Control, 84 (Ministry of Health of Russian Federation, Moscow, 2002)

21. GN 2.2.5.1315-03. Maximum allowable concentrations (MACs) of chemicals in the water of water objects used for drinking and domestic-recreation purposes, 100 (Ministry of Health of Russian Federation, Moscow, 2003)
22. P 2.1.10.1920-04. Human Health Risk Assessment from Environmental Chemicals, 143 (Federal Center for State Sanitary and Epidemiological Supervision of the Ministry of Health of the Russian Federation, Moscow, 2004)

Assessment of noise parameters in elements of urban transport infrastructure

Taisia Bolshanina[1,*], and *Sergei* Ovsiannikov[1]

[1]Tomsk State University of Architecture and Building, 634003 Tomsk, Russia

Abstract. The paper considers urban road network in combination with elements of transport infrastructure which are presented as noise sources. Acoustic analysis of the level of noise produced by transport accounts only railway track spans and highways as linear sources of noise. Within the study road network elements and objects of transport infrastructure were classified; traffic interchanges and objects of transport infrastructure were presented as spatial noise source, which provides relevance of this study. The purpose of the research is development of methods for analyzing noise parameters of road networks and objects of transport infrastructure, as well as analysis of noise level in the residential territories caused by the combination of linear and spatial noise sources. The results of analysis and measurements demonstrate that evaluation of noise level caused by transport of the near-highway territory without the account of traffic interchanges as spatial noise sources leads to downward bias of noise level up to 10 dBA.

Introduction

Transport is the main source of noise in cities. Acoustic analysis while evaluating noise pollution of urban territories in Russia is performed in compliance with regulatory documents: GOST 20444-2014 [1], GOST 23337-2014 [2], SP 51.13330.2011 [3], etc. These standard documents are based on the methods for defining the noise level in residential territories developed by the following authors: G. Osipova, [4, 5], P. Pospelova [6, 7], A. Faktorovich [8], etc. Drawback of the existing methods for noise level evaluation consists in lack of classification of elements of urban road network and objects of transport infrastructure with regard to noise parameters. Besides, traffic interchanges and objects of transport infrastructure are not accounted as spatial noise sources, whereas the measurements show that noise parameters of urban road network cannot be presented as a combination of linear noise sources only.

The accuracy of acoustic analysis depends on the correctly defined type and number of noise sources. The method of analysis applicable either for linear noise sources or for spatial ones is selected based on these parameters. This research is aimed at developing methods for analyzing noise parameters of road networks and objects of transport

* Corresponding author: BolshaninaTS@gmail.com

infrastructure, as well as analyzing noise level in the residential territories caused by the combination of linear and spatial noise sources.

Methods of research

The existing methods for evaluation of noise level in urban territories do not account objects of transport infrastructure and elements of road networks which occupy large areas. Road network elements include track spans and traffic interchanges which can be represented by intersections or multilevel transport interchanges. Elements of railway transport are single-track or double-track railway track spans, freight or passenger stations and passing loops.

Objects of transport infrastructure are represented by transport storage and maintenance facilities: parking lots, car washes, repair facilities and depot for railway transport.

According to GOST 23337-2014 [2], the following types of noise produced by transport are singled out: roadway noise, railway noise, waterborne noise and aircraft noise. Below, in Tables 1 and 2, the authors present own classification.

Table 1. Classification of urban transport and road network.

Urban transport and road network†	**road network**	• track spans: residential streets; main district highways; main city highways; cargo lines; • traffic interchanges and junctions: intersections; traffic circles; multilevel interchanges.
	railway transport	• railway hauls: single-track; double-track; • stations and loops: passing loops; passenger stations; freight stations.
	underground railway and street railway	• underground railway hauls; • street railway hauls; • open underground stations; • tram stations.
	waterborne transport	• water ways; • riverside stations; • cargo ports.

Table 2. Classification of urban transport infrastructure.

Urban transport infrastructure‡	**motor transport**	• passenger transportation companies; • goods transportation companies; • trolleybus fleet; • specialty vehicles companies; • car washes; • repair facilities; • tire services; • car parkings.
	railway transport	• railway depot.
	underground railway and street railway	• underground depot; • tram depot.
	waterborne transport	• water transport repair facilities.

† without the account of air transport
‡ without the account of air transport

Elements of road network and objects of transport infrastructure can be classified with regard to the noise source:

- track spans as linear noise sources;
- multilevel interchanges and transport maintenance enterprises as spatial noise sources.

Highways shall be considered as linear noise sources, and noise levels in residential territory shall be evaluated with the account of noise reduction factors [4]. With two and more highways, the levels of noise they produce are defined by energy summation.

To evaluate the impact of traffic interchanges on the acoustic mode of near-highway territories one shall define the integral value of noise level in the calculated point obtained from a combination of sources using the following method. According to Figure 1a, the calculated point is affected by noise produced by two linear sources and one spatial source represented by a traffic interchange. The noise parameters of the linear noise source were defined based on the results of measuring equivalent and maximum noise levels according to GOST 20444-2014 [1]. Sound power applied to acoustic center which is a geometrical center of the traffic interchange shall be considered a noise parameter for spatial noise source.

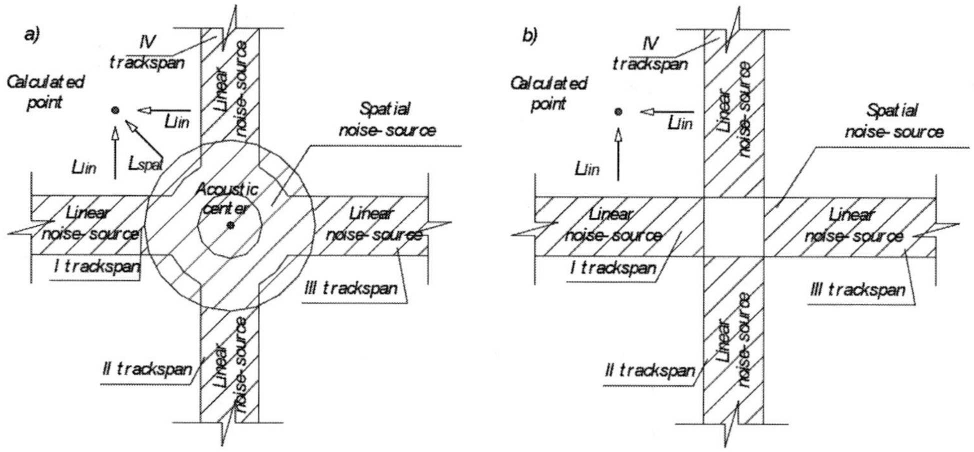

Fig. 1. Schemes of noise sources location in transport interchanges: a – single-level interchange - a combination of spatial and linear noise sources; b – intersection – crossing of linear noise sources.

Below is presented the algorithm for defining noise parameters in the calculated points located in proximity to transport interchanges, with the example of a cloverleaf interchange in Tomsk city (Russia):

1. Measure the noise levels in the calculated points. Location of the calculated points is presented in Figure 2.

Let us present a cloverleaf interchange, as well as a traffic circle interchange, in the form of an ellipse-shaped spatial source, similar to the scheme in Figure 1a. The track spans of the interchange are Pushkina Street in both ways, Komsomolsky Avenue and 79 Gvardeiskoy Divisii Street (Fig. 2). At each liner section of a noise source two points are selected being 7.5 meters away from the closest axis of motion for defining the mean value of noise level produced by the linear noise source (1, 1.1, 5, 5.1, 11, 11.1, 15, 15.1). To define the noise parameter of a spatial noise source two uniform points are selected in the track spans.

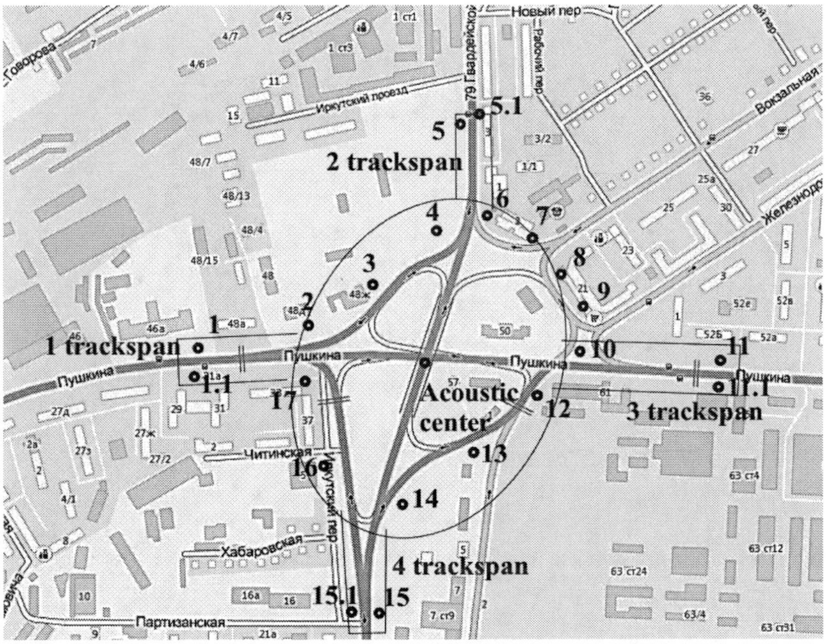

Fig. 2. The schematic map of a cloverleaf interchange with marked noise sources and calculated points.

2. Define the average noise level produced by the track spans, which is an arithmetic mean of the two measured values.

3. Find the average noise level of a spatial noise source in a cloverleaf interchange by the formula (1):

$$\overline{L_{eq}} = 10 \cdot \lg\left[\left(\frac{1}{n}\right)\sum_{i=1}^{n} 10^{0,1 \cdot L_i}\right], \quad dBA \tag{1}$$

where L_i – equivalent sound level at the i point, dBA; n – number of i points.

4. Track spans have the same traffic intensity, hence the directivity index of a nose source is defined by the formula (2):

$$\Phi_i = G_i = L_{eqi} - \overline{L_{eq}}, \quad dBA \tag{2}$$

where L_{eqi} - equivalent noise level at the i point, dBA.

5. Find the sound power of a spatial noise source in compliance with the "Recommendations on Measurement and Evaluation of External Noise Produced by Industrial Enterprises" by the formula (3):

$$L_p = \overline{L_{eq}} + 10 \cdot \log\frac{2S}{S_0}, \quad dBA \tag{3}$$

where S – area of the territory occupied by a spatial noise source, m^2; $S_0 = 1\ m^2$.

6. To perform the calculation of equivalent sound level in the territory adjacent to a traffic interchange two calculated points are selected (Fig. 2) located 2 meters away from the building facades and 1.5 meters above the road surface level. Figure 3 demonstrates the average sound levels.

7. Find the noise level in the calculated points located two meters away from the facades using the summability method of integrals (Fig. 4).

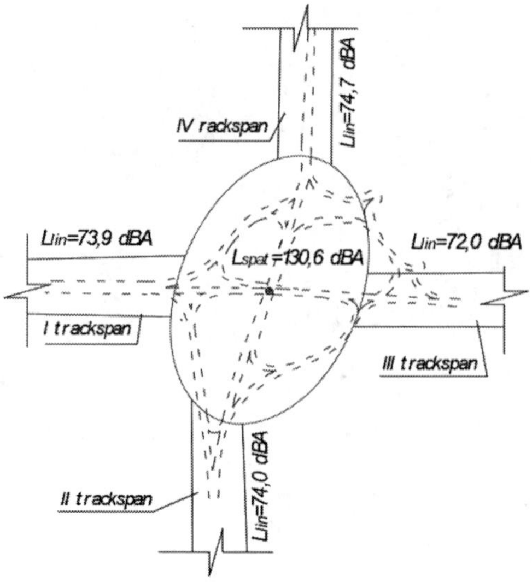

Fig. 3. Averaged parameters of noise levels (L) in the track spans of a traffic interchange.

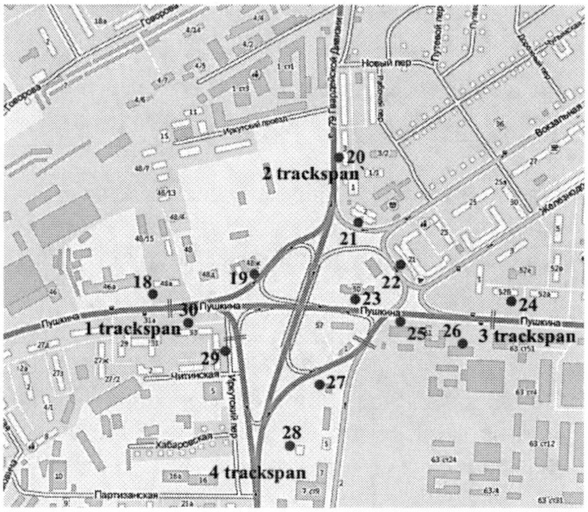

Fig. 4. Schematic map with marked calculated points in a transport interchange.

Results

The measurement results for the calculated points marked in Figure 2 are presented in Table 3. The average noise level of a spatial noise source of a cloverleaf interchange defined by the formula (1) equals 72.8 dBA.

Table 3. Measured parameters of a cloverleaf interchange (intersection of Komsomolsky Avenue and Pushkina Street) and the noise level in the calculated points.

Number of the calculated point	Velocity km/h	Intensity, vehicles/h	Traffic flow composition			Measured noise level, L_{eq} / L_{max}, dB	Directivity index $\Phi(G)$, dBA
			trucks	buses	cars		
1	50-60	4182	5	7	88	73.3/83.3	-
1.1	50-60	3588	3	7	90	74.5/87.7	-
2	30-40	1248	3	5	92	69.3/76.6	-3.5
3	30-40	1260	4	5	91	67.8/76.2	-5.0

4	30-50	2424	4	4	92	76.9/96.3	4.1
5	40-60	2400	4	3	93	75.4/82.8	-
5.1	50-60	2796	5	3	92	74.0/82.6	-
6	30-40	648	13	2	85	63.7/78.5	-9.1
7	30-40	1122	5	3	92	63.3/73.6	-9.5
8	30-40	1050	6	5	89	66.6/84.6	-6.2
9	30-40	894	5	5	90	66.5/80.4	-6.3
10	30-40	540	7	3	90	69.4/87.1	-3.4
11	50-60	4254	5	7	88	71.4/79.9	-
11.1	50-60	4182	5	6	89	72.6/85.7	-
12	30-40	1056	5	3	92	73.1/91.2	0.3
13	40-60	882	1	7	92	69.8/80.0	-3.0
14	40-60	558	6	6	88	72.4/87.2	-0.4
15	50-60	3168	5	2	93	74.2/83.5	-
15.1	50-60	3348	5	3	92	73.8/93.1	-
16	50-60	504	2	4	94	64.2/74.8	-8.6
17	30-40	498	6	0	94	66.8/80.6	-6.0

Averaged noise parameters of track spans of a transport interchange are demonstrated in Figure 3.

Sound power of noise of a spatial source is calculated by the formula (3) and equals:

$$L_p = 72{,}8 + 10 \cdot \log \frac{604136}{1} = 130{,}6 \quad dBA$$

$$S_{tr.interchange} = S_{ellipse} = \pi \cdot a \cdot b = 3{,}14 \cdot 370 \cdot 260 = 302068 \quad m^2$$

The traffic interchange area is defined in a horizontal plane as an ellipse.

All calculated and measured values of noise levels in the calculated points marked in Figure 4 are collected in Table 4.

Table 4. Calculated and measured values of noise levels in the near-highway territories for the calculated points, dBA.

Number of the calculated point	***Llin*, from linear sources**	***Lspat*, from a spatial source**	***Lspat±G*, from a spatial source with the account of directivity index**	***Lter*, the total of a linear and a spatial sources with the account of directivity index**	**$L_{изм}$, measured noise level**	**$\Delta = L_{изм} - Llin$, difference between the calculated and measured noise levels without the account of directivity index**	**$\Delta = L_{изм} - L_{Атер2}$, difference between the calculated and measured noise levels with the account of directivity index**
1	*2*	*3*	*4*	*5*	*6*	*7*	*8*
18	68.3	69.2	65.7	70.2	69.0	0.7	-1.2
19	57.1	75.0	70.0	70.2	67.8	10.7	-2.4
20	74.1	64.6	64.6	74.6	76.6	2.5	2.0
21	55.3	73.0	63.7	64.3	64.3	9.0	0.0
22	64.6	69.2	63.0	66.9	67.8	3.2	0.9
23	68.6	82.1	78.7	79.1	80.8	12.2	1.7
24	65.0	61.9	61.9	66.7	68.7	3.7	2.0
25	69.6	73.3	69.9	72.8	68.2	-1.4	-4.6
26	67.7	62.3	62.3	68.8	70.5	2.8	1.7
27	66.5	72.3	66.3	69.4	64.2	-2.3	-5.2
28	65.6	69.0	63.0	67.5	69.8	4.2	2.3

Table 4 demonstrates that when taking the integral sum for the calculated points 18, 20, 21, 22, 23, 24, 26 the difference between the calculated and measured noise levels with the account of spatial source lies in the range of error of a sound level meter which is ±1-2 dBA (Table 4). When disregarding the transport interchange as a spatial noise source, the values obtained with the account of linear sources only mark down the results by 2 to 12 dBA.

For the calculated points located close to the edge of spatial noise source (points number 9, 25 and 27) noise level values are observed to be up to 5 dBA higher. The results of other research in the field can be found in the paper [9, 10].

Conclusion

The paper presents classification of elements of road network and urban transport infrastructure which enables to apply the method for calculating noise pollution for a certain urban situation. The research shows that the model of acoustic sources of road network including not only linear sources (track spans), but also spatial sources (traffic interchanges and objects of transport infrastructure) realistically present noise parameters of all elements of urban road network. Error of the suggested method for noise level calculation in certain points with the combination of linear and spatial noise sources lies in the range of sound level meter error. This method is appropriate for predicting the noise parameters of built-up urban area. Noise profiles built on the basis of noise parameters of track spans only in certain cases show lower noise levels in the calculated points located in urban areas by up to 10 dBA.

Extension of this research requires detailed analysis of the reason for isolated errors in the obtained values for the calculated points located at the edge of acoustic center. The following hypothesis will be tested: while evaluating acoustic pollution objects of transport infrastructure shall be presented as spatial noise sources.

References

1. GOST 20444-2014 (ISO 1996-1:2003, NEQ; ISO 1996-2:2007, NEQ). Noise. Traffic flows. Methods of noise characteristic determination (Standartinform, Moscow, 2015) (in Russian)
2. GOST 23337-2014 (ISO 1996-1:2003, NEQ; ISO 1996-2:2007, NEQ). Noise. Methods of noise measurement in residential areas and in the rooms of residential, public and community buildings (Standartinform, Moscow, 2015) (in Russian)
3. SP 51.13330.2011. Noise protection. Revised edition of SNiP 23-03-2003 (Minregion Rossii, Moscow, 2010) (in Russian)
4. G.L. Osipov, V.E. Korobkov, A.A. Klimukhin, *Zashhita ot shuma v gradostroitel'stve* [Noise protection in urban development] (Strojizdat, Moscow, 1993) (in Russian)
5. G.L. Osipov, E.Yu. Yudin, G. Hubner, *Snizhenie shuma v zdanijah i zhilyh rajonah* [Reduction of noise in buildings and residential districts] (Strojizdat, Moscow, 1987) (in Russian)
6. P.I. Pospelov, *Bor'ba s shumom na avtomobil'nyh dorogah* [Noise control on highways] (Transport, Moscow, 1981) (in Russian)
7. P.I. Pospelov, V.I. Purkin, B.A. Shchit, *Metodicheskie rekomendacii po zashhite ot transportnogo shuma territorij, prilegajushhih k avtomobil'nym dorogam* [Recommendations on traffic noise protection of the territories adjacent to highways], (Informavtodor Publ., Moscow, 2011)
8. A.A. Faktorovich, G.I. Postnikov. *Zashhita gorodov ot transportnogo shuma* [Protection of cities from traffic noise], 144 (Budivelnik, Kiev, 1982)

9. S.N. Ovsiannikov, E.M. Kotova, Zhilishchnoe Stroitel'stvo [Housing construction], **6**, 6 (2013)
10. T.S. Bolshanina, S.N. Ovsiannikov, BST — Bulletin of Construction Equipment 2.0, **6,** 84 (2017)

Cost efficiency assessment of automated quality control of precast structures

Liudmila Kaverzina[1,*], *Galina* Kovalenko[1], *Irina* Dudina[1], and *Oleg* Belskii [1]

[1]Bratsk State University, 665709 Bratsk, Russia

Abstract. Relevance of the research is conditioned by the necessity to enhance the factory quality control of reinforced concrete structures based on integral assessment of their reliability. The current system of selective quality control of precast concrete structures does not provide reliability assurance of the whole lot of products. The present research aims to develop operational procedure and consider economic feasibility of automated quality control of precast RC structures. Quality control is performed each shift according to the developed software system based on probabilistic methods considering statistic variability of the controlled parameters. The critical criterion of operational integrity of structures is integral assessment of the reliability indicators. The following theoretical research methods were used in the study: probabilistic-statistical, methods of system and economic analysis. Validity of the obtained results and economic feasibility were proved by experimental studies including full-scale tests.

Introduction

Serious attention has always been paid to the issue of operational integrity assessment of building structures and enhancement of their quality [1-7]. This objective is especially relevant during implementation of quality control of precast reinforced concrete (RC) structures [1, 2, 8-14], which comprise the basis of modern construction in Russia due to complexity of climatic conditions. Production of precast RC structures requires providing the basic condition of their reliability at minimum expenses. The relevance of this engineering and economic problem is motivated by the necessity to enhance the system of in-process control and quality control of the output products to ensure that they are up to the required consumer properties [1, 2].

Currently, reinforced concrete factories perform operational integrity assessment of the output products in several stages as specified by the acting standard. First, current control of the separate quality indicators is performed (strength and deformation material properties, data on geometrical parameters and reinforcement). The stage of end products acceptance includes periodic proof tests of full-scale structures using strength load up to structural failure [8, 15]. As described in [1, 2, 11] the current system of selective factory control apart from economic inefficiency also almost does not consider process variability of each exact production and does not provide quality and reliability assurance of all the

* Corresponding author: Dekanfps@mail.ru

structures from the test batch. The necessity of transition from selective to full factory control was stated in the research of V.V. Sudakov, E.A. Sekhniashvili, G.V. Slyusarev, V.I. Korobko, etc. [9, 10, 12, 13]. The most rational and accurate solution of this objective could be implementation of integral reliability assessment of RC structures resulting from current control of certain parameters [1, 2, 11, 14]. Along with that, it should be noted that current control at RC factories is conducted every shift and expenses for it are much lower than those for proof tests of the structures themselves [16].

So, the objective of integral reliability assessment of precast RC structures based on using probabilistic calculation methods gains its research relevance. Developed software system based on probabilistic algorithms enables to generalize and take into account the influence of variability of process factors and the controlled parameters on the consumer properties of output products on a shift basis [2, 14, 17]. The present study aims to develop methodology and prove economic feasibility of automated quality control of precast RC structures at the production stage.

Materials and methods

During solving the objectives set in the research probabilistic-statistical methods were used [1, 2, 7, 18-27], methods of system and economic analysis [16], as well as experimental methods pertaining to testing of full-scale structures [8, 14, 15].

Relying on the analysis of development and enhancement of probabilistic methods of building structures calculation, studies of the following Russian and foreign scientists were distinguished: V.D. Raizer, A.G. Tamrazjan, M.B. Krakovskiy, V.P. Chirkov, G. Augusti, A. Baratta, F. Casciati, G. Spaethe, etc. [1-3, 20-25]. To assess reliability of RC structures probabilistic methods were used, as described in [1-3, 14, 18, 21-23, 26]: method of statistical testing; statistical modeling; linear approximation of functions.

Following from analysis in the works [1, 2, 14, 18] the following condition serves as the basic criterion of operational integrity control of load bearing RC structures:

$$R \geq [R_i], \tag{1}$$

where R – actual reliability indicator, obtained from probabilistic calculations [1, 2, 14, 18];

$[R_i]$ – standard value of reliability indicator, accepted in accordance with the considered limit state [1, 2].

Based on the chosen probabilistic methods software system for reliability assessment of the basic load bearing RC structures was developed (roof and floor slabs, cross-beams, beams, trusses, wall panels, columns).

While developing methodology and operational procedure of automated shift-based quality control of structures based on integral assessment of their reliability, results of experimental studies of concrete samples, reinforcement and the structures themselves were used. Testing was conducted at the test room of "Bratskzhelezobeton" integrated plant (Russia). Operational procedure of automated non-destructive testing of operational integrity of RC structures is described in application for an invention [17], flow chart presenting the algorithm of this control method is given in Figure 1.

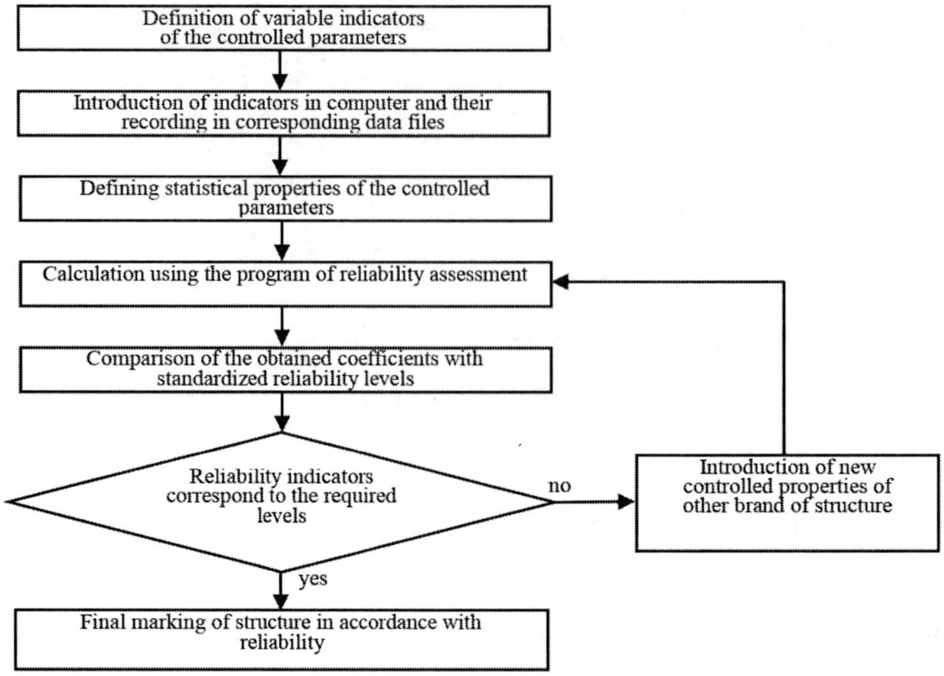

Fig. 1. Algorithm of automated shift-based reliability assessment of RC structures at the production stage.

Suggested non-destructive quality control of RC structures at the production stage begins with introduction of indicators of controlled parameters which are recorded into the corresponding data files after each shift [14, 18]. Statistical processing of initial data is performed on the basis of small sample obtained during every shift and general total of testing results obtained considering a priory information per definite period of observation [11, 14]. Further step is performing calculation with the basic software system for structural reliability assessment taking into account the worst statistical properties of controlled parameters. The obtained reliability indicators form the basis for final acceptance of the given lot of products and their certification according to requirements of strength, rigidity and crack resistance [2, 17, 19]. Structures are admitted as corresponding to the designed brand when following the condition (1). Final acceptance of structures is performed after their reliability assessment also considering check of flatness, diagonality and other parameters [8].

Checking the accuracy of operational procedure of the output products automated quality control requires conducting proof tests by strength loading at least once in 6 months [2, 17]. These tests can be replaced with effective methods of non-destructive testing [12, 13]. In case the results of proof tests are lower compared to the probability assessment of structural reliability the reasons of this noncompliance are revealed and the actual load which shift products can sustain is defined [2].

Automated shift-based assessment of operational integrity is especially relevant when it concerns structures with complex stress state [14, 28]. This refers to wall panels operating under biaxial bending and eccentric-compressed columns. Field tests of such structures (especially columns) in factory conditions are quite hard to implement.

Results

The suggested automated non-destructive quality control provides opportunity for RC factories to manufacture structures with the required reliability level. Inclusion of computer system into workflow contributes to solving the objectives of operational stability assessment and defining reserves to reduce material consumption [2, 11]. The options of structural optimization by reduction of their material consumption are implemented according to the developed software system taking into account variability of the calculated parameters and comparative economics [14, 16, 19].

Economic efficiency of the developed acceptance quality control of structures is defined by sharp decrease of the number of field tests (by 6-10 times). Following from the analysis of economic feasibility of implementing non-destructive testing for ribbed floor and roof slabs at "Bratskzhelezobeton" integrated plant, the output products showed production costs reduction by 2.4% [2]. The efficiency of the suggested automated non-destructive quality control for the basic list of precast structures is given in Table 1. Annual expenses for production and testing of the given structures according to Russian State Standard GOST 8829-94 and by automated method at "Bratskzhelezobeton" integrated plant are two-year-old prices.

Economic effect in this case will be formed due to cost reduction and can be defined the following way:

$$E=C_1-C_2, \tag{2}$$

where C_1 - expenses for conducting testing of structures in accordance with Russian State Standard GOST 8829-94 considering their production;
C_2 – expenses for production of RC structures and conducting testing using automated reliability assessment.

When the corresponding values from Table 1 are put into Formula (2), the economic effect is obtained, which is equal to 4069.05 thousand rubles. (4800.87-731.82).

Costs reduction due to implementation of automated quality control of RC structures under study is 6.7 times. Generally, implementation of the suggested shift-based non-destructive quality control of precast building structures allows improving performance indicators of construction industry enterprises which is shown on the case of "Bratskzhelezobeton" integrated plant (Table 2).

Table 1. Annual expenses for production and testing of different types of structures according to GOST 8829-94 requirements and by automated method.

Type of structure	Production volume, pcs. per year	Production costs of one structure, thousand rubles	Total production costs of structures, thousand rubles per year	Expenses for testing of one structure, thousand rubles	Testing according to GOST 8829-94		By suggested method	
					Annual demand for testing, pcs.	Total expenses for testing considering production, thousand rubles	The volume of annual tests according to new operational procedure, pcs.	Total expenses for automated control implementation, thousand rubles
Truss	124	27.24	3377.64	22.95	4	200.76	2	120.46
Beam	286	24.96	7137.70	17.01	8	335.76	2	125.90
Cross-beam	420	18.37	7715.82	14.27	36	1174.97	4	195.83
Wall panel	6800	16.70	113587.2	15.48	96	3089.38	6	289.63
Total	7630	-	131818.36	-	144	4800.87	14	731.82

Table 2. Variability of performance indicators of enterprise by using automated quality control of building structures.

Type of structure	Production volume, pcs. per year	Selling price excluding VAT, thousand rubles per piece	Indicators by different control ways Testing acc. to GOST 8829-94 By suggested method				
			Volume of sales, pcs. per year	Nett proceeds, thousand rubles per year	Production costs, thousand rubles per year	Sales profit, thousand rubles per year	Cost effectiveness of production, %
Truss	124	31.33	120 122	3759.6 3822.26	3469.40 3443.62	290.20 378.64	8.36 11.00
Beam	286	28.90	278 284	8034.2 8207.6	7273.70 7213.69	760.50 993.91	10.46 13.78
Cross-beam	420	21.68	384 416	8325.12 9018.88	8229.54 7838.16	95.59 1180.72	1.16 15.06
Wall panel	6800	19.71	6704 6794	132135.84 133909.74	115073.28 113776.61	17062.56 20133.13	14.8 17.7
Total	7630	-	7486 7616	152254.76 154958.48	134045.92 154958.48	18208.85 22686.40	13.58 17.50

The information presented above proves economic feasibility of implementing the suggested system of factory quality control. The income obtained from products sales of the factory under study was increased by more than 20%, both due to revenue growth and reduction of construction costs. It has positive effect on the indicator of production profitability which increases by 3.92%.

Conclusion

The objective of integral reliability assessment of precast RC structures using probabilistic calculation methods implemented in software system was solved as a result of the conducted study. Further, methodology of automated shift-based quality control of precast RC structures was developed on its basis with the account of the variability of process-dependent parameters. Thus, the aim of the study was accomplished. Practical approval was conducted and economic efficiency of the suggested automated quality control of precast RC structures was proved.

Conducting shift-based non-destructive quality control of structures allows reducing the number of expensive full-scale tests up to 10 times. It makes possible to reduce production costs and, therefore, provides growth of income for construction industry enterprises and, finally, will result in increasing performance ratio. Suggested control method is crucial for construction industry enterprises as it finally provides reduction of construction products cost along the whole process chain.

Further research in this direction will enable finding out the reserves for materials consumption reduction and solving production tasks on structural optimization using probabilistic methods on the basis of various models. This will contribute to developing the basis of automated control systems of the whole process procedure.

References

1. V.D. Raizer, *Teorija nadezhnosti sooruzhenij* [Theory of structural safety] (Association of Civil Engineering Universities, Moscow, 2010) (in Russian)
2. A.G. Tamrazjan, I.V. Dudina, Zhilishhnoe stroitel'stvo [Housing Construction] **3**, 8-10 (2001) (in Russian)
3. A.G. Tamrazjan, Vestnik MGSU **10**, 109-115 (2012) (in Russian)
4. O.G. Kumpyak, Z.R. Galyautdinov, O.R. Pakhmurin, Vestnik of TSUAB **3**, 88-95 (2014) (in Russian)
5. A.V. Filatov, European science **9(10),** 28-29 (2015) (in Russian)
6. I.A. Rubaratuka, Journal of Engineering and Technology **2**(1), 46-52 (2008)
7. E.I.G. Abdelouafi, K. Benaissa, K.Abdellatif, Procedia Engineering **114**, 650-657 (2015)
8. GOST 8829-94. Izdelija stroitel'nye zhelezobetonnye i betonnye zavodskogo izgotovlenija. Metody ispytanij nagruzheniem. Pravila ocenki prochnosti, zhestkosti i treshhinostojkosti [Reinforced concrete and prefabricated concrete building products. Loading test methods. Assesment of strength, rigidity and crack resistance] (Izdatel'stvo standartov, Moscow, 1994) (in Russian)
9. V.V. Sudakov, *Kontrol' kachestva i nadezhnost' zhelezobetonnyh konstrukcij* [Quality control and reliability of reinforced concrete structures] (Strojizdat, Leningrad, 1988) (in Russian)
10. E.A. Sekhniashvili, *Integral'naja ocenka kachestva i nadezhnosti predvaritel'no naprjazhennyh konstrukcij* [Integral assessment of quality and reliability of prestressed structures] (Nauka, Moscow, 1988) (in Russian)
11. Yu.A. Samarin, Stroitel'naja industrija v energetike [Building industry in power engineering] **3**, 2-60 (Informenergo, Moscow, 1989) (in Russian)
12. G.V. Slyusarev, News of higher educational institutions. Construction **5-6**, 122-125 (1995) (in Russian)
13. V.I. Korobko, A.V. Korobko, *Kontrol' kachestva stroitel'nyh konstrukcij: Vibroakusticheskie tehnologii* [Quality control of building structures: Vibroacoustic technologies] (Association of Civil Engineering Universities, Moscow, 2003) (in Russian)
14. G.V. Kovalenko, S.A. Zherdeva, I.V. Dudina, Systems. Methods. Technologies **3**, 161-167 (2014) (in Russian)
15. I.V. Dudina, E.V. Nester, Systems. Methods. Technologies **9**, 113-118 (2011) (in Russian)
16. L.A. Kaverzina, V.A. Kaverzin, *Vnutrifirmennoe planirovanie dejatel'nosti stroitel'nogo predprijatija: mehanizm realizacii, vlijanie na jeffektivnost' funkcionirovanija* [Corporate planning in construction company: implementation arrangements, influence on efficiency] (LAP Lambert Academic Publishing, 2011) (in Russian)
17. G.V. Kovalenko, V.A. Ljublinskij, I.V. Dudina, S.A. Zherdeva, E.V. Nester, Application for an invention no. 2015128114, **2** (Rospatent, Moscow, 2017) (in Russian)
18. G.V. Kovalenko, I.V. Dudina, E.V. Nester, *European Science and Technology: Materials of the International Research and Practice Conference* (Wiesbaden, Germany, January 31, 2012)
19. O.A. Kalash, Determinirovannaja i verojatnostnaja model' naprjazhenno-deformirovannogo sostojanija zhelezobetonnyh ferm s uchetom fizicheskoj nelinejnosti [Deterministic and probabilistic models of stress-strain state in reinforced

concrete trusses with regard to physical nonlinearity], PhD thesis abstract, Bratsk State University, 2009 (in Russian)
20. M.B. Krakovskiy, Beton i zhelezobeton [Concrete and Reinforced Concrete] **5**, 9-11 (1997) (in Russian)
21. G. Spaethe, *Die Sicherheit tragender Baukonstruktionen* (Springer-Verlag Wien, New York, 1992)
22. V.P. Chirkov, *Prikladnye metody teorii nadezhnosti v raschetah stroitel'nyh konstrukcij* [Applied methods of reliability theory in structural analysis] (Marshrut, Moscow, 2006) (in Russian)
23. G. Augusti, A. Baratta, F. Casciati, *Verojatnostnye modeli v stroitel'nom proektirovanii* [Probabilistic Methods in Structural Engineering] (Strojizdat, Moscow, 1988) (in Russian)
24. L. Hasofer, J. of the Engineering Mech. **100**, 111-121 (1974)
25. O. Ditlevsen, J. of Struct. Engineering **113**(1), 38-49 (ASCE, 1988)
26. F. Liang, L. Chu, R. J. Carroll, *Advanced Markov chain Monte Carlo methods* (Wiley, 2010)
27. J. Voss, *An Introduction to Statistical Computing: a simulation-based approach* (Wiley, 2013)
28. V.S. Plevkov, I.V. Baldin, S.V. Baldin, A.G. Laskovenko, G.A. Laskovenko, *Sovremennye problemy rascheta zhelezobetonnyh konstrukcij, zdanij i sooruzhenij na avarijnye vozdejstvija* [Modern calculation problems of reinforced concrete structures exposed to accidental impacts], 323-328 (MGSU, Moscow, 2016) (in Russian)

Nondestructive testing method for a new generation of electronics

Anton Azin[1,*], *Andrey* Zhukov[1], *Anton* Narikovich[2], *Sergey* Ponomarev[1], *Sergey* Rikkonen[1], and *Vladimir* Leitsin[2]

[1]National Research Tomsk State University, 634050 Tomsk, Russia
[2]Immanuel Kant Baltic Federal University, 236016 Kaliningrad, Russia

Abstract The implementation of the Smart City system needs reliable and smoothly operating electronic equipment. The study is aimed at developing a nondestructive testing method for electronic equipment and its components. This method can be used to identify critical design defects of printed circuit boards (PCB) and to predict their service life, taking into account the nature of probable operating loads. The study uses an acoustic emission method to identify and localize critical design defects of printed circuit boards. Geometric dimensions of detected critical defects can be determined by the X-ray tomography method. Based on the results of the study, a method combining acoustic emission and X-ray tomography was developed for nondestructive testing of printed circuit boards. The stress-strain state of solder joints containing detected defects was analyzed. This paper gives an example of using the developed method for estimating the degree of damage to joints between PCB components and predicting the service life of the entire PCB.

1. Introduction

The Smart City is the latest development in the field of urban planning. The key resources for implementing the Smart City strategy are information and intelligent processing systems designed to ensure a stable link between individuals and knowledge. This will become the driving force for the transition to a fundamentally new way of energy consumption [1]. The Smart City system integrates all information flow into a single center where data is accumulated, systematized and processed. The operational performance of this analytical center depends on smooth operations of all computer systems and servers. Since electronic devices and their components are becoming more and more compact, new nondestructive testing methods need to be developed.

Two nondestructive testing methods, optoelectrical and X-ray, are widely used to control the quality of electronic products [2, 3]. Optoelectrical systems can be used to monitor and analyze performance of automatic installation equipment and to measure qualitative parameters of solder joints. These systems, however, do not guarantee the

* Corresponding author: antonazin@mail.ru

detection of latent defects (hidden defects that have not been detected through standard tests of printed circuit boards (PCB) but can lead to PCB failure during operation).

At present, the X-ray method allows us to test a PCB with a resolution of 0.5 μm, to identify all discontinuities in its design, and to create a detailed 3D model of the test object. This analysis requires significant time and additional risk assessment of identified defects. With these methods, it is difficult to identify latent defects that will be growing in size during the operation of a device with PCB. This makes it impossible to accurately predict the service life of the entire device.

The analysis of the existing nondestructive testing methods demonstrates that the acoustic emission (AE) method can be adapted to identify growing design defects of printed circuit boards. Mistras, (USA), is one of the companies engaged in research in this field. The researchers detected latent defects in printed circuit boards with BGA chips. The defect localization accuracy was within a circle with a radius of 1 mm [4–6].

Based on the results of theoretical and experimental studies performed by the authors, the difficulty in using the AE method for defect identification in PCB can be explained by the following reasons. Firstly, solder joints (test object) have small geometric dimensions (on the surface of a 40 x 40 mm chip, there are from 600 to 1,200 solder joints). Secondly, the PCB design combines four materials with different technical characteristics (ceramics, copper, solder, and fiberglass).

The experimental studies on sample PCBs show that the AE method can detect a growing design defect of a PCB (the probability of defect detection is 85%). However, it is difficult to determine defect criticality in areas with high packing density on the basis of AE data [7].

In order to clarify the location of latent defects and determine their geometric dimensions through numerical analysis, the authors suggest combining the acoustic emission method with the X-ray tomography method. This study is the first to use such an integrated nondestructive testing approach to printed circuit boards.

2. Materials and Methods

Mechanical characteristics of the POS-61 solder material designed to attach a chip to a PCB were determined using the INSTRON 8801 (USA) fatigue testing system. The tests were carried out in compliance with GOST 1497-84 [8].

Cyclic tests of the POC-61 solder material were performed using the INSTRON 8801 fatigue test system (USA). The tests were carried out in compliance with GOST 25502-79 [9].

In order to localize latent design defects of a PCB, the AEWin software and Micro2 digital AE system (USA) were used [10].

Geometric dimensions of detected latent defects were determined by a Y. Cheetah 160kV microfocus X-ray inspection system [11].

In order to determine the maximum stress rate in the solder material under a given load, BGA solder joints were numerically modeled using the ANSYS Mechanical software package.

3. Experiment

3.1 Identification of Latent Design Defects in a Printed Circuit Board Using the Acoustic Emission Method

A video card containing a BGA chip on its surface was selected as a sample printed circuit board for the experiment. The geometric dimensions of the chip package are 0.032 x 0.032 m. It is located in the center of the video card (dimensions of the PCB: 0.07 x 0.14 m). The sample printed circuit board is shown in Figure 1*a*. Defects were found using a planar location test setup with four sensors. A diagram in Figure 1*b* shows the results of defect identification (in the figure, • means the AE sensor; ■ means the detected defect). In the diagram, the locations of sensors are indicated by numbers. The chip boundaries are shown by rectangular outline. The test results demonstrate that there are "active" defects in the top right corner of the chip that can lead to failure of the PCB during its operation.

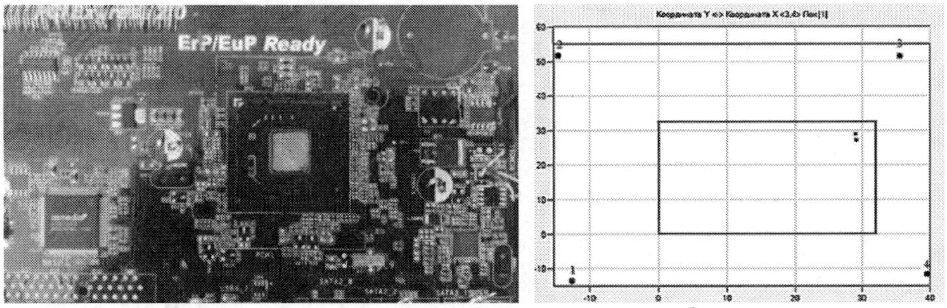

a *b*

Fig. 1. AE tests: a – sample PCB; b – diagram of defect identification results.

3.2 Determining Geometric Dimensions of PCB Latent Defects

In order to determine the geometric dimensions of detected growing defects, the PCB was tested by using a Y. Cheetah 160kV microfocus X-ray inspection system. The experimental results are given in Figure 2. Figure 2*a* presents a 3D reconstruction of solder joints through which geometric dimensions of any defects can be determined. Figure 2*b*–*e* shows sections of the solder joints at different heights from the board surface.

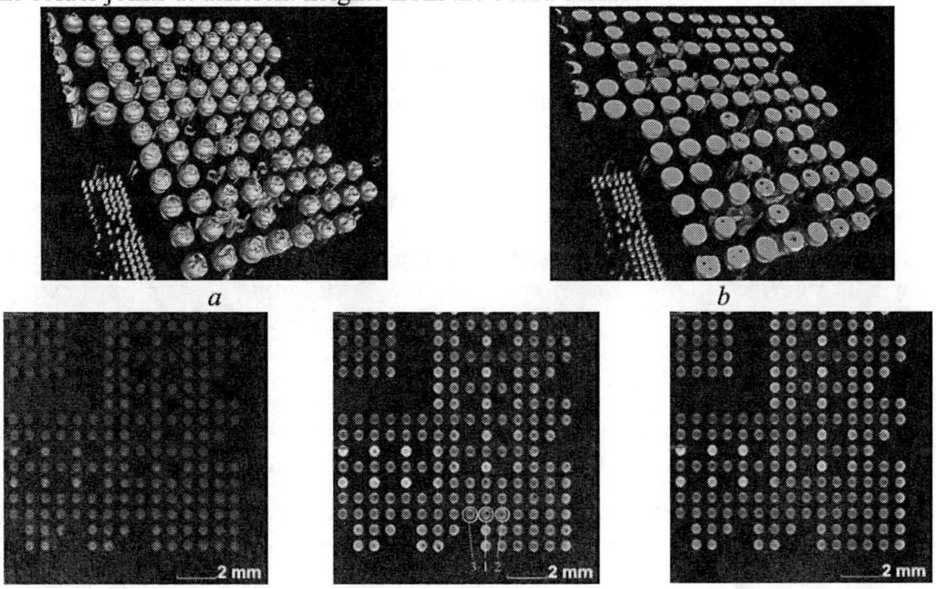

a *b* *c* *d* *e*

Fig. 2. Experimental results: *a* – 3D reconstruction of solder joints, *b* – section of solder joints, *c* – 0.37 mm, *d* – 0.20 mm, *e* – 0.06 mm.

Thus, by combining AE and X-ray tomography methods, it is possible to determine the geometry of a growing defect. The service life of the PCB can be determined by numerically analyzing the stress-strain state of solder joints under operating loads.

3.3 Numerical Analysis

3.3.1 Determining Mechanical Characteristics of POS-61 Solder Material

Uniaxial tensile and cyclic tests were carried out to determine mechanical characteristics of the solder. Based on the results of uniaxial tensile tests, the bilinear stress-strain curve for the solder material (the solid and dashed lines represent the experimental data and the bilinear approximation, respectively) (Fig. 3*a*). Parameters of the long-term strength model were determined on the basis of the Manson-Coffin equation:

$$N^m \Delta\varepsilon = C_N, \tag{1}$$

where N is the number of cycles for a given deformation; $\Delta\varepsilon$ is the strain range; C_N, m are parameters of the model (for this material $C_N = 0.1892$, $m = 0.53$). The results of theoretical calculations (dashed line) and experimental data (solid line) are reconciled in Figure 1*b*.

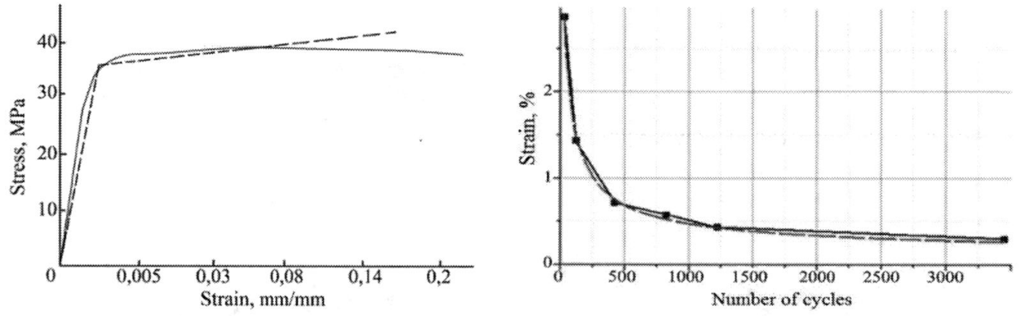

a *b*

Fig. 3. Test results: *a* is the stress-strain behavior in the sample; *b* is the dependence of strain rate on the number of cycles prior to fracture.

3.3.2 Damage Rate Function of the Material

To determine the current state of the material and predict the remaining service life, let us introduce the damage rate function proposed by V.V. Moskvitin [12] that must satisfy the following conditions:

$$\omega(0)=0, \quad \omega(t^*)=1, \tag{2}$$

where t_* is the time prior to the limit state (fracture).

The following expression can be introduced to assess the degree of material damage:

$$\omega(t,T) = \int_0^t \frac{(1+m_1)(t-\tau)^{m_1}}{t_0(\varepsilon,T)^{1+m_1}}\, d\tau \quad , \tag{3}$$

where $m_1 = -1 + 1/m$ (for the solder material $m = 0.53$); ε is the strain rate; t_0 is the durability of a given material determined experimentally at temperature T and not depending on deformation time. Let us use a special case to estimate the degree of damage of solder joints exposed to thermal cycling (3):

$$\omega(t,T) = \upsilon \int_0^t \frac{(1+m_1)(t-\tau)^{m_1}}{\left[C_1(a-bT)\Delta\varepsilon^{-m}\right]^{1+m_1}} d\tau, \tag{4}$$

where $\Delta\varepsilon$ is the time-independent strain range; ν is the loading rate, C_1 and m are, respectively, the parameters of the model determined for this material, where $C_N(T) = C_1 \cdot E(T)$ and $E(T) = a - bT$ (C_N = 18.92, m = 0.53 at 25°C); a and b are the parameters of the model that show the temperature's dependence of the material's elastic modulus (a = 6.4411, b = 0.0284).

After integrating the relation (4), we obtain an analytical dependence of material damage on time and at a given temperature T different from the initial one:

$$\omega(t,T) = \left[\frac{\upsilon}{C_1(a-bT)}\right]^{\frac{1}{m}} \Delta\varepsilon t^{\frac{1}{m}} \tag{5}$$

3.3.3 Estimating the Load-Bearing Capacity of Solder Joints

In order to predict the service life of a printed circuit board with detected defects, it is necessary to assess the stress-strain state of defective solder joints by means of numerical analysis. In our problem, we determined the stress-strain state of the printed circuit board at a temperature cycle from -55 to +85 ° C by exposing it to the maximum and minimum temperatures for 20 minutes. The rate of cooling and heating was 10 ° C/min [13].

In order to take into account all boundary conditions and loads to which the solder joint is exposed during thermal cycling, it is necessary to consider two types of numerical models: general and local ones [14].

The general model represents the entire printed circuit board (Fig. 4). The model contains a printed circuit board, contact pads on a printed circuit board and on a chip substrate, a substrate, a chip, and a ball grid array. Since the problem is axisymmetric, ¼ of the printed circuit board design is considered. The result of the solution is a displacement field for all model nodes. Next, this displacement field is applied to solder joints that have defects in the local model in order to determine the maximum stress rate.

The local model represents a defective solder joint. Three cases of defective contacts are considered (Fig. 5 and Fig. 2*d*). The model contains a printed circuit board, contact pads on a printed circuit board and on a chip substrate, a substrate, and a solder joint. The data obtained in the general model are used as a load in the local model. Since the problem is axisymmetric, ½ of the solder joint design is considered.

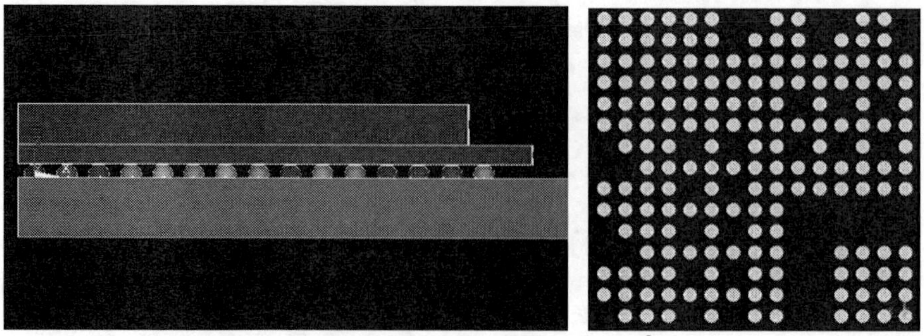

a *b*

Fig. 4. Global model: *a* is the general form; *b* is the BGA.

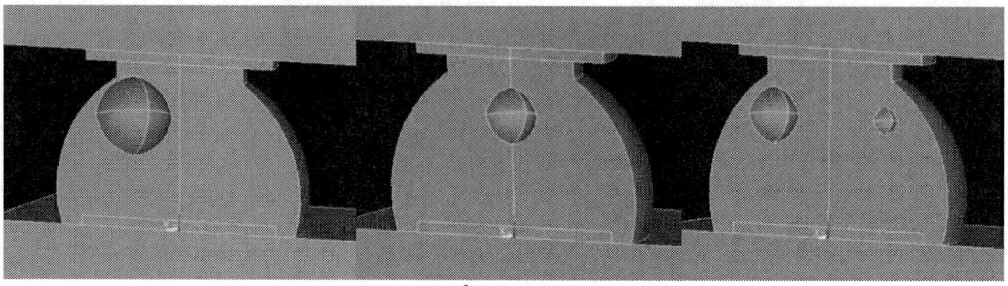

a *b* *c*

Fig. 5. Local model: *a* is the first; *b* is the second; c is the third cases.

Based on the results of numerical simulation, maximum stress rates were determined in each of the considered cases: 1 – $1.86 \cdot 10^7$ Pa; 2 – $3.07 \cdot 10^7$ Pa; 3 – $2.53 \cdot 10^7$ Pa. Shear strain can be calculated from the ratio (its form was determined from uniaxial tensile tests of solder samples):

$$\varepsilon(\sigma,t) = \left(\frac{\sigma}{E}\right) + A_1\left(\frac{\sigma}{E}\right)^{B_1}, \text{ where } A_1 = 1.17 \cdot 10^{57},\ B_1 = 22.26,\ E = 16.082 \cdot 10^9 \text{ Pa.}$$

In order to determine the number of cycles until the solder joint loses its load-bearing capacity under a given load (the duration of the temperature cycle is 68 min, the loading rate $\nu = 2.45 \cdot 10^{-4}$ cycles/s), the expression (5) is used.

Based on the results of the solution, we determined the degree of damage as a function of the operating time at a given loading rate for each case of a defective solder joint (Fig. 6). The curve shows that, at a given loading rate, the defective solder joint of the second case has the shortest time of failure. In this case, the service life of the PCB is 5,500 hours.

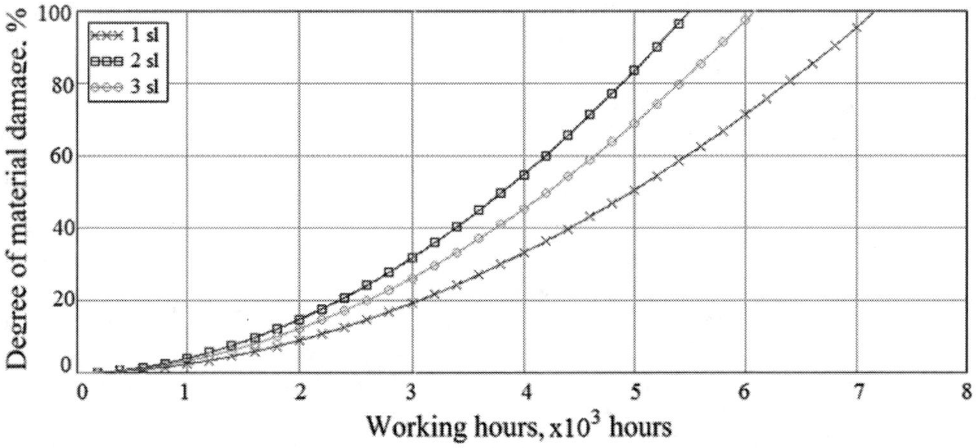

Fig. 6. The damage accumulation function for the three considered cases of damaged joints.

4. Results

By combining AE and X-ray tomography methods, we localized growing defects and determined their geometric dimensions. Using the developed numerical modeling algorithms and methods aimed at predicting the remaining service life based on loading history of the PCB, we predicted the service life of the PCB under given operating loads.

5. Conclusion

Based on the results of the studies, a nondestructive testing method for electronic equipment was developed. Using this method, it is possible to identify latent design defects of PCB, to analyze their criticality, and to determine the residual service life of PCB. The method can be used in the production of printed circuit boards and the testing of new solutions before commercialization. The method is also applicable to estimating the service life of equipment during operation. This innovation is necessary for the reliable functioning of Smart City systems.

The application of the developed method was demonstrated for PCBs containing BGA chips. Further research should be devoted to justifying the general applicability of this method to printed circuit boards of any configuration.

The reported study was funded by RFBR according to the research project No. 16-38-00743 мол_а.

References

1. Green Capacity, Umnye goroda [Smart Cities], available at: http://greencapacity.ru/ru/information/smart-cities (2017) (in Russian)
2. I.I. Kochegarov, I.V. Khanin, A.V. Lysenko, N.K. Yurkov, V.B. Almametov, University proceedings. Volga region. Technical sciences **3(27)**, 105–114 (2013) (in Russian)
3. Sovremennyj rentgenovskij kontrol' pechatnyh plat [Modern X-ray control of printed circuit boards], available at: http://www.rts-engineering.ru/Smounting/smArticle/smART_rentgen.html (2017)

4. A. Bansal, G. Ramakrishna, K. Liu, *A New Approach for Early Detection of PCB Pad Cratering Failures* (Cisco Systems Inc., San Jose, CA, 2011)
5. A. Bansal, G. Ramakrishna, K. Liu, *Investigation of Pad Cratering in Large Flip-Chip BGA using Acoustic Emission* (Cisco Systems Inc., San Jose, CA, 2011)
6. W.B. San, Proceedings of IPC APEX Expo, (San Diego, February 28–March 1, 2012)
7. S.V. Ponomarev, S. Rikkonen, A. Azin, A. Karavatskiy, N. Maritskiy, S.A. Ponomarev, *Conference Series: Materials Science and Engineering* **71,** 012056 (2015)
8. GOST 1497-84. Metals. Methods of tension test, available at: http://www.internet-law.ru/gosts/gost/4616/ (2008) (in Russian)
9. GOST 25502-79, Methods of metals mechanical testing. Methods of fatigue testing, available at: http://markmet.ru/gost/gost-25502-79/ (1979)
10. Physical acoustic, Micro-II – Compact PCI AE Chassis, available at: http://www.physicalacoustics.com/by-product/micro-ii/ (2017)
11. YXLON, YXLON Cheetah, available at: http://www.yxlon.com/products/x-ray-and-ct-inspection-systems/yxlon-cheetah (2017)
12. V.V. Moskvitin, *Soprotivlenie vjazkouprugih materialov* [Resistance of viscoelastic materials] (Nauka, Moscow, 1972) (in Russian)
13. K.C. Otiaba, R.S. Bhatti, N.N. Ekere, S. Mallik, M. Ekpu, Engineering Failure Analysis **28,** 192–207 (2013)
14. L.J. Ladani, A. Dasgupta, International J. of Fatigue **31**, 703–711 (2009)

Calculation of vehicle delay at signal-controlled intersections with adaptive traffic control algorithm

Roman Andronov[1,*], and *Evgeny* Leverents[1]

[1]Tyumen Industrial University, 625000 Tyumen, Russia

Abstract. By widely introducing information technology tools in the field of traffic control, it is possible to increase the capacity of hubs and reduce vehicle delays. Adaptive traffic light control is one of such tools. Its effectiveness can be assessed through traffic flow simulation. The aim of this study is to create a simulation model of a signal-controlled intersection that can be used to assess the effectiveness of adaptive control in various traffic situations, including the presence or absence of pedestrian traffic through an intersection. The model is based on a numerical experiment conducted using the Monte Carlo method. As a result of the study, vehicle delays, queue length and duration of traffic light cycles are calculated subject to different intensities of incoming traffic flows, and the presence or absence of pedestrian traffic.

Introduction

In recent years, there is a trend towards introduction of information technologies in the field of traffic control in cities with the aim of eliminating congestion and reducing vehicle delays. Transport navigation systems for navigating city streets and automated traffic control systems are becoming more popular; they are synchronized with GPS navigation data; the so-called smart traffic light technology is introduced, etc. [1]. This trend is consistent with the principle of sustainable urban development and should ultimately lead to implementation of the livable city concept [2].

One of acceptable solutions to the traffic congestion problem is the use of automated traffic control system (ATCS). Adaptive traffic light control adjustable to traffic intensity is a particularly attractive solution.

The latest generation of ATCS is developed with regard to data from navigation systems (GPS, Glonass, Galileo) [3, 4], with the priority given to vehicles of emergency services [5]. However, most of current adaptive control systems use data from passage and vehicle presence detectors [6, 7, 8]. Spectrum 2.0, SCOOT and UTOPIA belong to this type of ATCS.

Due to adaptive traffic control vehicle delays at signal-controlled intersections can be reduced by 10–60% [8]. In order to justify the use of adaptive traffic control at a signal-controlled intersection, it is necessary to know time of vehicle delays and a value by which

[*] Corresponding author: aroma77777@mail.ru

they will be reduced. Some software products, such as VISSIM and Anylogic, have add-on modules for adaptive traffic control, but software vendors protect methodology and mechanisms behind these modules as a trade secret.

The applicable reference document of the Russian Federation [9] contains recommended uses of adaptive control but this document is over 40 years old and, therefore, does not reflect contemporary realities. Also, the introduction of the road safety audit concept [10] requires assessing the quality of roads and streets in terms of traffic safety. This task is performed in urban conditions by singling out pedestrian traffic as a separate phase with a resulting decrease in the overall traffic capacity of intersections.

The aim of this study is to create a numerical model representing performance of an isolated intersection under adaptive control to obtain reliable average values of traffic light cycle, queue length and vehicle delays in a stationary stochastic flow.

The novelty of the study lies in the model of traffic flow through a signal-controlled intersection with adaptive control. Provided that a traffic situation is simulated multiple times with different initial conditions, this model makes it possible to calculate average values of queue parameters and vehicle delays.

Materials and Methods

Based on traffic flow patterns, this study proposes a simulation model of traffic flows at an isolated signal-controlled intersection with numerical Monte Carlo simulation [11] and subsequent calculation of vehicle delays and queue parameters. This method has proved itself in simulating both complex systems and traffic flows at micro- and macrolevels [12, 13].

Processes of approaching vehicles to the intersection and leaving it from the stop line at specific intervals are simulated. As a result, there is a queue of cars which should be completely resolved for the next green phase. There is a significant spread of data, including queue parameters, duration of phases and the final value of delays. The essence of the method is to simulate the traffic situation multiple times with the aim of achieving a final average result with required reliability and confidence.

Experiment

To avoid overloading of the model, a conditional intersection of two-lane one-way streets was selected. Activation of the enabling signal and beginning of movement of vehicles occurred alternately subject to the passage of the entire queue of vehicles on the intersected street and the appearance of a gap in the flow [8] (a 5-second interval during which no vehicle approaches the stop line).

The simulation was performed without regard to pedestrian traffic and with a pedestrian phase. A 20-second pedestrian phase was activated alternately in the final part of the traffic light cycle. These conditions satisfy the performance of the Spectrum 2.0 adaptive light control system.

Approaching vehicles for a short time interval (5 seconds) was emulated on the assumption of a Poisson distribution of intervals. The flow was assumed to be stationary for a one-hour interval. The passage of cars through the stop line in traffic conditions with a stable saturation flow was subject to the normal distribution law. The saturation flow was simulated in "ideal" conditions (absence of slopes, good visibility, summer season, sufficient coefficient of adhesion).

The general nature of the simulation is presented in formulas 1 and 2.

$$x_i = F^{-1}(x_{av}, \sigma_x, s)(1) \quad y_i = F^{-1}(y_{av}, \sigma_y, K_h, r)(2)$$

where r, s are uniformly distributed random numbers in the range of 0 to 1;
x_{av}, y_{av} are mathematical expectations for the traffic capacity of the stop line at the intersection (lane) and suitable number of vehicles per cycle, respectively, vehicles/5 s;
σ_x, σ_y are the standard deviations of the traffic capacity of the stop line at the intersection (lane) and suitable number of vehicles per cycle, respectively, vehicles/5 s.

In order to create a virtual model of intersection and carry out a numerical experiment, we used an Excel data systematization tool, which, despite its relative simplicity, offers a large set of mathematical formulas and computation techniques in its latest versions. Another advantage of this application is that it can be conveniently used with statistical models, since the Monte Carlo method is based on statistical modeling, one of types of simulation modeling [14].

Adaptive algorithms were implemented by switching through logical functions and links with given parameters of accumulated and resolving queues.

The methodology given in this study can be implemented in any programming language, including Python.

When the intersection is close to its capacity limit, the time of the traffic light cycle with adaptive control can grow infinitely. Some authors recommend limiting this parameter [8]. In this paper, however, the cycle time was not limited due the principle of comparative simplicity.

Results

The results of the numerical experiment are shown in Tables 1, 2 and Figures 1, 2. In addition, we compared the results with delay of one vehicle subject to fixed traffic light control. A technique based on the Webster's formula [15] is suitable for this purpose.

In total, the model was simulated 50 times for each traffic situation. This number was selected with the aim of reaching the required confidence level (95%) and permissible error of mean result (not more than 10%). The relevance of the obtained data is supported by the general theory of traffic flows and preliminary experiments, in particular those described in [16].

Table 1. Results of the numerical experiment, without regard to the pedestrian phase.

Traffic intensity for the 1st street	Traffic intensity for the 2nd street	Cycle duration, s			Delays, s		Deviation of average delay	
		Min	Max	Average	Fixed control	Adaptive control	Absolute	Relative
300	300	15	50	31.3	19.1	11.9	7.20	38%
500	500	20	55	39.4	20.7	12.3	8.44	41%
1,000	500	30	75	49.3	24.23	13.7	10.56	44%
1,000	1,000	40	125	78	26	18.2	7.81	30%
1,500	500	25	105	67.1	19.25	16.0	3.26	17%
1,200	1,200	40	215	128.9	29.7	25.8	3.93	13%
1,400	1,000	65	205	116.3	30.63	24.4	6.27	20%

Table 2. Results of the numerical experiment, with regard to the pedestrian phase.

Traffic intensity for the 1st street	Traffic intensity for the 2nd street	Cycle duration, s			Delays, s		Deviation of average delay	
		Min	Max	Average	Fixed control	Adaptive control	Absolute	Relative
300	300	20	75	57.8	27	22.4	4.60	17%
500	500	45	95	74.2	29.2	24.5	4.69	16%
1,000	500	70	130	95.4	30	27.69	2.31	8%
1,000	1,000	95	225	161.2	39.9	37.19	2.71	7%
1,500	500	85	185	134	34.4	32.87	1.53	4%
1,200	1,200	170	370	244.9	54.3	52.97	1.33	2%
1,400	1,000	140	315	236.9	58.8	52.0	6.80	12%

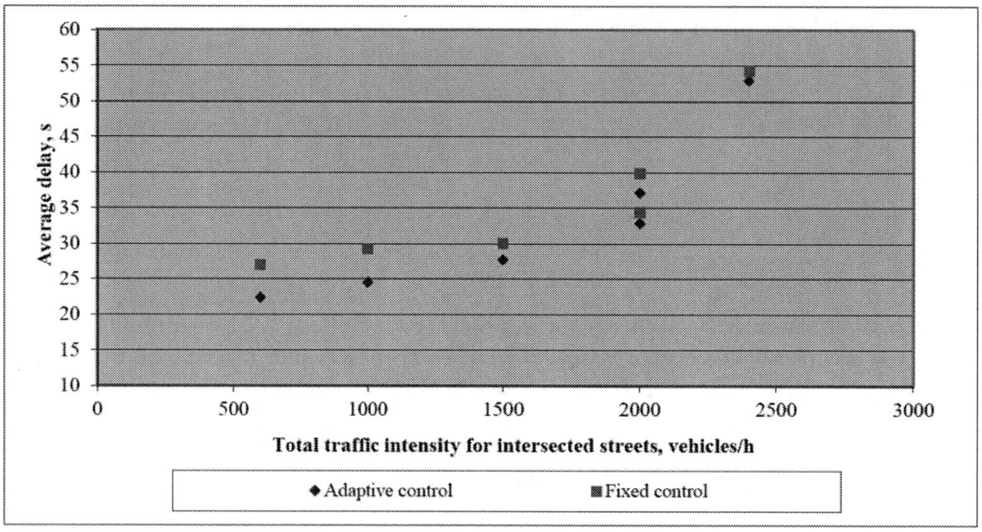

Fig. 1. Average vehicle delay, without regard to the pedestrian phase at the intersection.

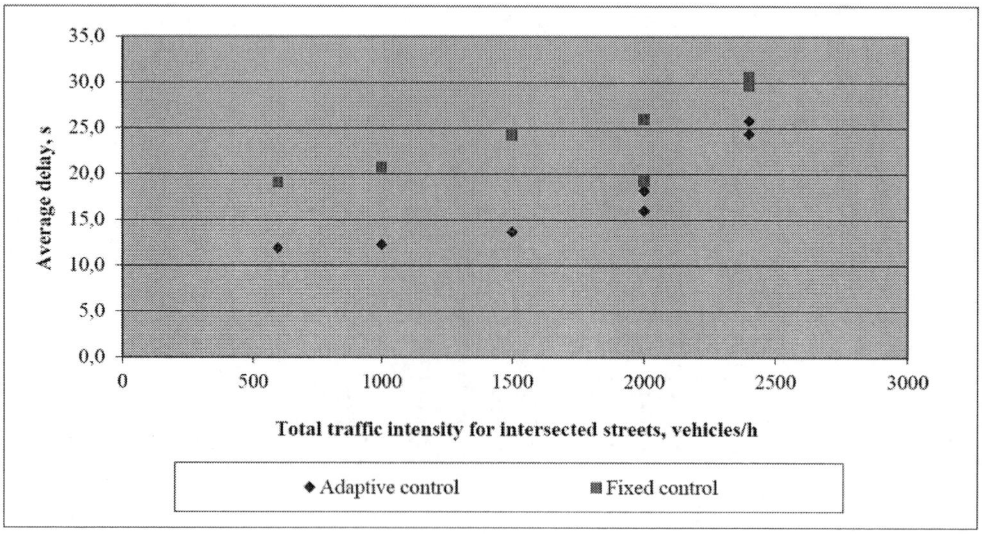

Fig. 2. Average vehicle delay, with regard to the active pedestrian phase at the intersection.

As can be seen from the tables, adaptive control is very effective for intersections with relatively low load levels (delays decreased by 30–44%). But it remains significant even at high levels if absolute values of delays and increased traffic intensity are taken into account. The total time loss at the intersection will be significant as well.

In the presence of pedestrian phase, adaptive regulation loses its effectiveness due to many actuations of this phase. Therefore, it is necessary to consider limiting the number of its actuations or using pedestrian-controlled traffic lights, if there is light pedestrian traffic at the intersection.

Based on the developed model, it is possible to simulate a real-life traffic flow throughout the day where a traffic flow is unevenly distributed across the day, and then to check the convergence of results. This model is especially relevant when it comes to combination of adaptive regulation with reorganization of intersections (allocation of lanes for turning, additional phases), which is particularly important in central parts of dense cities where large-scale reconstruction is a challenge.

Further, it is advisable to identify a recommended field of application for adaptive control in the same way as in [9], as related to various types of intersections, traffic situation at them and possible reorganization.

Conclusion

In this study, the simulation model of a signal-controlled intersection with adaptive control was developed using the Monte Carlo method; final average values of vehicle delays, duration of traffic light cycle and queue length were calculated. This model can be used to simulate the performance of crossings in various traffic situations with or without pedestrian traffic.

It is necessary to determine a recommended field of application for adaptive control as related to various types of intersections and traffic situations at them. In addition, knowing recommended applications of adaptive traffic light control will make it possible to optimize road traffic and to reconstruct road networks in a more purposeful and effective manner.

References

1. A.K. Debnath, Cities **37**, 47-56 (2014)
2. V.R. Vuchic, *Transportation for Livable Cities* (Territorija budushhego, 2011) (in Russian)
3. V. Astarita, V.P. Giofrè, A.A. Vitale, ASRJETS **19**, 1 (2016)
4. A.O. Evdokimov, A.V. Gorohov, D.Ja. Lugov, I.R.J. **5**, 55 (2015)
5. S. Djahel, *Adaptive traffic management for secure and efficient emergency services in smart cities* (PERCOM Workshops, IEEE, 2013)
6. E.A. Petrov, Transportnaja strategija XXI vek [Transport strategy – XXI century] **12** (2011) (in Russian)
7. V. Erisov, V. Trushevski, EEJET **4**, 9 (2014)
8. Ju.A. Kremenec, M.P. Pecherskij, M.B. Afanas'ev, *Tehnicheskie sredstva organizacii dorozhnogo dvizhenija* [Technical facilities for traffic organization] (Akademkniga, 2005) (in Russian)
9. V.M. Polukarov, L.A. Jakushin, *Rukovodstvo po proektirovaniju i vnedreniju avtomatizirovannyh sistem upravlenija dorozhnym dvizheniem na baze ASSUD* [A Manual on the Design and Implementation of Automated Traffic Control Systems] (MIA USSR, 1981) (in Russian)
10. ODM 218.6.010-2013. Metodicheskie rekomendacii po organizacii audita bezopasnosti dorozhnogo dvizhenija pri proektirovanii i jekspluatacii avtomobil'nyh

dorog [Methodical recommendations on the organization of road safety audit in the design and operation of highways] (Informavtodor, 2014) (in Russian)

11. N. Metropolis, S. Ulam, JASA **44**, 247 (1949)
12. S. Sinha, *Dynamics of urban traffic congestion: A kinetic Monte Carlo approach to simulating collective vehicular dynamics* (COMSNETS, IEEE, 2016)
13. S. Tsiakmakis, TRR Journals **2572**, 66-77 (2016)
14. V.V. Silyanov, *Teorija transportnyh potokov v proektirovanii dorog i organizacii dvizhenija* [Theory of traffic flows in the design of roads and traffic organization] (Transport, Moscow, 1977)
15. ODM 218.6.003-2011. Metodicheskie rekomendacii po proektirovaniju svetofornyh objektov na avtomobil'nyh dorogah [Methodical recommendations on the design of traffic lights on highways] (Росавтодор, 2013)
16. R.V. Andronov, *Modelirovanie ocheredej na reguliruemyh peresechenijah ulichno-dorozhnoj seti krupnogo goroda v uslovijah plotnogo transportnogo potoka* [Modeling of queues at regulated street intersections in a large city within the dense traffic] (Tyumen, 2007)

MATEC Web of Conferences **143**, 04009 (2018) https://doi.org/10.1051/matecconf/201814304009
YSSIP-2017

Development of multiparameter equations for satellite monitoring analysis of traffic flow

Alexander Testeshev[1], VeraTimohovetz[1,*], and *Tatyana* Mikeladze[1]

[1]Tyumen Industrial University, 625000 Tyumen, Russia

Abstract. The paper is dedicated to the development of multiparameter equations of traffic flows for satellite monitoring analysis of the road networks of major cities. Use of multiparameter dependences will allow interpreting the road situation by transforming a static image into dynamic characteristics of traffic and refuse from multiple mono-dependences which account road and weather conditions. Multiparameter dependences are developed based on mathematical methods for traffic flow modeling. The issue of using cartographic resources with updatable databases are investigated, sufficient amount of observation samples was determined for development of reliable functional dependences. The suggested method allows minimizing resource expenses on creation of calculation framework for analysis of network monitoring results for traffic flows on road networks of modern metropolises.

Introduction

Multifunctionality of urban territories and abundance of routes are the main factors that make selective control of traffic flow in modern metropolises ineffective for obtaining the full vision of traffic situation. Subject of the research is development of method for network monitoring of roads in major cities.

Representative data on traffic flows characteristics are required for making managerial decisions on urban infrastructure development. It is difficult to obtain such data using the known methods due to high labor intensity and economic costs.

Remote monitoring method based on analysis of satellite images from open sources is one of the most promising directions of network research [1]. Functional dependences of traffic flow theory serve for decoding and recognition of dynamic parameters of the flow (traffic intensity and speed) using a static image (flow density) [2-5]. Review of the topical research [6-8] showed incomplete correspondence of the existing deterministic model with the modern requirements to mathematical methods of traffic modeling.

This research is aimed at improving mathematical tools for decoding satellite data with the purpose to solve operational tasks of transport infrastructure management.

*Corresponding author: verochka1987@mail.ru

Calculation methods

One of the most labor-intensive and important stages of the studied method is development of mathematical dependences of traffic flow which account traffic (T_i), road (M_j) and weather conditions (R_k) [9].

The need in account of these variable factors consists in the necessity to develop a set of monodependences which accumulate a set of certain conditions for each scenario (Fig. 1).

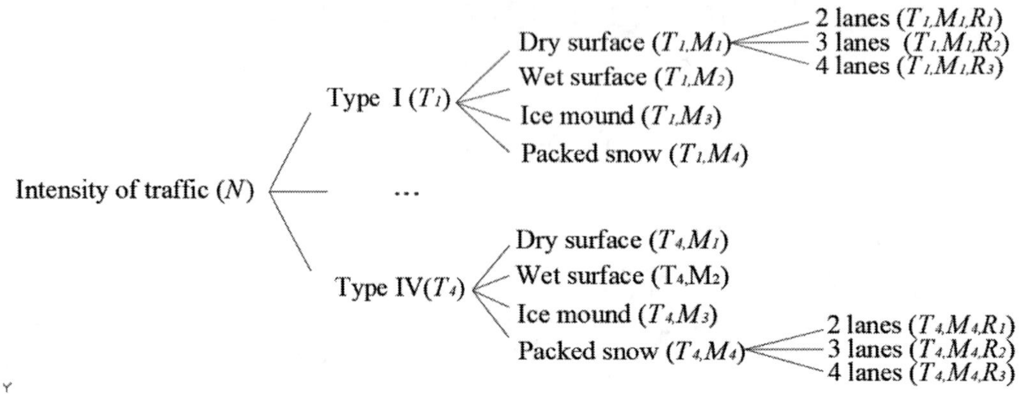

Fig. 1. Structural scheme of rows of variable factors for different monodependences.

Systematic study of traffic in mesoscale of 4 major cities of a single Federal district of the Russian Federation (RF) requires development of at least 64 monodependences [1], which considerably complicates the use of mathematical tool for satellite image decoding.

Analysis in macroscale of the territory of the country requires essentially different concept for equation development. It shall have open architecture and allow defining the unknown parameters through correction factors of variable conditions for universal dependences with regard to the reference conditions.

This criterion defines the necessity of transition to mathematical model represented by graphical dependence of the type $N = f(\rho_{ijn})$ [2] to the following multiparameter function that includes quantitative account of the variable factors:

$$N(\rho)= f\left(\rho, S_T^M, P_{T,R}^M\right), \tag{1}$$

where S_T^M – reduction factor of actual road surface condition (M_j), type of street (T_i) to the reference road surface (M_1);

$P_{T,R}^M$ – reduction factor of design number of same-direction lanes (R_j), road surface (M_j), type of street (T_i) to the reference number of lanes (R_1).

Among the advantages of the suggested method is the ability to adjust correction factors with the account of variable conditions without variation of the basic dependence. In prospective development of urban territories, this allows extending the nomenclature of design conditions of the road surface and variability of lane number, as well as improving precision of calculations due to increase in the volume of statistical sampling. Besides, analogously with [10], in case of the need to detect traffic in non-reference period, the suggested algorithm will enable to indirectly define the traffic flow parameters due to correction factors.

Development of dependences according to the suggested method required to account regional peculiarities of the major cities of the RF, such as Moscow, Saint Petersburg,

Novosibirsk, Yekaterinburg, Nizhniy Novgorod, Rostov-on-Don, Irkutsk, Kemerovo, Krasnoyarsk, Lipetsk, Omsk, Ryazan, Saratov, Tolyatti and Yaroslavl.

The above-mentioned peculiarities include search of constantly updated electronic sources of satellite images, justification of sufficient amount of observation samples, beginning of citizens' daily activity in each city.

One of the main advantages of the method is the ability to use free and regularly updated content [11], for this purpose cartographic services were carefully selected.

During this selection process frequency and dates of the database updates were determined (Table 1).

Table 1. Frequency of updates of free Russian language online sources in the RF territory for the last 5 years.

Year	Bing Maps	Yahoo! Maps	Yandex.Maps	Google Maps	Rambler-Maps	Maps @Mail.Ru
2012	-	-	-	20.10; 17.12.	-	-
2013	-	-	-	-	-	Terminated its operation
2014	-	6.10.	-	07.03; 29.09; 06.10.	Terminated its operation	-
2015	-	-	-	03.06; 06.08.	-	-
2016	-	-	15.04.	08.11; 24.11.	-	-
2017	-	-	-	01.05.	-	-

Additionally, foreign paid cartographic services were analyzed: TomTom (the Netherlands), HEREmaps (Germany), Paikkatietoikkuna (Finland), Applemaps (USA), geo.admin.ch (Switzerland), Allgemeng – Geoportal Luxembourg (Luxemburg) and Gisserver.nic.in (India), the last one has limited access. Comparative analysis proved Googlemaps to be the most appropriate website with the most relevant and update information.

Necessary condition for development of objective and accurate decoding algorithm was sufficient amount of observation samples within the research [6].

Amount of observation samples (n) was calculated based on mean traffic intensity($\overline{N}$); average values of relative accuracy of records (Δ) and relative probability function ($t\alpha$) are confirmed for all metropolises. Then, root mean square error for a certain value of intensity was found with regard to average value (σ) and variation coefficient (ν). In order to verify validity of the obtained data, absolute error (ξ) was defined, the value of which lies within the specification range (Table 2).

Table 2. Fragment of the table for minimum amount of measurements.

Cities	Δ	ξ	σ	ν	t_α	$\overline{N}$	n
Moscow	0.11	10865.9	33585.5	0.264	1.65	98781	16
Saint Petersburg	0.11	4842.97	14528.91	0.256	1.65	44027	15
Novosibirsk	0.11	9113.94	26513.28	0.249	1.65	82854	14

While specifying typology of roads and streets of the metropolises under study the beginning of the life cycle of the city was accounted. Thus, for types I and II various temporary transport demand was observed, due to this fact division of the RF territory on time zones was accounted while remote monitoring, in the part of displacement of traffic investigation intervals for various cities (Table 3).

Table 3. Fragment of table for displacement of intensity peak values for various cities.

City	Time zone	Type of roads and streets					
		I			II		
		Street	Charact-eristics	Peak hours, h	Street	Charact-eristics	Peak hours, h
Moscow	GMT+3	Volgogradsky Ave.	Periphery	6:00-8:00 15:00-19:00	Eniseyskaya	Periphery	6:00-10:00 11:00-14:00 19:00-21:00
		Yunykh Lenintsev	Periphery		Nizhegorodskaya	Periphery	
		Akademika Skryabina	Periphery		Skobelevskaya	Periphery	
Saint Petersburg	GMT+3	Nevsky Ave.	Center	8:00-11:00 18:00-20:00	Liteynyi Ave.	Center	8:00-11:00 11:00-14:00 19:00-21:00
		Glavnaya	Periphery		Sadovaya	Center	
		Stolyarny Lane	Center		Zarechnaya	Periphery	
Novosibirsk	GMT+7	Planovaya	Periphery	6:00-10:00 17:00-20:00	Lenina Sq.	Center	6:00-9:00 12:00-15:00 16:00-19:00
		Kirova	Center		Oktyabrskaya	Center	
		Mostovaya	Center		Dimitrova Ave.	Center	

Remote monitoring of sufficient amount of observation samples was performed for major RF cities with the account of defined peculiarities, which allowed presenting a set of statistical observations through graphical dependences (Fig. 2) for various traffic, road and weather conditions.

a) b)

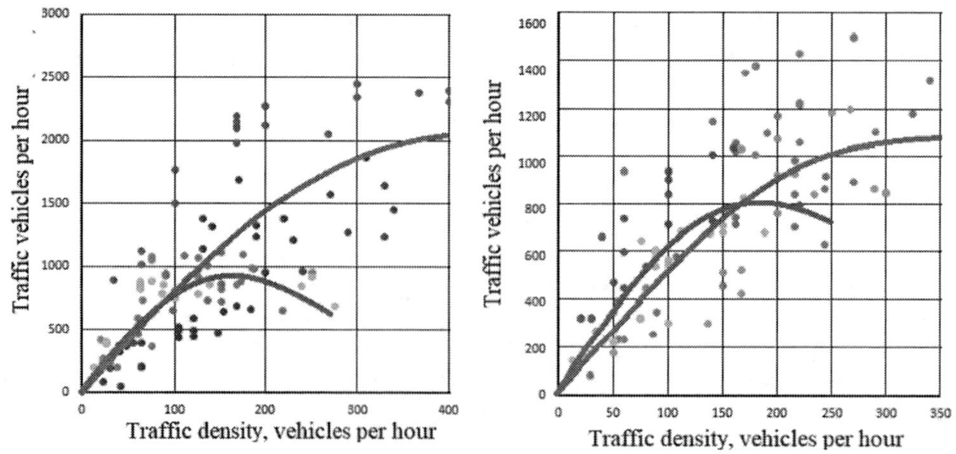

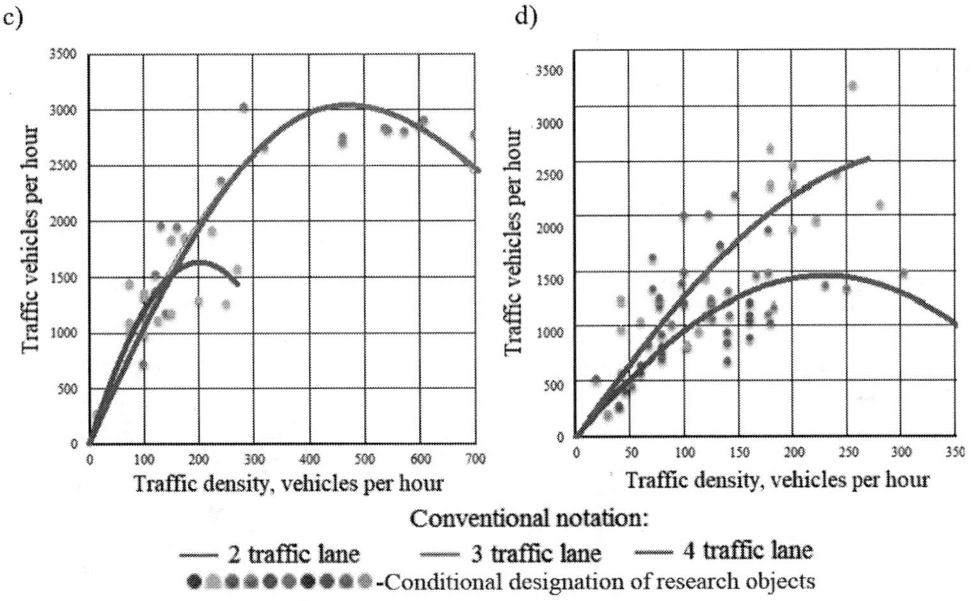

Fig. 2 Examples of dependences of traffic intensity on traffic flow density for various cities and streets with different number of lanes and for dry surface: a) 1 type; b) 2 type; c) 3 type; d) 4 type.

Due to similar behavior of graphical dependences for each type basic multiparametric equations were developed with regard to reference conditions, which can then be reduced to the design conditions by using correction coefficients.

According to the method described in [12], the reference conditions are represented by a single vehicle movement on a dry surface along straight horizontal road; for rough surface the adhesion coefficient in wet state is no lower than 0.6; evenness is no higher than 3.0 m/km; coefficient of rolling resistance is 0.01-0.02.

Cumulative account of the above factors is represented by a system of coefficients for variable parameters which are put in the basic equation. The system of coefficients and the basic equation were defined using methods of mathematical treatment and statistics.

Common structure of basic multiparametric equations is presented as follows:

a) for type I: $$N(\rho)=\frac{-0{,}0289\cdot\left(s_T^M\cdot p_{T,R}^M\cdot\rho\right)^2+9{,}6731\cdot s_T^M\cdot p_{T,R}^M\cdot\rho}{s_T^M\cdot p_{T,R}^M} \quad (2)$$

b) for type II: $$N(\rho)=\frac{-0{,}0285\cdot\left(s_T^M\cdot p_{T,R}^M\cdot\rho\right)^2+13{,}316\cdot s_T^M\cdot p_{T,R}^M\cdot\rho}{s_T^M\cdot p_{T,R}^M} \quad (3)$$

c) for type III: $$N(\rho\ )=\frac{-0{,}0415\cdot\left(s_T^M\cdot p_{T,R}^M\cdot\rho\right)^2+16{,}494\cdot s_T^M\cdot p_{T,R}^M\cdot\rho}{s_T^M * p_{TR}^M} \quad (4)$$

d) for type IV: $$N(\rho)=\frac{-\text{0,03ë38}\cdot\left(s_T^M\cdot p_{T,R}^M\cdot\rho\right)^2+11{,}457\cdot s_T^M\cdot p_{T,R}^M\cdot\rho}{s_T^M\cdot p_{T,R}^M} \quad (5)$$

Results

Based on the results of iterative calculations reduction factors for actual monitoring conditions were developed for each type of basic equation (formulae 2-5).

The obtained coefficients of street lane pattern and urban roads account for various cities are given in the Table 4.

Table 4. Reduction factors for lane number.

Type	Lanes	Surface condition			
		Dry, M_1	Wet, M_2	Ice mound, M_3	Packed snow, M_4
T_1	2, n_1	$P_{1,1}^{1} = 1,0000$ $p_{1,1}^{1} = 1,0000$	$P_{1,1}^{2} = 1,0000$ $p_{1,1}^{2} = 1,0000$	$P_{1,1}^{3} = 1,0000$ $p_{1,1}^{3} = 1,0000$	$P_{1,1}^{4} = 1,0000$ $p_{1,1}^{4} = 1,0000$
	3, n_2	$P_{1,2}^{1} = 0,7346$ $p_{1,2}^{1} = 0,6963$	$P_{1,2}^{2} = 1,1218$ $p_{1,2}^{2} = 0,8140$	$P_{1,2}^{3} = 0,3036$ $p_{1,2}^{3} = 0,2756$	$P_{1,2}^{4} = 0,5234$ $p_{1,2}^{4} = 0,5110$
	4, n_3	$P_{1,3}^{1} = 0,3746$ $p_{1,3}^{1} = 0,3686$	$P_{1,3}^{2} = 0,2974$ $p_{1,3}^{2} = 0,3461$	$P_{1,3}^{3} = 3,1766$ $p_{1,3}^{3} = 3,3636$	$P_{1,3}^{4} = 0,4299$ $p_{1,3}^{4} = 0,5221$
T_2	2, n_1	$P_{2,1}^{1} = 1,0000$ $p_{2,1}^{1} = 1,0000$	$P_{2,1}^{2} = 1,0000$ $p_{2,1}^{2} = 1,0000$	$P_{2,1}^{3} = 1,0000$ $p_{2,1}^{3} = 1,0000$	$P_{2,1}^{4} = 1,0000$ $p_{2,1}^{4} = 1,0000$
	3, n_2	$P_{2,2}^{1} = 1,6333$ $p_{2,2}^{1} = 1,8065$	$P_{2,2}^{2} = 0,7798$ $p_{2,2}^{2} = 0,3389$	$P_{2,2}^{3} = 0,3059$ $p_{2,2}^{3} = 0,2168$	$P_{2,2}^{4} = 0,4194$ $p_{2,2}^{4} = 0,6174$
	4, n_3	$P_{2,3}^{1} = 0,4686$ $p_{2,3}^{1} = 0,5129$	$P_{2,3}^{2} = 0,4149$ $p_{2,3}^{2} = 0,6519$	$P_{2,3}^{3} = 0,3787$ $p_{2,3}^{3} = 0,3140$	$P_{2,3}^{4} = 0,4929$ $p_{2,3}^{4} = 1,4737$
T_3	2, n_1	$P_{3,1}^{1} = 1,0000$ $p_{3,1}^{1} = 1,0000$	$P_{3,1}^{2} = 1,0000$ $p_{3,1}^{2} = 1,0000$	$P_{3,1}^{3} = 1,0000$ $p_{3,1}^{3} = 1,0000$	$P_{3,1}^{4} = 1,0000$ $p_{3,1}^{4} = 1,0000$
	3, n_2	$P_{3,2}^{1} = 0,4771$ $p_{3,2}^{1} = 1,3121$	$P_{3,2}^{2} = 0,5510$ $p_{3,2}^{2} = 0,1732$	$P_{3,2}^{3} = 0,5665$ $p_{3,2}^{3} = 0,2478$	$P_{3,2}^{4} = 0,3633$ $p_{3,2}^{4} = 0,3637$
	4, n_3	$P_{3,3}^{1} = 0,4241$ $p_{3,3}^{1} = 0,5610$	$P_{3,3}^{2} = 0,3001$ $p_{3,3}^{2} = 0,3601$	$P_{3,3}^{3} = 0,5591$ $p_{3,3}^{3} = 0,3035$	$P_{3,3}^{4} = 0,4056$ $p_{3,3}^{4} = 0,7835$
T_4	2, n_1	$P_{4,1}^{1} = 1,0000$ $p_{4,1}^{1} = 1,0000$	$P_{4,1}^{2} = 1,0000$ $p_{4,1}^{2} = 1,0000$	$P_{4,1}^{3} = 1,0000$ $p_{4,1}^{3} = 1,0000$	$P_{4,1}^{4} = 1,0000$ $p_{4,1}^{4} = 1,0000$
	3, n_2	$P_{4,2}^{1} = 0,3402$ $p_{4,2}^{1} = 0,2122$	$P_{4,2}^{2} = 1,2609$ $p_{4,2}^{2} = 0,8856$	$P_{4,2}^{3} = 0,2941$ $P_{4,3}^{3} = 0,1834$	$P_{4,2}^{4} = 0,2615$ $P_{4,2}^{4} = 0,1862$
	4, n_3	$P_{4,3}^{1} = 0,4026$ $p_{4,3}^{1} = 0,2838$	$P_{4,3}^{2} = 0,3421$ $p_{4,3}^{2} = 0,4060$	$P_{4,3}^{3} = 0,3719$ $P_{4,3}^{3} = 0,3221$	$P_{4,3}^{4} = 0,1819$ $P_{4,3}^{4} = 0,1408$

Note: * above fraction - value of the factor put in the numerator of the basic equation; below fraction - value of the factor put in the denominator of the basic equation.

The developed reduction factors for surface condition to the type of street are presented in Table 5.

Table 5. Reduction factors for surface conditions.

Type	Surface condition			
	Dry, M_1	Wet, M_2	Ice mound, M_3	Packed snow, M_4
T_1	$S_1^1 = 1,0000$ $s_1^1 = 1,0000$	$S_1^2 = 1,2338$ $s_1^2 = 0,7061$	$S_1^3 = 1,7627$ $s_1^3 = 0,834$	$S_1^4 = 1,3609$ $s_1^4 = 0,6915$
T_2	$S_2^1 = 1,0000$ $s_2^1 = 1,0000$	$S_2^2 = 1,1005$ $s_2^2 = 0,8850$	$S_2^3 = 2,1564$ $s_2^3 = 2,2089$	$S_2^4 = 2,0239$ $s_2^4 = 0,9163$

T_3	$\frac{S_3^1 = 1{,}0000}{s_3^1 = 1{,}0000}$	$\frac{S_3^3 = 1{,}3248}{s_4^3 = 1{,}8210}$	$\frac{S_3^3 = 0{,}3093}{s_4^3 = 0{,}4671}$	$\frac{S_3^3 = 2{,}0918}{s_3^3 = 1{,}8159}$
T_4	$\frac{S_4^1 = 1{,}0000}{s_4^1 = 1{,}0000}$	$\frac{S_3^2 = 0{,}9909}{s_3^2 = 0{,}9564}$	$\frac{S_4^3 = 1{,}5928}{s_4^3 = 1{,}3441}$	$\frac{S_4^4 = 2{,}9980}{s_4^4 = 2{,}4740}$

Validity of the obtained values when compared to the actual traffic intensity was assessed using mean squared error (Table 6).

Table 6. Fragment of values for deviation of actual intensity from design intensity.

No.	Legend	Variable factors			Deviations		
		Type	Surface condition	Number of lanes	between design and actual σ, %	Mean error of approximation $\bar{A}$,%	Coefficient of variation c_N,%
1	$T_4M_1R_1$	4	Dry	2	9.17	21.62	31.47
2	$T_4M_1R_2$	4	Dry	3	5.95	18.24	33.69
3	$T_4M_1R_3$	4	Dry	4	6.25	12.10	32.96
4	$T_4M_2R_1$	4	Wet	2	7.44	18.23	20.24
5	$T_4M_2R_2$	4	Wet	3	8.18	10.18	16.79
6	$T_4M_2R_3$	4	Wet	4	10.92	22.24	33.17
7	$T_4M_3R_1$	4	Ice mound	2	12.50	24.12	30.85
8	$T_4M_3R_2$	4	Ice mound	3	14.28	20.16	22.76
9	$T_4M_3R_3$	4	Ice mound	4	4.62	17.89	30.47
10	$T_4M_4R_1$	4	Packed snow	2	3.72	26.36	20.40
11	$T_4M_4R_2$	4	Packed snow	3	11.00	21.26	30.11
12	$T_4M_4R_3$	4	Packed snow	4	11.24	7.89	28.96

Conclusion

The assessment results showed that the the deviation of actual values from the design ones for each condition lies within the range from 3 to 14 %, which characterizes representation of the obtained data while solving spatial problems.

These positive results of the suggested methodology serve as a basis for verification of mathematical tools universality for the world metropolises scale. Further improvement of the methodology is possible through extension of geography of the studied objects and use of paid content. Increase in resolving power of satellite images enables to verify differentiation of the studied traffic flows by content.

At the moment, documentation for solving actual operational tasks is under development, it will be further tested in urban bodies and enterprises.

References

1. A.A. Testeshev, V.D. Timohovetz, AIP Conference Proceedings, **1800**, 050006 (2017)
2. V.V. Silyanov, *Teorija transportnyh potokov v proektirovanii dorog i organizacii dvizhenija* [Theory of traffic flows in the design of roads and traffic organization] (Transport, Moscow, 1979) (in Russian)
3. M. Treiber, A. Kesting, *Traffic Flow Dynamics* (Springer, 2013)
4. Yu.V. Trofimenko, M.R. Yakimov, *Transportnoe planirovanie: formirovanie effektivnyh transportnyh system krupnyh gorodov* [Transport planning: establishment of efficient transport systems of big cities] (Lotos, Moscow, 2013) (in Russian)

5. A.E. Gorev, K. Bettger, A.V. Prokhorov, R.R. Gizatullin, *Transportnoe modelirovanie* [Transport Simulation], 1st ed. (SPSUACE, St. Petersburg, 2015) (in Russian)
6. Y.V. Khomyak, *Organizacija dorozhnogo dvizhenija* [Organization of road traffic] (Vysshaja shkola, Kiev,1986) (in Russian)
7. V. Markuc, Transportnye potoki avtomobil'nyh dorog i gorodskih ulic [Transport flows of highways and the city streets], 48 (Tyumen, 2008) (in Russian)
8. V.M. Starozhilets, Yu.V. Chekhovich, Journal of Machine Learning and Data Analysis, **2**(3) (2016) (in Russian)
9. A.A. Testeshev, V.D. Timohovetz, Transportation Research Procedia, **20**, 643-648 (2017)
10. ODM 218.4.005-2010. Rekomendacii po obespecheniju bezopasnosti dvizhenija na avtomobil'nyh dorogah [Recommendations for road safety ensuring on roads] (Rosavtodor, Moscow, 2010) (in Russian)
11. H. Lieu, Traffic-Flow Theory. Public Roads, **62**(4) (1999)
12. F. Heit, *Matematicheskaja teorija transportnyh potokov* [Mathematical theory of traffic-flows] (Mir, Moscow, 1966) (in Russian)

Integration of urban developers with regard to social and environmental responsibility

Irina Nuzhina[1, 2,*], *Maria* Zolotareva[1], and *Iuliia* Vasileva[1]

[1]Tomsk State University of Architecture and Building, 634003 Tomsk, Russia
[2]Tomsk State University of Control Systems and Radioelectronics, 634050 Tomsk, Russia

Abstract. The aim of this study is to develop a model for institutionalizing the social and environmental responsibility of the construction business as an important factor in the modern urban development. Methods of systems, logical and comparative analysis, as well as a questionnaire survey are used in the research. Based on the situation in Tomsk (Russia), public preferences and needs for construction products have been examined taking into account environmental friendliness and the quality of the urban environment. The results of the examination are presented in this paper. The study reveals the public need for building products and shows how these needs are related to greening the construction industry and developing the social and environmental responsibility of the construction business. The green urban development needs an environmentally sustainable regulation mechanism based on appropriate methods, tools and organizational forms of interactions between stakeholders in the construction industry. A concept for managing interactions between stakeholders in urban development is proposed on the basis of the ECOURBANBUILD platform. The integration based on this platform will create additional opportunities to implement the social and environmental responsibility of the construction business in the interests of sustainable urban development.

Introduction

Throughout the world, the absolute priority in urban development is to create a safe and comfortable environment for human beings, which is closely tied to environmental conditions and quality of life [1, 2]. Therefore, it is a matter of urgency to search for effective mechanisms for regulating urban development that will ensure socially and environmentally acceptable territorial development, and increase the environmental performance of the entire economic system. According to the Environmental Performance Index 2016, Finland is the greenest country in the world. The other top ten countries are Iceland, Sweden, Denmark, Slovenia, Spain, Portugal, Estonia, Malta, and France. Russia ranks 32nd among 180 countries in the index and has improved its ranking by 24% in the previous two years [3].

* Corresponding author: irinanuzhina@yandex.ru

There are also annual environmental ratings for various Russian regions. As of May 1, 2017, the Tomsk Region has ranked 22nd among 86 regions of the Russian Federation. In the Siberian Federal District (SFO), the Tomsk Region is second only to the Altai Republic and the Altai Krai [4].

Urban planning has a direct bearing on environmental rankings of territories. Socio-environmental aspects of urban planning are studied on an interdisciplinary basis. Problems of environmental support for urban planning are discussed in Plotnikova (2008) and Krygina (2015). Environmental safety of construction and management is considered in Telichenkoetal (2016). Greening of industries and environmental innovations are addressed in Morgunovetal. (2017), Chuanqi (2012), Rudneva et al. (2016), Zakharova et al. (2015), Kasianenko et al. (2015). Sustainable development problems in construction and architecture, as well as the development of environmental infrastructure are investigated in Tetior (2008). Social aspects of urban development are considered in Grabovyy (2016), Pakhomova et al. (2008) [5-11].

However, institutionalization of social and environmental responsibility of the construction business has yet to be fully explored both in theory and in practice. Methods and tools for regulating the construction industry should be aimed at achieving socially and environmentally significant results.

This paper concerns institutionalization of the social and environmental responsibility of the construction business, and aims to determine the nature of this process and structure it, as well as to develop a conceptual model for managing the integrated interactions between stakeholders in urban development as a key factor in greening the construction regulation mechanism. For these purposes, the paper demonstrates the importance of the environmental factor for society and the urban environment, examines the nature and content of the social and environmental responsibility of the construction business, identifies public preferences and the need for building products.

Research Methods

The theoretical and methodological framework of the research is based on provisions of the institutional theory [12–14], works by Russian and foreign scientists and specialists addressing problems of regulation in the field of urban development, taking into account environmental factors, environmental and economic support for construction activities [7–9, 11].

Data from Russian and foreign statistical systems serves as a basis for the research. These are Eurostat, the Unified Interdepartmental Information and Statistical System (UISIS), Rosstat (Federal State Statistics Service of Russia), Tomskstat (Territorial Body of the Federal State Statistics Service in the Tomsk Region), UN reports and publications, Cities Alliance, All-Russia Public Opinion Research Center, and other Russian and foreign organizations. [4, 15–24].

When elaborating the problematic issues stated in the paper, we relied on opinion polls about the comfort level of the urban environment [16, 18, 20], a questionnaire survey to identify public preferences and the need for building products [21], the method of systematic and comparative economic analysis [25], as well as statistical data processing methods [26].

Research Materials

The human environment is a result of transformations in the field of urban development and closely tied to social (socio-psychological and socio-economic) factors of societal development. The awareness of the fact that consequences of construction activities are relevant to the environment determines the priorities of the modern urban development in territorial entities of Russia. These priorities include the availability of housing and housing services; availability and accessibility of social and infrastructural facilities, the creation of a comfortable, accessible and safe environment; availability of facilities that contribute to socialization, communication and recreation; preservation of natural landscapes; preservation of cultural, historical and architectural monuments; architectural expression and aesthetic appearance of facilities; application of environmentally friendly materials, technologies, engineering equipment and systems; energy efficiency and resource saving in construction and maintenance of facilities; use of "inconvenient" and "unattractive" areas; integrated housing development [23].

The fully-fledged environmentally oriented development of the construction sector is possible only by institutionalizing and developing the social and environmental responsibility mechanism of the construction business. The social and environmental responsibility mechanism for the construction business is a deliberate initiative of construction companies aimed at adopting additional, stricter social and environmental requirements for investment and construction projects, as well as at complying with mandatory environmental requirements, regulations and standards in the construction sector [23]. Construction companies go beyond the regulatory and legal framework, and initiate the adoption of additional, stricter social and environmental requirements to production activities and their results (products, services, work).

It is the institutional environment that creates opportunities for developing the construction business in an environmentally and socially sustainable manner. The institutional environment provides conditions for the functioning of construction companies and, therefore, affects the economic, environmental and social performance [23].

Public opinion polls make an important contribution to the promotion of green building [16, 18]. For example, since 2004, Eurobarometer has been conducting *Quality of Life in European Cities* surveys every three years [20]. Similar surveys are conducted in Russia. On April 20, 2017, the All-Russia Public Opinion Research Center presented the results of monitoring in the field of environment and landscaping. According to the survey, citizens are primarily concerned about problems of landscaping in the urban environment and areas adjacent to their houses. These problems include garbage generation (23%); condition of roads and traffic jams (22%); a large number of cars and a shortage of parking lots (12%); children's playgrounds (8%), and lack of green spaces (6%) [22].

In order to identify preferences and needs of citizens for eco-friendly building products, the authors conducted a questionnaire survey of Tomsk residents (200 respondents). Respondents were asked to assess the social and ecological environment of the city by identifying the three most significant factors that determined their dissatisfaction with the entire urban environment and the environment of their residential district (Table 1).

According to the survey, 57% of respondents are not satisfied with the quality of the environment in their districts and would like to move to another district with a more developed social infrastructure and a more favorable environmental situation. 62% of respondents are willing to make a choice in favor of a more eco-friendly and expensive house. This group is mainly represented by employees and managers aged 25 to 40 years. 79% of respondents would prefer to choose a house made of eco-friendly building materials.

Table 1. Analysis of Survey Results.

	Socio-demographic characteristics of respondents													Total proportion of respondents, %
	Sex		Age				Occupation				Household income level			
	Male	Female	under 25 years	25–40 years	40–60 years	over 60 years	Unemployed	Students	Pensioners	Employees	Low	Medium	High	
Proportion of respondents, %	43	57	33	42	20	5	6	43	11	40	34	43	13	
Dissatisfaction with the city's social and ecological environment.														
Sanitary condition of streets	47	53	42	26	8	8	18	37	5	39	29	53	18	63
Landscaping and gardening of the urban environment	37	63	39	29	32	0	0	34	21	45	39	37	24	57
Architectural appearance of buildings	50	50	0	75	25	0	0	25	38	38	0	50	50	50
Condition of tourist zones	43	57	30	57	13	0	11	28	21	40	24	58	18	53
Too much outdoor advertising	69	31	8	54	38	0	15	19	27	38	23	58	19	78
Dissatisfaction with the social and ecological environment in the residential district.														
Sanitary condition of yards	53	47	42	24	29	5	24	16	32	29	26	71	3	86
Landscaping and gardening of areas adjacent to the house	48	52	26	38	33	2	24	43	29	5	26	45	17	63
Architectural appearance of buildings	38	31	22	53	19	0	13	31	41	16	22	44	22	36
Lack of organized parking areas	53	47	28	47	25	0	0	14	22	64	19	56	25	73
Lack of social infrastructure facilities in walking distance	63	37	35	40	25	0	13	27	38	21	17	67	15	42
Choice of respondents (subject to the availability of financial resources)														
Moving to another city with a more developed social and environmental infrastructure	50	50	25	47	28	0	9	9	56	25	13	50	38	16
Moving to a district with more developed social and environmental infrastructure	49	51	37	42	21	2	8	22	29	41	31	57	9	43
Investing money in developing my district through modern forms of management	27	73	27	29	44	2	13	15	52	21	6	67	27	24
I am satisfied with everything	54	46	75	13	13	0	33	25	29	13	8	75	17	12
I wish a lot of things were different but I don't see a chance to change something.	58	42	50	33	17	0	17	17	58	8	75	25	0	6

Certainly, the ability to meet the public needs for eco-friendly building products depends on the level of household income. The price (five points) is the determining factor for respondents with low household income (up to $1,000 per month), while high-income respondents give priority to the environmental situation in their districts, comfortable environment, and availability of recreational zones (Table 2).

Table 2. Importance of factors that influence the choice of housing (points), based on the survey results.

Factors	Average household income		
	low	medium	high
Environmental situation in the microdistrict	2.5	3.9	5
Use of eco-friendly materials and energy-saving technologies	3.8	3.8	4
Proximity to the city center	3.8	2.6	2
Infrastructure, transport accessibility	3.8	2.6	2
Price	5	3	1.6
Comfortable environment, proximity to the recreational area	2.4	4.3	5

Thus, in order to create a modern environmentally sustainable space in urban areas, an adequate system of methods and tools for regulating urban development must be available. These methods and tools are aimed at institutionalizing the social and environmental responsibility mechanism for the construction business, as well as at developing appropriate legal regulations and voluntary initiatives that would be proposed by the government and the business, and supported by civil society.

The institutionalization of the social and environmental responsibility mechanism for the construction business is the process of defining formal rules and informal restrictions, establishing institutions and organizational forms of interaction between stakeholders in construction investment, environmental and economic regulation, and developing interrelations that ensure the achievement of objectives subject to social and environmental priorities of territorial development. Introduction of an environmentally sustainable mechanism for regulating construction investment activities is the key element of the institutionalization process [23].

Results

In the context of greening, the mechanism for regulating construction investment activities should be considered as a system of organizational forms of interaction between stakeholders in construction investment activities, methods and tools to influence the economic behavior of these stakeholders. The purpose of this mechanism is to harmonize economic interests with social and environmental requirements for territorial development. One of the most important current tasks is to develop a model for integrated management of interactions between stakeholders in urban development aimed at coordinating efforts of scientific institutes, the government, business and society in addressing social and environmental problems of urban areas. The methodological principles of integrated interaction include goal orientation, sufficiency, readily available, consistent information and analytical support, competence in decision-making, flexibility and adaptability, a variety of solutions to choose from, and recognition of the life cycle of building products.

In implementing these principles of integrated interaction, one of the most acceptable options is to create "platforms" for solving urgent problems, independent and professional expert reviews, cooperation of commercial and non-profit organizations and the government in various fields.

The creation of the ECOURBANBUILD platform is considered to be a reasonable solution for regulating the economic behavior of stakeholders in construction investment activities with the aim of ensuring environmentally acceptable and socially oriented development of the construction investment sector. This platform is a mechanism aimed at mobilizing initiatives of the construction business, the government, civil society institutions, and other stakeholders in developing an architectural and urban environment that is both eco-friendly and favorable for the human and social development. The platform brings together enterprises, organizations and associations from various sectors that share goals and objectives of the platform, and contribute to their achievement.

The conceptual model of the proposed ECOURBANBUILD platform is presented in Figure 1. The ECOURBANBUILD platform designed for environmentally sustainable urban development should be implemented in all regions. The institutional component includes location, an organizational management structure, working and expert groups, a strategic research and implementation program, a roadmap for achieving short-term and long-term goals.

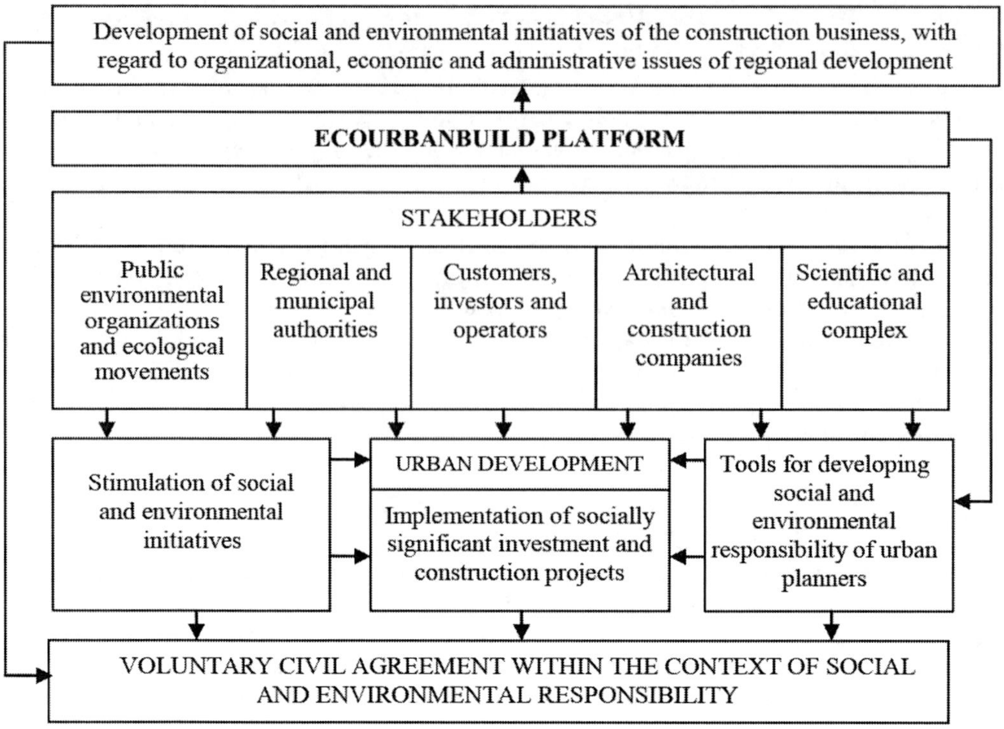

Fig. 1. The conceptual model for managing integrated interactions between stakeholders in construction investment activities, based on the ECOURBANBUILD platform (developed by the authors).

This model for managing integrated interactions between stakeholders in construction investment activities will create additional opportunities for promoting the development of social and environmental initiatives for the benefit of individuals and society as a whole.

Conclusion

A modern urban development model should be created under the dominant influence of social and environmental factors of territorial development reflecting the needs of citizens for more eco-friendly building products. The fully-fledged environmentally sustainable development of the construction business is possible only by introducing green methods and tools to regulate construction investment activities, as well as by improving organizational forms of interactions between urban development stakeholders, including the government, business and society.

Using ECOURBANBUILD as a platform for interactions, it is possible to reduce transaction costs for coordinating interests and business negotiations of stakeholders in investment and construction projects, exchange of information and disseminating experience in the field of green building, as well as to prevent possible consequences of decision-making without agreeing on controversial issues with potential consumers and the public, to exclude models of construction processes that negatively affect the environment and society.

The practical significance of this study is associated with the possibility and feasibility of using the proposed model for managing integrated interactions between stakeholders in construction and investment activities on the basis of the ECOURBANBUILD platform in the field of urban development regulation, while implementing socially significant investment projects.

The paper was funded by the Russian Foundation for Basic Research (RFBR) under the research project No. 16-32-00005a: Modeling of Management Techniques Aimed at Implementing Social and Environmental Responsibility of the Construction Business.

References

1. V.V. Buzyrev, I.P. Nuzhina, *Ekologo-ekonomicheskie aspekty investicionno-stroitel'noj dejatel'nosti* [Ecological and economic aspects of investment and construction activities] (Saint Petersburg State University of Engineering and Economics, Saint Petersburg, 2012) (in Russian)
2. WWW-LAFAGERHOLCIM-FOUNDATION, Understanding sustainable construction, available at: https://www.lafargeholcim-foundation.org/AboutPages/what-is-sustainable-construction/ (2014)
3. WWW-GTMARKET, Rating of countries in the world in terms of environmental performance in 2016, available at: http://gtmarket.ru/news/2016/01/29/7292/(2016) (in Russian)
4. WWW-GREENPATROL, Environmental rating of the subjects of the Russian Federation, available at: http://www.greenpatrol.ru/ru/stranica-dlya-obshchego-reytinga/ekologicheskiy-reyting-subektov-rf?tid=291 (2016) (in Russian)
5. L.V. Plotnikova, *Ekologija megapolisa* [Ecology of a metropolis] (Association of Civil Engineering Universities, Moscow, 2008) (in Russian)
6. A.M. Krygina Zhilishchnoe Stroitelstvo [Housing Construction], **6**, 57 (2015) (in Russian)
7. V.I. Telichenko, A.A. Benuzh, Vestnik of MGSU, **1**, 118 (2016)
8. H. Chuanqi, Modernization science, 472-476 (2012)
9. A.N. Tetior, Prirodoobustrojstvo [Environmental Engineering], **2**, 57-67 (2008) (in Russian)
10. P.G. Graboviy, Real estate: economics, management, **4**, 12-19 (2016) (in Russian)
11. N.V. Pahomova, G.B. Malyshkov, Problems of modern Economics, **2**, 310-317 (2008) (in Russian)
12. D. North, Questions of Economics, **3**, 6-17 (1997)
13. N.A. Asaul, *Institucional'noe vzaimodejstvie subjektov investicionno-stroitel'nogo kompleksa* [Institutional interaction of the subjects of the investment and construction complex] (Gumanistika, Saint Petersburg, 2005) (in Russian)
14. E.G. Matjugina, *Institucionalizacija nacional'nyh jekologo-jekonomicheskih otnoshenij kak faktor obespechenija ustojchivosti razvitija hozjajstvennoj sistemy* [Institutionalization of national ecological and economic relations as a factor in ensuring the development sustainability of the economic system] (TSUAB Pabl., Tomsk, 2006) (in Russian)
15. WWW-CONSTRUCTINGEXCELLENCE, Corporate Social Responsibility, available at:http://constructingexcellence.org.uk/wp-content/uploads/2015/03/social_responsibility.pdf/ (2016)
16. WWW-EC.EUROPA,"Quality of Life in European Cities," report, European Commission, Directorate-General for Regional and Urban Policy, available at: http://ec.europa.eu/regional_policy/sources/docgener/studies/pdf/urban/survey2015_en.pdf/ (2015)
17. M. Lowe, C. Whitzman, H. Badland, M. Davern, L. Aye, D. Hes, I. Butterworth, B. Giles-Corti, Urban Policy and Research, **33(2)**, 131-144 (2015)

18. WWW-CITIESALLIANCE, "Liveable cities. The benefits of urban environmental planning", report, The Cities Alliance, available at: http://www.citiesalliance.org/sites/citiesalliance.org/files/CA_Docs/resources/cds/liveable/liveablecities_web_7dec07.pdf/ (2007)
19. WWW-UN-DOCUMENT, "The urban challenge in developing countries", report, World Commission on Environment and Development, United Nations, available at: http://www.un-documents.net/ocf-09.htm (2015)
20. WWW-WCIOM, "Ekologicheskaja situacija: monitoring" [Ecological situation: monitoring] available at: https://wciom.ru/index.php?id=236&uid=116167 (2017) (in Russian)
21. WWW-WCRC, "Proekt realizacii tehnologicheskoj platformy "Tehnologii ekologicheskogo razvitija" [Project for the implementation of the technological platform "Technologies for Environmental Development, Russian Geographical Society, available at: http://www.wcrc.ru/TP.pdf/ (2011) (in Russian)
22. I.P. Nuzhina, M.V. Zolotareva, Yu.V. Vasileva, Basic research, **12(2)**, 447-452 (2016) (in Russian)
23. I.P. Nuzhina, M.V. Zolotareva, Journal of Economy and entrepreneurship, **12-1**, 945-953 (2016) (in Russian)
24. I.S. Godenov, *Evropejskie tehnologicheskie platformy* [European technological platforms] (TSU, Tomsk, 2011) (in Russian)
25. F.I. Peregudov, F.P. Tarasenko, Osnovy sistemnogo analiza [Fundamentals of system analysis], **3**, 396 (2001) (in Russian)
26. N.A. Yarushkina, *Optimizacija sistemy gradostroitel'stva na osnove informacionno-analiticheskogo podhoda* [Optimization of the urban planning system based on the information-analytical approach] (TSUAB Pabl., Tomsk, 2013) (in Russian)

Expansion of urban recreation areas as a factor for human capital renovation

Tatiana Ovsiannikova[1,*], *Mariya* Nikolaenko[1], and *Mergesh* Chilbakova[1]

[1]Tomsk State University of Architecture and Building, 2, Solyanaya Sq., 634003 Tomsk, Russia

Abstract. The paper is focused on the development of urban space on the basis of the modern urban development concepts. The role of urban environment in the development of individuals, their physical, intellectual and spiritual potential is described with the use of human capital theory. The City Recreation concept is suggested to be accounted while planning the urban territory development. The role of recreation areas in urban environment is described, classification of urban recreation areas is given. The paper recites the results of a social study devoted to evaluation of satisfaction of typical urban district residents with the number and quality of recreation areas. Recommendations are given on public spaces expansion to provide daily recreation.

Introduction

Human capital becomes one of the main production factors in post-industrial economy. Economic growth of a state and the level of population well-being are largely defined by its accumulated human capital. As Klaus Schwab, the Founder and Executive Chairman of World Economic Forum, said: "The key for the future of any country and any institution lies in the talent, skills and capabilities of its people" [1], "Talent, not capital, will be the key factor linking innovation, competitiveness and growth in the 21st century" [2]. That is why global scientific community pays close attention to the issue of providing socio-economic conditions for creation and development of human capital.

One of the consequences of global urbanization processes is concentration of human capital on urban territories. Multiple studies, including those performed by the authors of the paper, show that modern cities become attraction centers both for investment capital and human capital. The most economically developed territories get a strong incentive for economic growth and become ever more successful by attracting investment and human capital. At the same time, territories where human capital flight is observed switch to negative dynamics and become more depressive and stagnant. For that reason, attractiveness of urban areas for human capital becomes the major competitive advantage of the territory in the global labor and capital markets.

Due to the desire to increase the territory attractiveness for human capital local authorities and urban development stakeholders focus local urban policy on an individual. Analysis of theory and practice of urban territories development in various countries

* Corresponding author: tatovs@mail.ru

showed that the most successful are cities where socio-economic relations “city-person” are based on the modern concepts of urban development, which are *“Liveable City”, “Creative City”, “City of Opportunities”* [3]. According to the authors' opinion, these concepts can be complemented by the *“Recreation City”* concept. This paper is aimed to justify this urban development concept.

Methodology

Human capital theory has won strong positions in economic science since the second half of the XXth century, its authors being awarded with a Nobel Prize. This theory has become a methodological framework for numerous studies on the role of a person as the key development factor for social production. The theory is focused on a person as a source of knowledge, skills and competences. Accumulation of critical mass of human capital is considered as a condition for transit of the national economy from one technological frameworks to another.

Due to complexity and multidimensionality of the Human Capital concept multiple approaches and methodologies are used for its investigation. The majority of national and foreign researchers use investment approach as a basis, which considers assessment of human capital with regard to expenditures on its formation as a key issue. Such expenditures include investments in education, public health, etc. (Th. Schultz, 1963, 1971; G. Becker, 1975; R. Kapelyushnikov, 1993, S. Dyatlov, 1993, etc.). Later on, the reproduction approach was formed within the human capital theory, according to which human capital was considered throughout its whole life cycle. During this time, human capital undergoes the stages of its creation (formation), distribution, exchange and use. This is reflected in The Human Capital Reports where Human Capital Index is evaluated with the account of not only the stage of human capital formation, i.e. learning, but also the stage of use of human capital, i.e. employment [1, 2].

Human capital is a long-term resource; in the process of socio-economic activity this capital is depreciated, loses its quality and deserves renovation, just like physical capital. Reproduction approach to human capital study allows introducing the notion of human capital renovation. This category denotes the process of renovation and recovery of physical, intellectual and spiritual potential of a human which was spent while performing socio-economic activities. Unlike the renovation of fixed capital which is performed discretely and periodically throughout the life cycle of fixed assets, human capital renovation shall be performed almost permanently throughout the whole stage of its socio-economic activity.

Human capital being non-material wealth of a person, and collectively representing non-material wealth of a city, region and state, requires material assets for its formation, maintenance and renovation. These assets primarily include material costs for creation of high-quality living environment which provides formation of high-quality human capital and its renovation.

The great impact of living environment on formation of human personality, intellectual, physical and spiritual potential, as well as on human identity, is accepted by many researchers from various fields, including philosophy, sociology, psychology, architecture, urban planning, economics, etc. The special term of “place-identity” which denotes a substructure of self-identity, much like gender and social class, and is comprised of perceptions and comprehensions regarding the environment” has become almost interdisciplinary [4].

Thus, competitiveness of a city in global markets greatly depends on the level of conditions the city can provide for formation and renovation of human capital.

This function within the city is performed by public recreation areas meant for recovery and development of physical, intellectual and spiritual assets of a person spent in the process of his socio-economic activity. As a rule, such areas are multifunctional and perform communicative, educational, information, developing, entertaining, sportive, recreational, emotional, aesthetic, and other functions which contribute to human capital recovery. The importance of this task in Russia is acknowledged on the governmental level. The priority project "Formation of the comfortable urban environment" [5] was launched, its purpose being creation of new or improvement of the existing public spaces, including public recreation areas and courtyard spaces.

In this regard, investigation of the issue of sufficiency, availability and appropriateness of public recreation areas in urban environment is an important interdisciplinary task for economists, sociologists, urban planners, urban designers.

Methodologies for studying and understanding such problems usually involve qualitative techniques, such as interviewing, participant observation and mapping. Such studies are presented in works [6, 7]. This research was performed with the application of traditional methodologies, along with the method of correlation analysis and mapping.

Data & Research

The previous research demonstrate close correlation between the level of human development and the level of urban environment development. Thus, correlation analysis performed by the authors of this paper with the use of data presented in [1, 8] show that a definite relation exists between the City Prosperity Index (CPI) and the Human Capital Index (HCI) (Fig.1).

Obviously, this relation has a reciprocal nature, which means that a successful city creates a successful personality, and a person with high potential makes a contribution to the city prosperity and development. However, the city shall not only provide the opportunity for a person to function as a production factor, but also provide the renovation opportunities, since this directly affects the human capital efficiency.

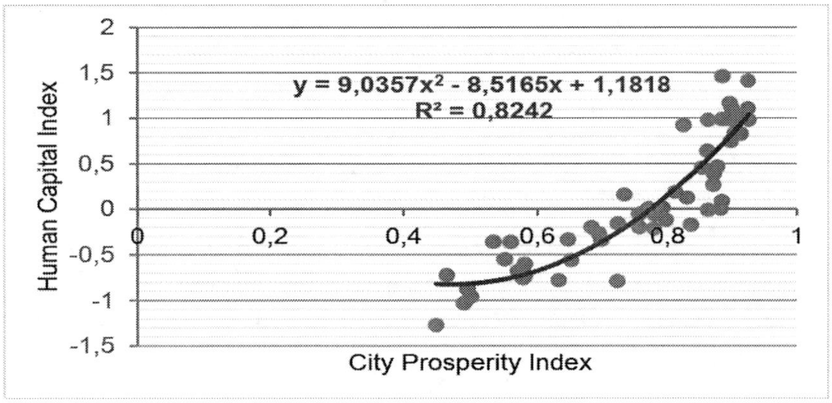

Fig. 1. The relation between City Prosperity Index and the Human Capital Index (performed by the authors).

As already mentioned, renovation is provided by recreation areas in the urban environment. Recreation as a specific type of human activity is performed with certain intervals - daily, weekly, annually, etc. Consequently, urban environment shall include public recreation areas of different types (Table 1).

Table 1. Classification of public recreation areas.

Type of public recreation area	Type of recreation activity	Frequency	Elements of public recreation areas
Micro-area (courtyard)	Recreation Communication	Daily	Playgrounds and sports grounds in courtyards
Local (quarter, district)	Recreation Communication Sports	Weekly Monthly	Parks, playgrounds and sports grounds
City	Recreation Communication Sports Entertainment Education	Weekly Monthly	Parks, squares, pavements, embankment, rest places in business centers, stadiums, groves, water objects.
Meso-area (regional)	Recreation Communication Sports Entertainment Education Rehabilitation	Monthly Annually	Parks, multifunctional complexes (sports, tourism, culture)
Macro-area (national)	Recreation Rehabilitation Communication Education Entertainment Aesthetic pleasure	Periodically, once in several years	National parks, national cultural complexes, national rehabilitation centers

The authors of the study surveyed residents of one of Tomsk districts (Russia). The purpose of the survey was to analyze provision of urban environment with public recreation areas, to assess the quality of recreation areas and sufficiency of functions they perform. Tomsk represent a typical Russian city, its population is 594 thousand citizens [9]. One of special features of Tomsk is its University City status, which results in a high share of young people in its demographic structure. This fact affected the structure of respondents, the majority of whom are represented by young people aged between 21 and 30. Opinion polling and remote polling methods, the latter with the help of Survio software (https://www.survio.com/ru), were used in the study. Around 400 people were surveyed within the study.

Results

Based on the survey results the authors obtained the data that enabled to analyze the quality and quantity of public recreation areas in Tomsk city. Some of the results are given in Figures 2-5 and Tables 2-3.

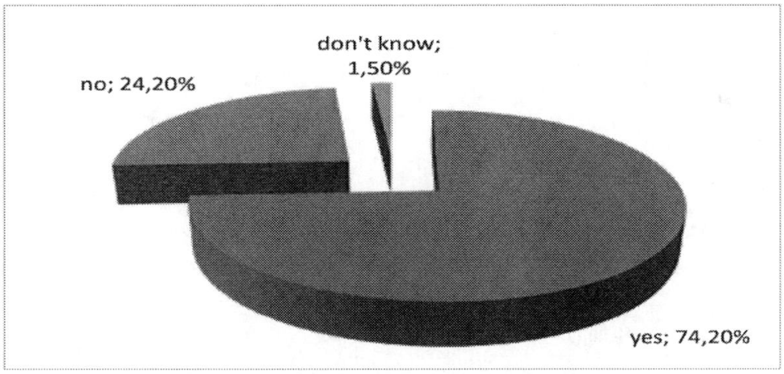

Fig. 2. Answers to the question: is your residential district comfortable for living?

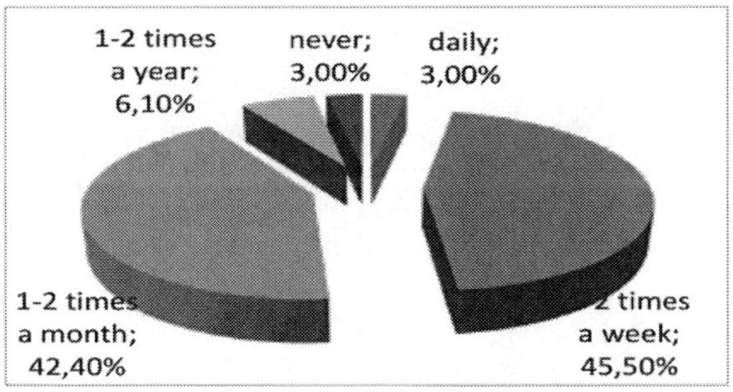

Fig. 3. Answers to the question: how often do you visit public recreation areas: parks, squares, embankments, etc.?

In their responses whether they are satisfied with the recreation areas, only 28.8% of respondents expressed their satisfaction. Those who were not completely satisfied with the existing recreation areas specified the following inconveniences (Table 2).

Table 2. Distribution of responses on what is particularly unsatisfactory in the existing recreation areas.

Answer	Share of those who specified what they are unsatisfied with
Lack of hygienic conditions (public facilities, toilets)	43.9 %
Lack of or insufficient entertainment services (games, roller skating, bicycles, amusements)	40.9 %
Lack of parking spots	19.7 %
Lack of space (small territory)	19.7 %
Lack of catering services	16.7 %
Noise	15.2 %
Lack of transport accessibility	6.1 %
Other (not specified)	18.2 %

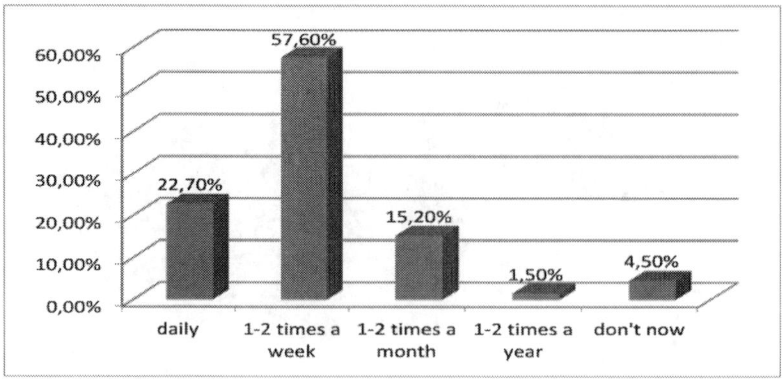

Fig. 4. Answers to the question: if the specified drawbacks are improved, how often would you visit the recreation areas?

Table 3. Change in demand of recreation services after improvement of drawbacks.

Attendance of recreation areas	Share of those who responded	
	Before improvement	After improvement
Daily	3 %	22.7 %
Once or twice a week	45.5 %	57.6 %
Once or twice a month	42.4 %	15.2 %
Once or twice a year	6.1 %	1.5 %
Never	3 %	-
I don't know	-	4.5 %

According to Table 3, in case the existing recreation areas are improved, the demand in recreation services among the local residents will be much higher.

In order to analyze the level of satisfaction of local residents' need in recreation the respondents were asked what kind of recreation areas lack in the city. The results of the survey are presented in Figure 5. The diagram demonstrates that the majority of respondents emphasized the lack of multifunctional objects, which indicates a variety of recreation preferences of the locals.

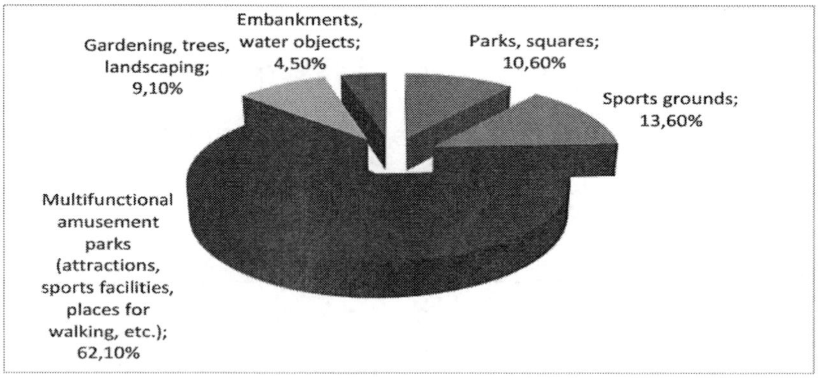

Fig. 5. Answers to the question: what kind of recreation areas lack in Tomsk?

Public micro-areas have special importance in the urban environment, since they provide daily recreation to all groups of residents. This problem is mostly relevant for

Russian cities with high share of multi-family residential buildings. This category of housing represents 75.8% of all residential housing in Russia, 65.2% of population reside here (calculated by the authors based on [10, 11]). Whereas in European countries a smaller share of population lives in apartments. For example, in Norway the share of apartment housing is 13.4%, in the UK - 14.5%, in Belgium - 25.1%, in the Netherlands - 29% [12]. In these countries daily recreation is performed in micro-areas - household territories. Those Russian citizens who live in apartments carry out their daily recreation in public micro-areas located between the buildings in residential areas.

As a result of the conducted sociological survey the authors found that more that 75% of the respondents are unsatisfied with the quality of public micro-areas, whereas 62.5% emphasized the lack of playgrounds, sports facilities and recreation areas for adults in their courtyards. And only 25% replied that they are completely satisfied with their neighborhood areas.

Visual inspection of courtyard areas in one of Tomsk residential districts was performed for more detailed analysis of the reasons for poor satisfaction level among local residents. The analyzed recreation micro-areas are located on a map and presented in Figure 6. The study showed that out of the 21 analyzed areas in the studied residential district, only one third of courtyard zones meet the needs of the residents in daily recreation.

Fig. 6. The analyzed recreation micro-areas located in a residential district of Tomsk.

Conclusion

The performed research confirms the importance of public recreation areas in the urban environment as a factor for human capital renovation. At the same time, the results of the study show that the recreation needs of the residents, particularly need in daily recreation, are not fully satisfied.

On the basis of visual inspection of the existing recreation micro-areas in one of the residential districts the complex of recommendations was developed for their improvement. It also allowed defining the volume and scope of work and assessing the required investments for creation of appropriate recreation micro-areas in the urban environment. With regard to different technical condition of equipment and street furniture in the courtyard areas, the amount of investments required for these courtyards is about $250,000, with the average of $12,000 per courtyard. The results were obtained on the example of one residential district, but they can be extrapolated on all urban space. This will enable to perform urban planning with the account of "City Recreation" concept.

References

1. The Human Capital Report (World Economic Forum, Cologny/Geneva 2013)
2. The Human Capital Report (World Economic Forum, Cologny/Geneva 2015)
3. T. Ovsiannikova, M. Nikolaenko, AIP Conf. Proc. **1800** (2016)
4. A. Hauge, Architectural Science Review (2007)
5. Ministerstvo stroitel'stva i zhilishchno-kommunal'nogo khozyaistva Rossiiskoi Federatsii, Available at: http://www.minstroyrf.ru/trades/zhilishno-kommunalnoe-hozyajstvo/strategicheskoe-napravlenie-razvitiya-zhkkh-i-gorodskaya-sreda (2017)
6. V. Yanchenkov, N. Shagov, Vestnik of TSUAB, **1,** 100 (2011) (in Russian)
7. V. Yanchenkov, N. Shagov, Vestnik of TSUAB. English version appendix, **1** (2012)
8. State of the world's cities 2012/2013. Prosperity of cities, United Nations Human Settlements Programme (UN-HABITAT, Nairobi, 2015)
9. Official data on population of Tomskstat, available at: http://tmsk.gks.ru/wps/wcm/connect/rosstat_ts/tmsk/ru/statistics/population/ (2017) (in Russian)
10. *Zhilishchnoe khozyaistvo v Rossii* [Housing in Russia] (Rosstat, 2013) (in Russian)
11. Results of the All-Russian population census 2010, Book 9. Living conditions of the population (Rosstat, 2017) (in Russian)
12. Living Conditions in Europe (Eurostat, 2017)

AUTHOR INDEX

AUTHOR INDEX

EDP Sciences – Web of Conferences
17, Avenue du Hoggar
Parc d'Activité de Courtabouf
BP 112
F-91944 Les Ulis Cedex A
France

ISBN 978-1-5108-5580-9

CURRAN ASSOCIATES INC.
proceedings.com